Advances in Comparative and Environmental Physiology 2

Volumes already published

Vol. 1: NaCl Transport in Epithelia
Edited by *R. Greger* (1988)

Vol. 2: From the Contents: Hibernation – Vapor Absorption
Nutrient Transport – Stable Pollutants (1988)

Volumes in preparation

Vol. 3: Molecular and Cellular Basis of Social Behavior in Vertebrates
Edited by *J. Balthazart*

Vol. 4: Animal Adaptation to Cold
Edited by *L. C. H. Wang*

Vol. 5: Vertebrate Gas Exchange
Edited by *R. G. Boutilier*

Advances in

Comparative and Environmental Physiology 2

With Contributions by

G. A. Ahearn · J. M. Bouquegneau · C. Joiris
W. H. Karasov · J. Machin · M. J. O'Donnell
L. C. H. Wang · S. H. Wright

With 60 Figures

Springer-Verlag
Berlin Heidelberg GmbH

ISBN 978-3-642-73377-2 ISBN 978-3-642-73375-8 (eBook)
DOI 10.1007/978-3-642-73375-8

Typesetting: K+V Fotosatz GmbH, Beerfelden
Printing and bookbinding: Brühlsche Universitätsdruckerei, Giessen
2131/3130-543210

Foreword to the Series

The aim of the series is to provide comprehensive, integrated reviews giving sound, critical, and provocative summaries of our present knowledge in environmental and comparative physiology, from the molecular to the organismic level.

Living organisms have evolved a widespread range of basic solutions to cope with the different problems, both organismal and environmental, with which they are faced. A clear understanding of these solutions is of course of fundamental interest for all biologists, zoologically or medically oriented. They can be best comprehended in the framework of the environmental and/or comparative approaches. These approaches demand either wide surveys of animal forms or a knowledge of the specific adaptive features of the species considered. This diversity of requirements, both at the conceptual and technological level, together with the fact that physiology and biochemistry have long been mainly devoted to the service of medicine, can account for the fact that these approaches emerged only slowly amongst the other new, more rapidly growing disciplines of the biological sciences.

The field has now gained the international status it deserves and the organization of a series devoted to it appeared timely to me in view of its actual rapid development and of the interest it arouses for a growing number of biologists, physiologists, and biochemists, independently of their basic, major orientation.

Liège, Belgium, Summer 1988 *Raymond Gilles*

List of Editors

Editor-in-Chief

R. Gilles
Laboratory of Animal Physiology
University of Liège
22, quai Van Beneden
B-4020 Liège, Belgium

Editorial Board

E. Florey
Fakultät für Biologie
Universität Konstanz
Universitätsstraße 10
D-7756 Konstanz 1, FRG

Ch. P. Mangum
Department of Biology
The College of William & Mary
Williamsburg, VA 23185, USA

T. J. Shuttleworth
Department of Physiology
University of Rochester Medical Center
601 Elmwood Avenue
Rochester, NY 14642, USA

G. N. Somero
Scripps Institution of Oceanography
University of California
La Jolla, CA 92093, USA

R. E. Weber
Biologisk Institut
Odense Universiteit
Campusvej 55
DK-5230 Odense M, Denmark

Contents

Chapter 4

Nutrient Transport Across Vertebrate Intestine
W.H. Karasov (With 3 Figures)

Chapter 5

Nutrient Transport Across the Integument of Marine Invertebrates
S.H. Wright (With 6 Figures)

Chapter 6

The Fate of Stable Pollutants – Heavy Metals and Organochlorines – in Marine Organisms
J.M. Bouquegneau and *C. Joiris* (With 14 Figures)

Chapter 1

Mammalian Hibernation: An Escape from the Cold

L. C. H. Wang

Contents

1 Introduction

In response to seasonal cold and shortages of food and/or water, many endotherms have evolved the ability to temporarily escape such environmental stresses by exhibition of torpor. During torpor, profound reductions in body temperature and other physiological functions occur which allow the sustainment of life with minimum energy requirement. Unlike torpor in the ectotherms, however, torpid endotherms are capable of spontaneous recovery to euthermia using heat generated exclusively from within by shivering and nonshivering thermogenesis. Depending on the timing, duration, and depth, torpor may be seasonal (e.g., estivation, hibernation) of nonseasonal (e.g., daily torpor, starvation-, and cold-induced torpor). Such diversities in torpor pattern no doubt represent the heterogeneity in selection pressure which different ecological niches exemplify. Extensive studies on the evolutionary, biochemical, physiological, neurophysiological, and neuroendocrinological aspects have firmly established that torpor in mammals is polyphyletic in origin and is an advanced form of thermoregulation

Department of Zoology, University of Alberta, Edmonton, Alberta, Canada T6G 2E1

Advances in Comparative and Environmental Physiology. Vol. 2

rather than a reversion to primitive poikilothermy (see Hudson 1973; Lyman et al. 1982, for reviews).

The very long (up to 76 days per bout of torpor; Twente et al. 1985) and very deep state of torpor (down to 2 °C; Wang 1979) in mammalian hibernation have attracted the most attention in studies of torpor because it epitomizes the thermal limits life may be traversed in a single organism among the higher vertebrates. Although members of at least six mammalian orders are capable of exhibiting torpor, i.e., Monotremata, Marsupialia, Primates, Insectivora, Chiroptera, and Rodentia (Lyman 1982), hibernators are much more commonly found in the last three orders. Some examples are the hedgehogs (*Erinaceus*), the bats (*Eptesicus* and *Myotis*), and the marmots (*Marmota*) and ground squirrels [*Spermophilus* (*Citellus*)]. In these animals, two clear physiological states exist annually: a nonhibernating and a hibernating phase. During the nonhibernating phase, the body weight is relatively constant and exposure to cold results in increased heat production and maintenance of euthermia, typical of other nonhibernating species. During the hibernating phase, the body weight may increase drastically due to hyperphagia and this often leads to a doubling of weight within 4 to 6 weeks. Once a weight plateau is reached, the animal becomes anorexic and a gradual loss of weight follows. Exposure to cold at this time results in the exhibition of hibernation. The physiological mechanisms underlying the transition from the nonhibernating to the hibernating phase are currently unknown; however, the alternation of the two phases is apparently regulated by an endogenous circannual rhythm. The free-running periodicity of this rhythm is species-specific and may range from 7 to 11 months under constant laboratory conditions (Pengelley and Asmundson 1974; Mrosovsky 1978). The generation of this endogenous rhythm is apparently not due to frequency demultiplication of the circadian rhythms (Mrosovsky 1978) since lesioning of the suprachiasmatic nuclei, although disrupting circadian rhythms for wheel-running activity, has no apparent effect on the circannual weight rhythm (Zucker and Dark 1986). The involvement of the pineal body in the regulation of circannual rhythm is also presently uncertain: pinealectomy does not affect the circannual cycles for body weight, gonodal activity, or hibernation (Zucker and Dark 1986) but shortens the periodicity for hibernation and testicular recrudescence by approximately 6 weeks in the second year (Phillips and Harlow 1982). It is suggestive that although the pineal gland is not necessary for the expression of endogenous circannual rhythm, it does exert a regulatory influence on the periodicity of the rhythm. Regardless of these negative findings in trying to decipher the mechanisms underlying the circannual rhythm, it is evident that the very precise schedules of immergence and emergence of hibernators under field conditions suggest that this endogenous rhythm may be entrained to synchronize with changes in nature. Some of the possible zeitgebers which have been investigated to show such entraining effects are photoperiod and temperature (Mrosovsky 1986) and the availability of food may also serve in this capacity (Collins and Cameron 1984).

The subject of mammalian hibernation has been reviewed many times in the last 3 decades. Earlier literature has been summarized by Lyman and Chatfield (1955), Kayser (1961), Hoffman (1964), Hudson (1973) and more recently, Raths and Kulzer (1976), Davis (1976), Lyman et al. (1982), Lyman (1984), and Wang (1985, 1986a). Specific aspects of hibernation have also been reviewed, for instance on neural (Beckman 1978; Beckman and Stanton 1982; Heller 1979), endocrinological (Hudson and Wang 1979; Wang 1982, 1987b), ionic (Willis 1979), and membrane aspects (Willis et al. 1981). In addition, proceedings from the seven international symposia on hibernation and related topics (Lyman and Dawe 1960; Suomalainen 1964; Fisher et al. 1967; South et al. 1972; Wang and Hudson 1978; Musacchia and Jansky 1981; Heller et al. 1986) have also provided extensive coverages on specific topics. The major thrust of this review is to provide a synthetic overview on the subject of mammalian hibernation incorporating recent advances in ecological, physiological, biochemical, and some neuroendocrinological aspects. Due to space constraint, selection of coverage is necessarily biased. Those topics which are conspicuously missing may be found elsewhere (e.g., endocrinological aspects in Wang 1987a, b; Heller et al. 1986). Priorities have been given to provide "patterns and generalizations" wherever possible rather than to reiterate detailed individual investigations. Speculations and directions for future research whenever suggested, are personal prejudices and must be viewed as conjectural.

2 Ecological Significance of Mammalian Hibernation

A hibernation bout consists of entry into, maintenance of, and arousal from hibernation (Fig. 1). Following arousal, a period of euthermia lasting from a few hours to a few days ensues before the animal reenters another bout of hibernation. Thus, in a hibernation season, torpor is not continuous but is composed of many individual hibernation bouts and interhibernation euthermic periods (Fig. 1). To quantify the energy savings derived from natural hibernation, it is necessary to conduct both field and laboratory studies. In the field, the timing of the hibernation season and the duration and depth of each hibernation bout and its seasonal variation must be first documented. In the laboratory, the energy costs associated with the different phases of a hibernation bout at comparable body and ambient (burrow) temperatures as those observed in the field must then be determined. Using this approach and employing implantable temperature-sensitive radiotransmitters, Wang (1979) has estimated that the proportional energy costs associated with different phases of hibernation in a natural population of the Richardson's ground squirrel (*Spermophilus richardsonii*) is as follows: entry into hibernation = 12.8%, maintenance of hibernation = 16.6%, arousal from hibernation = 19.0%, and interhibernation euthermia = 51.6%. The monthly energy savings range from 38% in July and 51.2% in March, but are 81.6 to 96% between August and February. For the hibernation season as a whole, which lasts some 8.5 months in adult Richardson's ground squirrels, the energy saving is 87.8% as compared to that if the same animals were resting but maintaining euthermia for the same duration (Wang 1979).

The very high energy costs associated with the interhibernation euthermic periods are often overlooked in the calculation of energetics of hibernation. In

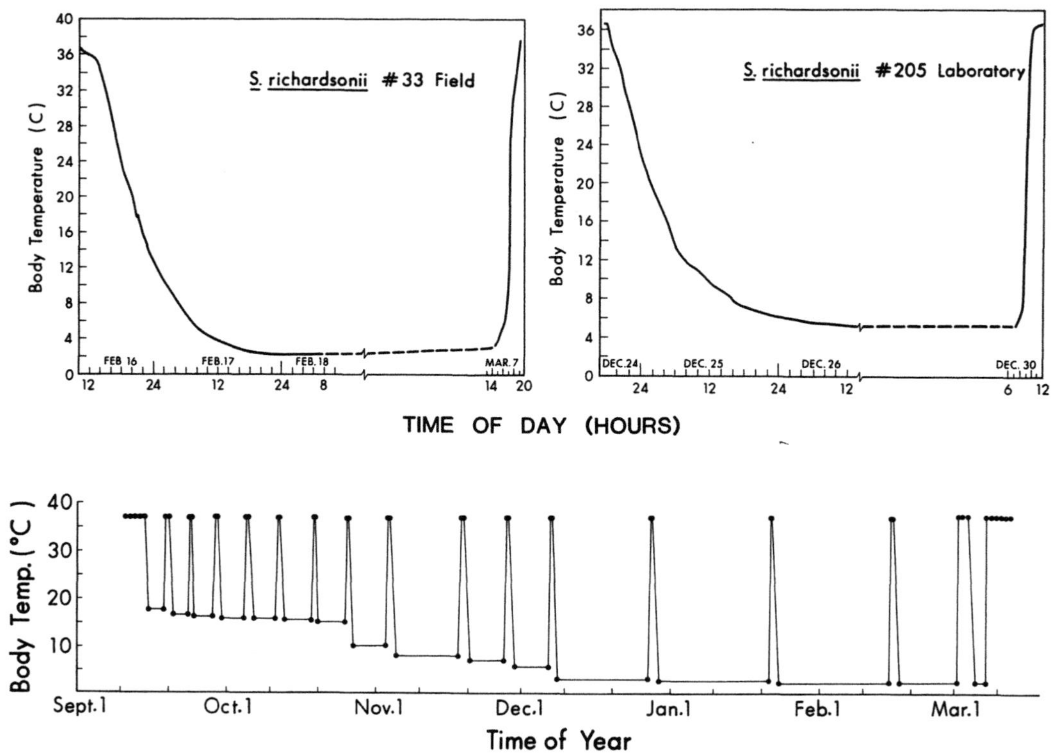

Fig. 1. Upper panel: Typical dynamic change of body temperature (via radiotelemetry) in a hibernation bout in the Richardson's ground squirrel under field (#33) and laboratory (#205) conditions. Note the qualitative and quantitative similarities between the two recordings indicating little disturbance under laboratory conditions in hibernation behavior. *Lower panel*: A complete hibernation season in a juvenile male Richardson's ground squirrel under field conditions. Note the seasonal variations in duration of hibernation and the declining minimum body temperature with progression of winter. The interhibernation euthermic period is too short to be represented accurately except in early March when it exceeds 24 h (Wang 1978)

the Richardson's ground squirrel the interhibernation period rages from 5 to 25 h in field animals depending on the season, being shortest in November-December than any other months (Wang 1979). In laboratory populations, the Uinta Basin ground squirrel (*S. armatus*) exhibits an interhibernation period from 7.5 – 21.5 h, with the shortest durations occurring between days 91 – 210 after the onset of hibernation (Cranford 1986). In the golden-mantled ground squirrel (*S. lateralis*), this duration averages 10.3 and 11.0 h for females and males, respectively (Torke and Twente 1977). There is apparently a positive correlation in the duration of interhibernation euthermia with (body mass)$^{0.38}$. This is interpreted to mean that the duration of euthermia is a direct consequence of the rate of euthermic metabolism which is also scaled to (body mass)$^{0.38}$ and that a fixed amount of metabolism is required during the interhibernation period before another bout of hibernation can commence (French 1985). It has been further suggested (French 1985, 1986) that the frequency of arousal is also proportional to body size, being less frequent in the smaller hibernators than it is in the larger. The reason is that due to the absolutely less energy expenditure in the smaller animals, the metabolic imbalance that may develop during hibernation would take longer to manifest than it is in the larger animals (French 1986). These observations and conjectures raise two important questions: (1) why do hibernating animals arouse periodically in spite of the very large energy expenditure associated with this process especially when the animals are in the midst of an energy conservation campaign (Wang 1978) and (2) why don't the aroused animals return to hibernation immediately thereafter since after urination and defecation, which may take less than a few minutes to complete, the animals are basically in a resting position in their nests with minimal activities (Torke and Twente 1977)?

At this time, there is no clear answer to either of these questions. With regard to the need of periodic arousal, there have been many suggestions (Willis 1982). To summarize, no definitive conclusion may be drawn with respect to predicting frequency and timing of arousal based on circadian rhythms of body temperature and sleep-wakefulness as observed in euthermia (Stanton et al. 1986). The need to eliminate metabolic wastes accumulated during torpor has also be discounted as a possible trigger for arousal. The need for replenishing carbohydrate supplies diminished during hibernation is also unlikely to be a universal mechanism triggering arousal (Zimmerman 1982), whereas the need to reestablish ionic balance perturbed during hibernation could serve as a trigger for arousal although this is difficult to test experimentally (Willis 1982). With regard to the regulator for duration of interhibernation euthermia, the resynthesis of certain key metabolic molecules which may have been depleted during hibernation but are needed for the reinitiation of hibernation could be a possibility. However, as in the search for the regulators of torpor duration, there is also no definitive evidence as to the possible identity of these mysterious molecules.

3 Organ System and Organ/Tissue Adaptations

3.1 Central Nervous System (CNS) Regulation of Body Temperature

As hibernation is a modified form of thermoregulation encompassing a wide range of body temperatures, it is anticipated that the CNS neurons involved in thermoregulation would have a broader thermosensitivity than that of the nonhibernating species. This is indeed true. For instance, single units in the preoptic and anterior hypothalamic areas, which are known to be involved in thermoregulation, show cessation of activity between 28° – 30 °C in the nonhibernators, near 15 °C in animals exhibiting shallow torpor, but close to 5 °C in the hibernators (Wünnenberg et al. 1986). Further, many neuronal models have been proposed to account for the strong seasonal difference in thermoregulatory responses to cold between the nonhibernating and the hibernating phase (South et al. 1978; Heller 1979; Beckman and Stanton 1982). One of these suggests that at least on theoretical grounds, two additional interactive neuronal pools which are active only during the hibernating phase may be responsible in exerting active inhibition or facilitation on thermogenesis in the cold, allowing the occurrence and termination, respectively, of hibernation (South et al. 1978). Although this supposition is weakened by the current lack of a precise definition on the nature and anatomical specificity of these neurons, it is nevertheless consistent with the observation that in the nonhibernating phase, euthermic hibernators show equivalent hypothalamic thermosensitivity to that seen in the nonhibernating species (Wünnenberg et al. 1978) but are capable of exhibiting a wider range of thermosensitivity during the hibernating phase. It is therefore apparent that seasonal alterations in CNS thermoregulatory sensitivity and responsiveness to thermal and chemical challenges may occur and such alterations may provide clues as to the selective activation or depression of CNS neuronal mechanisms governing thermoregulation.

It has been shown electroencephalographic evidence (i.e., EEG) that ground squirrels and marmots enter hibernation through the extension of sleep, in particular, the stage of slow wave sleep (Heller et al. 1978). During slow wave sleep, the body temperature is regulated at a lower level (lowered set-point temperature, Tset) than during wakefulness and the thermosensitivity (magnitude of thermoregulatory response to thermal stimulation) of the hypothalamus is decreased. Consequently, metabolic rate during slow wave sleep is reduced. During entry into hibernation, there is a further, progressive decline in Tset and thermosensitivity, resulting in corresponding decreases of metabolic rate and body temperature (Heller et al. 1978). Electrophysiological recordings from the diencephalon in hedgehogs, *Erinaceus europaeus* (Wünnenberg et al. 1978), indicate suppression of neuronal activities prior to any significant decrease in body temperature during entry into hibernation. Other studies have also shown that during entry into hibernation, suppression of electrical activities occurs first in the neocortex, followed by those in the midbrain reticular formation, and lastly in areas of the limbic system (Wünnenberg et al. 1978; Beckman and Stanton 1982). During hibernation, at a brain temperature of 6.1 °C (Strumwasser 1960), spontaneous electrical activity remains in most, if not all brain areas, but especially in the motor and sensory cortex, medial preoptic area, septum, and the ventromedial nucleus of the hypothalamus although the magnitude of EEG has decreased to about 10% of that observed in euthermia. Of particular interest is the typical electrical silence in the midbrain reticular formation during hibernation; however, upon touching the animal with a glass rod, bursts of activities immediately appear, indicating that this area may be inhibited during hibernation but is capable of responding when stimulated. Although at profoundly depressed body temperature, Tset is still maintained during hibernation: cooling below the lowest Tset or Talarm which may be as low as 1.7 °C in the golden-mantled ground squirrel (Heller et al. 1978) and 1.0° – 3.5 °C in the eastern chipmunk, *Tamias striatus* (Pivorun 1986), elicits increased heat production in a manner qualitatively similar to that seen in the euthermic state. Summarizing the electrophysiological evidence, Heller (1979) and Beckman and Stanton (1982) proposed that increased inhibitory influence on midbrain reticular formation from the hippocampus facilitates sleep and this may be ex-

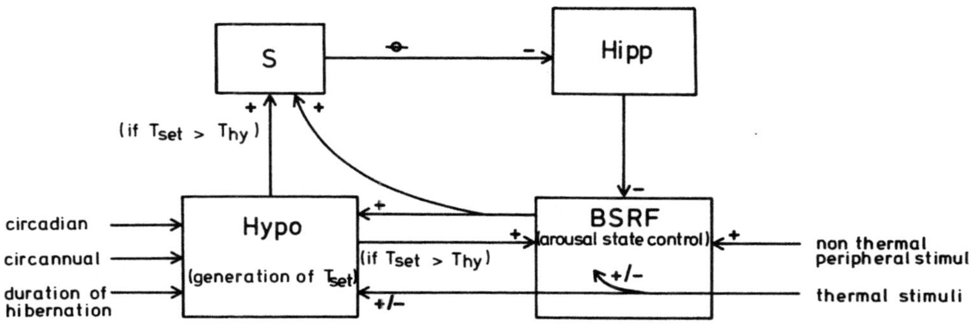

Fig. 2. Possible interactions between the brain stem reticular formation (BSRF), the hypothalamus (Hypo), the limbic system as represented by the septum (S), and the hippocampus (Hipp) in the control of the sleep/wakefulness cycle and the hibernation/euthermia cycle. Excitatory influences are indicated by "+" and inhibitory influences by "−". T_{set} refers to the hypothalamic temperature threshold (set point) for the metabolic heat production response. T_{hy} is the actual hypothalamic temperature. $T_{set} > T_{hy}$ indicates a thermogenic drive. θ represents the activity of theta-rhythm pacemaker cells in the septum. The inputs to the hypothalamic compartment represent influences on the generation of T_{set}. At the onset of sleep, excitatory input form BSRF to Hypo is reduced, resulting in decreased T_{set} and decreased error signal of $T_{set} > T_{hy}$ which further decreases the excitatory influence on BSRF from Hypo (an internal positive feedback loop). Declining activity also decreases excitation to the septum, curtailing O cell activity and disinhibition of the hippocampus. Increasing hippocampal inhibition on the BRSF maintains the sleep state. For entry into hibernation, the T_{set} is lowered further than that for sleep but how this is accomplished remains unknown. Awaking from sleep or hibernation may be initiated by peripheral stimuli or internal influences sufficient to overcome the tonic hippocampal inhibition during sleep or hibernation (Heller 1979)

tended to initiate entry into and the maintenance of deep hibernation. The release of hippocampal inhibition or an activation of the midbrain reticular formation reinstates wakefulness or arousal from hibernation. A schematic representation of this thesis is depicted in Fig. 2.

Although EEG measurements provide evidence that hibernation is an extension of slow wave sleep and that sleep and hibernation are functionally homologous serving to conserve energy through decreased body temperature and metabolism (Walker and Berger 1980; Berger 1984), there is currently no biochemical evidence to suggest that sleep and hibernation are being promoted by the same set of chemical messengers (Krueger and Shoham 1986). Sleep-promoting factors have been chemically defined and include several peptides (e.g., delta sleep-inducing peptide, vasoactive intestinal peptide, arginine vasotocin, and Factor S), polypeptides (e.g., interleukin-1 and interferon), and lipid soluble substances (e.g., prostaglandon D2, exogenous endotoxin, and melatonin) (Krueger and Shoham 1986). In contrast, the hibernation-promoting factors (e.g., antabolone and hibernation induction trigger, HIT; see Sects. 5.1 and 5.2) are still chemically undefined and cannot be directly compared to the sleep promoters. Further, whether the sleep-promoting factors promote hibernation in the hibernators and vice versa are also unknown (Krueger and Shoham 1986). Thus, until such molecular and mechanistic interrelationships are resolved it is difficult to ascertain whether there is true homology (i.e., derived from the same ancestral molecular structures) or simply analogy (i.e., similarity of function but not origin) between sleep and hibernation.

Neurochemically, the brain monoamines are known to be involved in thermoregulation in mammals. Of particular importance is the action of norepinephrine (NE) and serotonin (5-HT) which exert their thermoregulatory effects via the preoptic/anterior hypothalamic neurons (Myers 1980). The possible involvement of these amines plus other neurochemicals in the regulation of hibernation has been documented in several studies. For instance, microinjections of NE, 5-HT, and acetylcholine into the preoptic/anterior hypothalamic area trigger premature arousal in the golden-mantled ground squirrel, whereas only acetylcholine injected into the midbrain reticular formation elicits the same effect (Beckman 1978; Beckman and Stanton 1982). These studies indicate two general aspects: (1) that differential chemical sensitivity of the CNS thermoregulatory neurons is retained during hibernation despite the profoundly depressed body temperature; and (2) that monoaminergic and cholinergic activities are involved in the reactivation of thermogenesis for arousal although the anatomic sites involved may be very specific.

Among the monoamines, the possible role of 5-HT in the induction and maintenance of hibernation has attracted particular attention perhaps because of its possible involvement in the regulation of sleep (Jouvet 1972). In the Syrian (golden) hamster, *Mesocricetus auratus*, intraventricular injection of 5-HT elicits a decrease in body temperature (Jansky and Novotona 1976). In Richardson's ground squirrel, intraventricular 5-HT elicits differential thermoregulatory effects depending on the physiological state of the animal (Glass and Wang 1979a). In the nonhibernating phase, 5-HT increases heat loss in the cold resulting in a slight decrease of body temperature (0.4 °C) before compensatory increase of heat production restores normal body temperature. In the hibernating phase, however, similar injection not only increases heat loss but also suppresses heat production resulting in a much greater decrease in body temperature (1.6 °C). The implication is that an enhanced serotonergic activity during the hibernating phase may facilitate the transition from euthermia to hibernation through inhibition of heat production and enhancement of heat loss. In the red-cheeked ground squirrel, *S. erythrogenys* (Popova and Voitenko 1981), an increased 5-HT level has been found in the hypothalamus, hippocampus, and midbrain prior to entry into hibernation; central injection of 5-HT causes a decrease in heat production and an increase in heat loss possibly via the 5-HT2 receptor (Popova 1986) and may facilitate entry into hibernation. In the Syrian hamster (Novotona et al. 1975), the brain 5-HT turnover rate is some 24 times higher during torpor than it is when active, suggesting increased serotonergic activity associated with the onset of hibernation. In the golden-mantled ground squirrel (Spafford and Pengelley 1971), injection (i.p.) of parachlorophenylalanine or lesioning of the medial raphe nucleus, both serve to reduce the endogenous synthesis of brain 5-HT and result in an interruption of the animal's ability to enter hibernation. This observation has been further amplified in a recent study on the European hamster, *Cricetus cricetus* (Canguilhem et al. 1986), in which only electrolytic lesion of a small area in the anterior part of the medial raphe nucleus prevents the onset of hibernation, whereas electrolytic lesion of other parts of the same nucleus or a general reduction of brain 5-HT by 5,7-dihydroxytryptamine has no effect. This study emphasizes the fact that it is not the absolute brain 5-HT concentration but the intactness of certain specific 5-HT pathways that critically dictates the occurrence or absence of hibernation. Finally, feeding of a tryptophan-rich diet to Syrian hamsters appears to facilitate the occurrence of hibernation, presumably through an increase of substrates for 5-HT synthesis in the brain (Jansky 1978). Taken together, these studies indicate that an increased brain serotonergic activity, especially in a circumscribed area of the medial raphe nucleus, promotes the onset of hibernation. How this heightened serotonergic activity is transmitted into action is currently unknown. In view of the ascending inputs from the medial raphe nucleus to the septum, hippocampus, neocortex, thalamus, and hypothalamus, it may be speculated that the enhanced serotonergic activity may either directly or indirectly affect the hypothalamo-hippocampal inhibition of the midbrain reticular formation (see above) and thus the initiation of sleep and entry into hibernation.

In contrast to 5-HT, brain turnover of NE is essentially nil in the 13-lined ground squirrel, *S. tridecemlineatus* (Draskoczy and Lyman 1967), and the European hedgehog (Sauerbier and Lemmer 1977) during hibernation. In the Richardson's ground squirrel (Glass and Wang 1979b), intraventricular injections of NE increases thermogenesis in the cold in euthermic animals regardless of whether they are in the hibernating or nonhibernating state. Further, intraventricular injections of NE at body temperatures of 10°, 20°, and 30 °C during arousal from hibernation also increases the magnitude of thermogenesis and the rate of rewarming (Glass and Wang 1979b). It therefore appears that brain NE exerts mainly a thermogenic action in these species and the noradrenergic activity is minimal during hibernation but is increased to promote the exit from hibernation at the time of arousal.

3.2 Cardiovascular Functions

During deep hibernation at a body temperature near 5 °C, the heart rate may be 5 – 10 beats min^{-1}, or less than 1/30 of normal and cardiac output may be 1/60 of the euthermic level. The blood pressure is typically maintained at 90/30 mm Hg in comparison to the 130/80 mmHg seen in euthermia. These and other impressive cardiovascular modifications related to hibernation have been extensively reviewed (e.g., South et al. 1972; Hudson 1973; Lyman 1984) and only selected highlights will be presented here.

Studies on the isolated, perfused heart indicate that hearts from species which hibernate (body temperature = 2° – 7 °C typically, e.g., woodchuck, ground squirrel, chipmunk, and hamster) continue to beat at – 0.5° to 7 °C, whereas hearts from those which do not hibernate (tree squirrel, white rat, cotton rat, and mountain beaver) stop beating at 10° to 16 °C (Lyman and Blinks 1959). The left ventricular pressure is maintained between 70 and 110 mmHg and between 22° and 6 °C and the autoregulated coronary flow remains constant down to 6 °C in the isolated, perfused heart of the golden hamster (Jones and Romano 1984). In the 13-lined ground squirrel, the autoregulation of coronary flow and the cardiac output of isolated working heart persist down to at least 7 °C but both are abolished in the rat below 17 °C (Burlington et al. 1986). Similarly, the arotic smooth muscle from the 13-lined ground squirrels can maintain a normal transmembrane K^+ and Na^+ gradient down to 7 °C for 48 h, whereas in the rat, half-times for loss of intracellular K+ and gain of Na+ are 14 and 11 h, respectively, at the same temperature (Kamm et al. 1979). In the hedgehog, the cardiac action potential lacks a plateau phase in contrast to the prominent plateau phase found in the guinea pig (Duker et al. 1986), indicating a difference in trans-sarcolemmal Ca^{2+} flux during depolarization. The hedgehog's heart also shows a negative staircase (treppe) in its force-frequency response in contrast to the positive staircase observed in the guinea pig (Duker et al. 1986). A negative staircase phenomenon has also been found in the hearts of other mammalian hibernators at least during certain times of the year, for instance, the Richardson's ground squirrel in December – January (Charnock et al. 1983; Zhou et al. 1987) and the Asiatic chipmunk during the hibernating phase (Kondo and Shibata 1984). Interestingly, it is also found in a nonhibernator, the rat. Since a negative staircase is considered to be anomalous among mammals and is a reflection of greater reliance on intracellular Ca^{2+} stores than extracellular Ca^{2+} for contraction (Langer 1980), it remains to be investigated whether a reduced dependence on extracellular Ca^{2+} in excitation-contraction coupling is of adaptive significance for maintenance of cardiac function under prolonged and profound depression in body temperature. In the ground squirrel atrial preparation, both the membrane potential and action potential are maintained between 37° and 6 °C (Marshall and Willis 1962), whereas in the nonhibernating species both potentials decline rapidly below 20 °C (Marshall 1957). These species differences between the hibernators and nonhibernators on cardiovascular performance at low temperature point to the existence of cellular and subcellular adaptations in the hibernators. Although the specific mechanisms are not yet fully elucidated, they are likely to involve aspects on ionic regulation, excitation-contraction coupling, mitochondrial bioenergetics, and membrane lipid-protein interaction. These will be discussed in detail in later sections (Sects. 4.2 and 4.3).

In addition to species differences, there also exists a seasonal difference in cardiac performance within the same species (presumably, in the same individual, too). For instance, in the hibernating hamster, the resting membrane potential is maintained at 12 °C but is significantly decreased below 20 °C in the nonhibernating hamster (Jacobs and South 1976). The cardiac trabeculae also show greater excitability, peak tension, and intensity of active state between 38° – 5 °C in the hibernating hamster than it is in the nonhibernating hamster (South and Jacobs 1973). These observations indicate that transitional changes occur at the cellular level in cardiac functions during the cold-acclimating period which is typically required by the hamsters prior to their exhibition of hibernation. In the Asiatic chipmunk (Kondo and Shibata 1984), hibernation is associated with a reduced reliance on extracellular Ca^{2+} for myocardiac contraction. This is evidenced by (1) a reduced, slow inward Ca^{2+} current; (2) a shift from a positive to negative staircase in a force-frequency response; and (3) ryanodine, an inhibitor of sarcoplasmic reticulum (SR) Ca^{2+} release, abolishes tension development in papillary muscle from the hibernating animal but only partially reduces that in the nonhibernating animal. In the Richardson's ground squirrel, myocardial contractility to external Ca^{2+} is diminished during hibernation

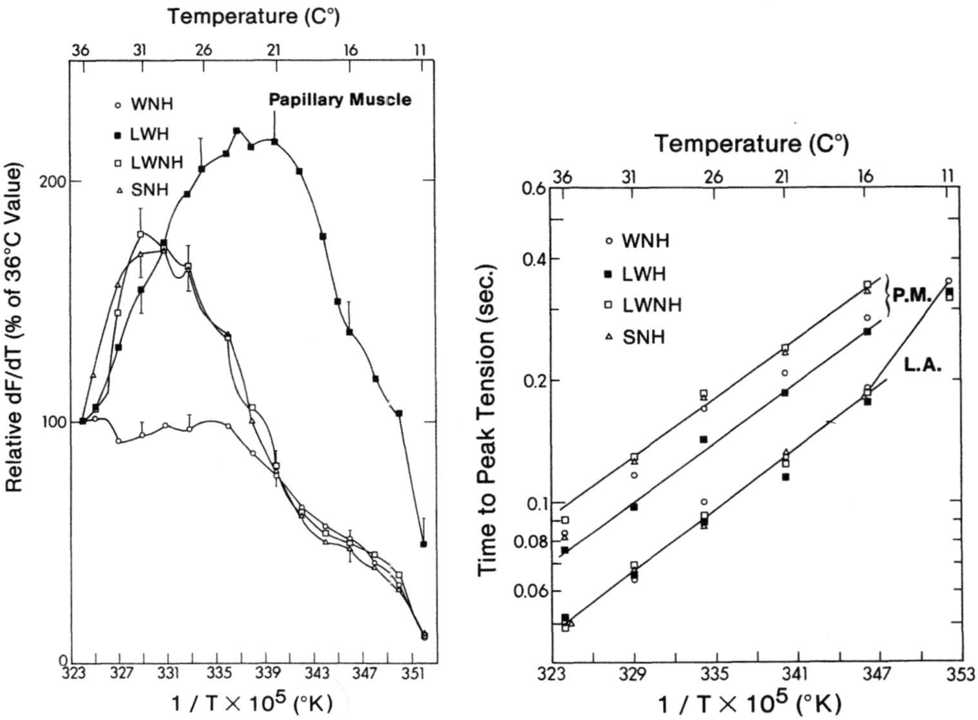

Fig. 3. Seasonal difference in the intensity of the active state (dF/dT; *left panel*) and duration of active state (*right panel*, P.M.) in the papillary muscle of the Richardson's ground squirrel. LWH (■) denotes late winter hibernating animals (sacrificed in mid-February to April; n = 7); WNH (○) are winter nonhibernating animals (December to mid-February; n = 10); LWNH (□) are late winter nonhibernating animals (mid-February to April; n = 12); and SNH (△) are spring nonhibernating animals (June; n = 8). Note the significantly greater intensity (below 26 °C) but shorter duration of the active state in the hibernating state (LWH) vs those in the nonhibernating state (LWNH) even though both groups of animals were sacrificed at the same time of year. This indicates manifestation of physiological differences at different phases of the endogenous circannual rhythm for hibernation.

(Charnock et al. 1980a), also indicating a reduced trans-sarcolemmal Ca^{2+} influx during excitation-contraction coupling. However, since in the same species, the electrically stimulated twitch tension of papillary muscle is significantly elevated during hibernation over that seen in the nonhibernating state (Fig. 3; Zhou, Dryden and Wang, unpubl.), it is apparent that a greater reliance on the intracellular Ca^{2+} source must have also occurred similar to that seen in the hibernating Asiatic chipmunk. Significantly, the enhanced myocardial contractility during hibernation, especially that at the lower temperatures, is achieved through a shorter duration but greater intensity of the active state than those seen in the nonhibernating state. This indicates a state-dependent change in intracellular Ca^{2+} regulation (i. e., sequestration, binding, and uptake) in association with hibernation (Zhou, Dryden, and Wang, unpubl.). The simplest explanation would be that a greater rate and amount of Ca^{2+} release from the SR are responsible for the enhanced contractility seen during hibernation. If so, and in order to quickly restore the normal intracellular Ca^{2+} level after a high, transient Ca^{2+} release, the rate of ATP-dependent Ca^{2+} uptake by the SR should also be enhanced during hibernation. This is indeed true as a 1.8 – 2.7-fold increase in SR Ca^{2+} uptake between 37° and 4 °C has been observed in the hibernating state over that in the nonhibernating state (Belke et al. 1987). Whether this seasonally increased rate is due to a greater amount of the same or variant Ca^{2+} pump protein, greater Ca^{2+}/ATP coupling ratio in Ca^{2+} transport, or SR membrane lipid-protein interaction remains to be investigated.

A similar seasonal difference in vascular smooth muscle function has also been observed. For example, in the euthermic woodchuck, *Marmota monax* (Miller et al. 1986), the sensitivity to NE-stimulated tension generation in strips of the thoracic aorta and the portal vein is highest in the spring as compared to that of summer and winter, although the maximum response is constant year-round. In the hibernating state, tissue-specific differences in sensitivity to NE stimulation are evident: the renal arterial strip shows an increased sensitivity, whereas the aortic and femoral strips show no difference as compared to those observed in the warm- and cold-acclimated woodchucks. A decrease in temperature from 37° to 28 °C increases tension development to NE in the aortic and renal strips but not that of the femoral strip. Whether these differential sensitivity and magnitude changes reflect the specific perfusion requirements during entry, maintenance, and arousal from hibernation remains to be evaluated.

3.3 Respiratory Functions

During entry into hibernation and in deep hibernation, there is significant reduction in ventilation due to the reduction of oxygen demand. The pattern of breathing during hibernation is different in different species but is typically one of two types (Fig. 4, upper part). The first type, as seen in the marmot (*M. marmota*) and the Columbian ground squirrel (*S. columbianus*), is represented by single breaths separated by periods of apnea lasting from 1 – 6 min. The second type, as seen in the hedgehog and the golden-mantled ground squirrel, is characterized by bursts of continuous breathing separated by longer periods of apnea (Cheyne-Stoke breathing) lasting sometimes up to 45 to 150 min (Malan 1982; Steffen and Riedesel 1982; Milsom et al. 1986; Riedesel et al. 1986). However, in certain species, such as the golden-mantled ground squirrel (Milsom et al. 1986), Cheyne-Stoke breathing may be switched to continuous single breathing simply by decreasing the ambient temperature for hibernation from above 5° to below

Fig. 4. Upper part: Ventilatory patterns of Columbian and golden-mantled ground squirrels as measured by plethysmograph pressure differentials (awake at ambient temperature of 20 °C) and pneumotachograph air flow (hibernating at ambient temperature of 5 °C). Note the bursting pattern of ventilation (Cheyne-Stoke) in the hibernating golden-mantled ground squirrel. *Middle part*: Effects of changing ambient temperature from 7° to 4 °C and back to 6.5 °C on the ventilatory pattern of a hibernating golden-mantled ground squirrel. *Lower part:* Absolute (left) and relative (right) ventilatory responses of euthermic and hibernating ground squirrels to varying levels of inspiratory CO_2 (F_{ICO_2}) (Milsom et al. 1986)

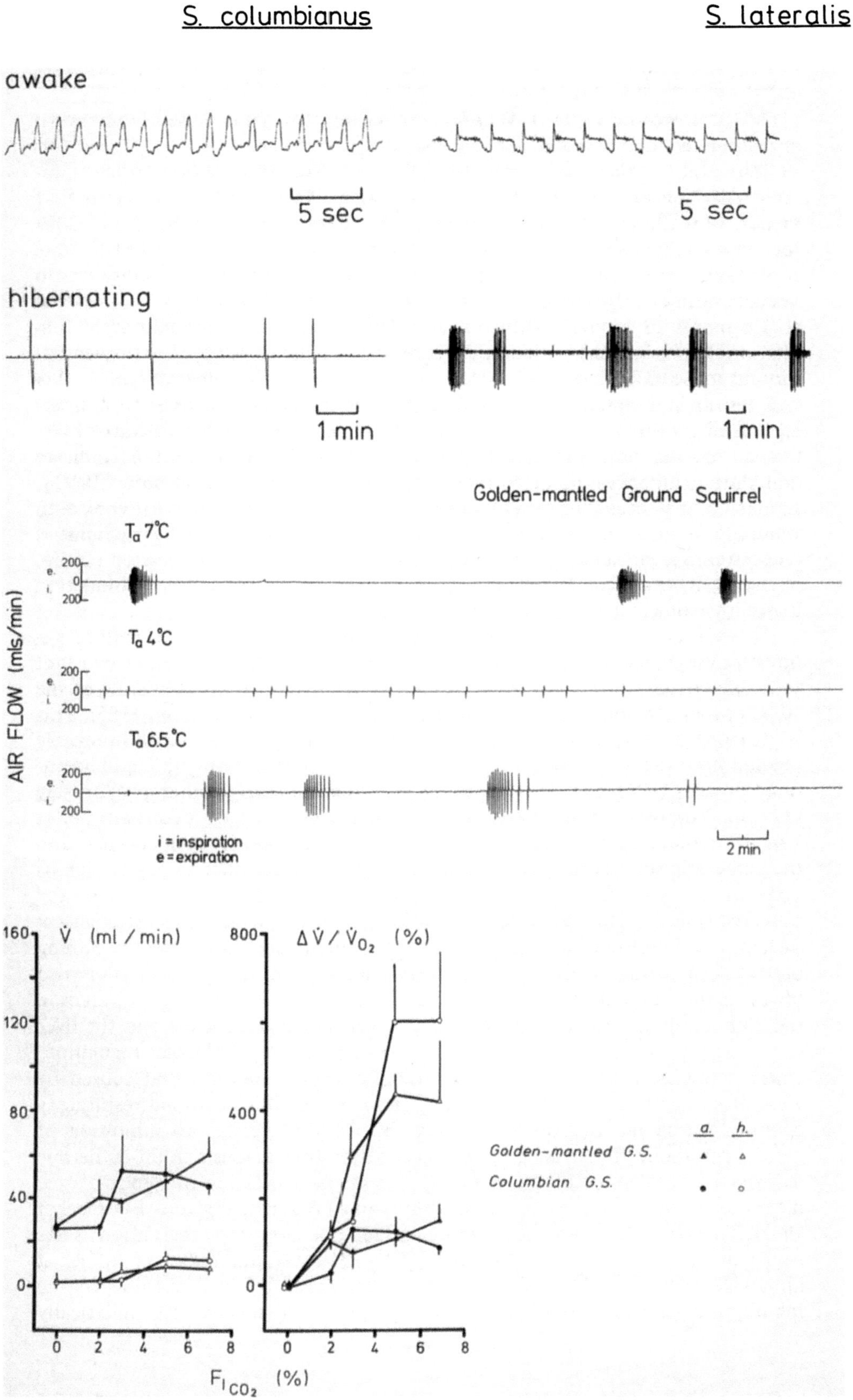
S. columbianus
S. lateralis
awake
5 sec
5 sec
hibernating
1 min
1 min
Golden-mantled Ground Squirrel
T_a 7°C
T_a 4°C
T_a 6.5°C
AIR FLOW (mls/min)
200
0
200
e.
i.
i = inspiration
e = expiration
2 min
$\dot{V}$ (ml / min)
$\Delta \dot{V} / \dot{V}_{O_2}$ (%)
160
120
80
40
0
800
400
0
2
4
6
8
$F_{I_{CO_2}}$ (%)
a.
h.
Golden-mantled G.S.
Columbian G.S.

5 °C (Fig. 4, middle section). What determines this switch is presently unknown but is unlikely due to an increased hypoxia drive (see below).

In the hibernating golden-mantled ground squirrel, the end tidal oxygen concentration is near 15% and that of CO_2 near 3 – 4% in the first three to four breaths of a Cheyne-Stoke breathing episode. These values change to 19 – 20% for oxygen and less than 1% for CO_2 in the subsequent 10 – 12 breaths (Riedesel et al. 1986). These values indicate a very competent ventilatory exchange. As to the respiratory "drives" for ventilation, breathing of 100% nitrogen (Riedesel et al. 1986) or 1 – 3% oxygen (Milsom et al. 1986) results in no stimulation of ventilation. Similarly, denervation of the carotid sinus also does not alter the ventilatory response to hypoxia (Milsom et al. 1986). In contrast, breathing of 5 – 7% CO_2 results in elevated ventilation (Fig. 4, lower part) and a change from intermittent to continuous breathing (Milsom et al. 1986) although breathing of 4% CO_2 has no stimulatory effect (Riedesel et al. 1986). These observations indicate that during hibernation, the primary "drive" for ventilation is blood PCO_2, similar to that seen during euthermia. The inability of systemic hypoxia to stimulate ventilation indicates that either the effect of this sensory input is reduced or absent during hibernation or that central hypoxic depression occurs, rendering it incapable of responding to the peripheral ventilatory stimulation under hypoxic challenge (Milsom et al. 1986).

In addition to the typical reductions in ventilation during hibernation, another physiologically unique trait is the maintenance of a relatively constant extracellular pH (near 7.40) between euthermia and hibernation in spite of the 30 °C or more change in body temperature (Malan 1982; Malan et al. 1985). This is in strong contrast to the maintenance of constant alkalinity (alpha imidazole regulation) rather than constant pH in the ectotherms under fluctuating temperatures (Reeves 1977). Since the neutrality of water increases from approximately pH = 6.90 at 37 °C to pH = 7.40 at 5 °C, the maintenance of a constant pH at 7.40 during hibernation represents a strong, relative acidosis when compared to the euthermic state. The acidosis is respiratory in nature since PCO_2 as high as 150 torr (Vs 40 – 60 torr in euthermia, both corrected to 37 °C) is typically observed (Fig. 5). This respiratory acidosis is likely due to an accumulated effect of hypoventilation during entry into hibernation and the intermittent breathing pattern seen during hibernation. The intracellular pH during hibernation is also low, ranging from 6.8 in the brain to 7.3 in the liver, with intermediate values in the diaphragm and skeletal muscle (Malan 1982). Only the heart and the liver show partial metabolic compensation toward alpha imidazole regulation, whereas the brain shows a further metabolic acidosis beyond that caused by respiratory acidosis. The physiological significance of this profound respiratory acidosis during hibernation is uncertain but may be related to inhibition of certain metabolic functions. For instance, respiratory acidosis in an euthermic hamster induced by breathing 10% CO_2 results in depression of hypothalamic neuronal activity involved in thermoregulation (Wünnenberg and Baltruschat 1982). This could facilitate the lowering of Tset and entry into hibernation as has been demonstrated in hamsters under 5% CO_2, 21% O_2, and 74% N_2 (Kuhnen et al. 1983). Further, glycolysis is severely depressed by acidosis due to the inhibition of phosphofructokinase by low pH, which dissociates the catalytically

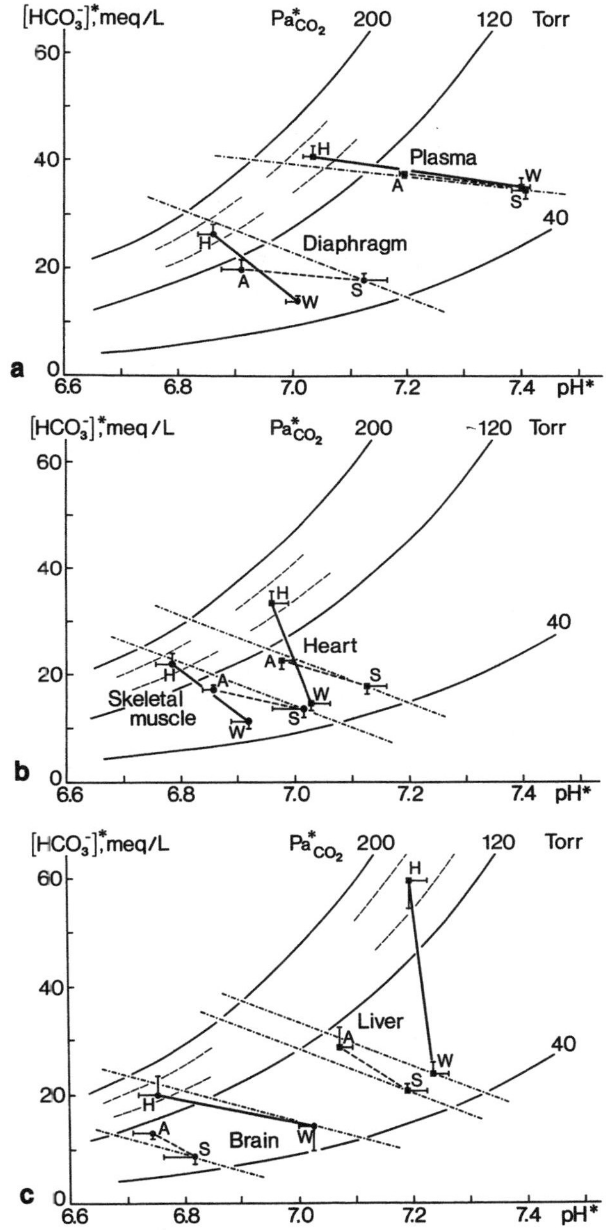

Fig. 5 a – c. Bicarbonate pH diagrams for blood plasma and intracellular pH for diaphragm, heart, skeletal muscle, liver, and brain in the European hamster in spring under normocapnia (*S*) and respiratory acidosis (*A*) and in winter under euthermia (*W*) and hibernation (*H*). *Asterisks* indicate that all values have been corrected to 37 °C. A respiratory acidosis is indicated by the high $Pa*CO_2$ and low pH. Uncompensated respiratory acidosis is seen in plasma in both hibernation and euthermic hypercapnia and in the brain during hibernation. This is evidenced by their predicted changes in $Pa*CO_2$ and pH following the extra- and intracellular fluid buffer lines. Partial metabolic compensation of respiratory acidosis is seen in the diaphragm, skeletal muscle, and heart during hibernation. Nearly perfect metabolic compensation is seen in the liver during hibernation (Malan et al. 1985)

active tetrameric form into the catalytically inactive dimeric form (Hand and Somero 1983). This may conserve liver and muscle carbohydrate reserves while undergoing long-term fasting during hibernation. The norepinephrine-stimulated nonshivering thermogenesis in the brown adipose tissue is also significantly depressed by acidosis perhaps due to the increased binding affinity by low pH of the mitochondrial GDP-binding protein to its ligands (Malan 1986). Because of these inhibitory effects by low pH, it has been suggested that a metabolic suppression beyond the level dictated by low temperature alone may exist during hibernation (Malan 1986). Since respiratory acidosis is easily correctable by hyperventilation, the inhibition of neuronal and metabolic functions can be quickly reversed. In this regard, it is significant that during the early stages of arousal, when thermogenesis is required, hyperventilation has been observed in the marmot resulting in a decrease in blood PCO_2 and an increase in pH (Malan 1982). It is currently unknown, however, how hyperventilation is suddenly initiated at the time of arousal while prior to this, the hibernating animal has been quite tolerant to the prevailing acidotic state without ventilatory compensation.

3.4 Heat Production

During hibernation, the metabolic rate may decrease to 1/100 or less of the euthermic level measured at the same ambient temperature. In addition to the possibility of acidosis suppressing metabolism as described above, certain other cellular features may also contribute to the reduction in metabolism beyond that dictated by low body temperature alone. In the Richardson's ground squirrel (Pehowich and Wang 1984), the absolute rate of succinate-supported mitochondrial respiration is significantly depressed by 50 – 70% during hibernation at all assay temperatures between 4° and 37 °C as compared to the values found in summer or winter euthermic animals. Interestingly, this depression is completely released after arousal resulting in indistinguishable or even higher rates of respiration than those found in summer and winter euthermic animals. Similar depressions of mitochondrial respiration during hibernation have also been reported in hamsters and other species of ground squirrels (see Pehowich and Wang 1984; Fedotcheva et al. 1985; Gehnrich and Aprille 1986). It is apparent that the suppression of mitochondrial respiration during hibernation is not due to the prevailing acidotic pH since all preparations are assayed at the same pH of 7.40 for all temperatures (Pehowich and Wang 1984). Many other possible mechanisms have been suggested but no definitive consensus has been reached to account for the depression of mitochondrial respiration during hibernation. These include the limiting role of mitochondrial nucleotide permeation (see Pehowich and Wang 1984), inhibition of succinate oxidation by oxaloacetate and a decreased mitochondrial K^+ transport (Fedotcheva et al. 1985), and a decreased respiratory chain activity in the region of the coenzyme Q-cytochrome bC1 (Gehnrich and Aprille 1986).

Two types of heat production are involved during arousal from hibernation: shivering and nonshivering thermogenesis (NST). Shivering is involuntary and is manifested by rhythmic, random, and uncoordinated but simultaneous contractions of the flexors and extensors. The primary center for regulating shivering is in the dorsomedial area of the hypothalamus (Hemingway 1963). The activation of shivering appears to be determined by both the skin and the spinal cord temperature in nonhibernators such as the guinea pig (Brück and Wünnenberg 1970), but in the golden hamster, spinal cord temperature plays no role in enhancing or depressing shivering (Wünnenberg et al. 1986). This latter observation may have functional significance during arousal from hibernation since the warm blood from brown fat, although warming the spinal cord, does not lead to inhibition of shivering so that heat produced by shivering will continue to contribute until rewarming is complete (Wünnenberg et al. 1986).

Nonshivering thermogenesis is a heat-producing mechanism without muscle contraction. The major site of NST in mammals is the brown adipose tissue (BAT) which is well developed in neonates, many small mammals after cold-acclimation or winter acclimatization and in particular, adults of species which exhibit hibernation (Jansky 1973). The BAT is characterized by its multilocular lipid droplets, numerous mitochondria and high cytochrome oxidase activity, extensive vascularization, and rich sympathetic innervation. In the cold-acclimated rat although the total amount of BAT is less than 2% of body weight, it can account for 65 – 80% of the total thermogenesis during cold exposure (Foster and Frydman 1979). This extraordinary thermogenic capability of the BAT is due to its biochemical uniqueness which allows extensive uncoupling of oxidative phosphorylation resulting in the production of heat instead of ATP. A BAT-specific inner mitochondrial membrane protein of 32,000 daltons, which has specific binding sites for purine nucleotides (e.g., GDP, GTP, ADP, ATP; the GDP-binding protein), has been identified as the key molecule which forms a part of the natural protonophore of the inner mitochondrial membrane (Cannon and Nedergaard 1983; Nicholls and Locke 1984). When bound by GDP, the proton channel is impermeable to proton flow and the proton gradient generated by the electron transport chain provides the energy needed for ATP synthesis (the coupled state). When the BAT is stimulated by sympathetic discharge, NE, via the beta-receptor, adenylate cyclase, cAMP pathway activates the lipase and leads to lipolysis and the formation of fatty acids and acyl-CoA. The fatty acids or acyl-CoA competitively binds to the GDP-binding protein and displaces GDP. Consequently, the proton channel becomes permeable to proton flow and a proton short-circuit is formed which dissipates the energy gradient into heat instead of ATP synthesis (the uncoupled state). The metabolism of fatty acids sustains this uncoupled state.

The proportional contribution of shivering and NST during arousal from torpor is different among different species. In the little brown bat (*Myotis lucifugus*), 80% of heat production comes from NST (Hayward 1968), whereas in the hamster and dormouse (*Glis glis*), approximately 40 and 20%, respectively, are derived from this source (Hayward and Lyman 1967). In a recent study (Lyman and O'Brien 1986), surgical removal of up to 81% of their normal BAT does not affect the rate of rewarming in the Turkish hamster (*M. brandti*), 13-lined and the golden-mantled ground squirrels. This implies that the contribution of heat production by BAT is probably minor in these rodents for arousal from hibernation. However, since peak heat production during arousal from hibernation in some ground squirrels is only about 60 – 70% of their maximum thermogenic capability elicited by severe cold (Wang and Abbotts 1981), there remains the possibility that shivering thermogenesis could have increased to compensate for the heat deficit created by the removal of the BAT. In this regard it is interesting to note that in the Richardson's ground squirrel, the maximum thermogenic capacity does not vary seasonally but the NST is, however, significantly elevated from 20.6% of total thermogenesis in animals sampled in June to 45.6% sampled in the hibernating phase (Abbotts and Wang 1980). The significance of this seasonal shift is presumably to aid heat production during the early stages of arousal when conditions for shivering thermogenesis may not have been fully optimized due to low muscle temperature and restriction of blood flow to some of the large muscle masses in the posterior part of the body. If one assumes 60 – 80% of NST is from the BAT (Foster and Frydman 1979) and 80% of BAT is removed from the ground squirrels prior to hibernation as is done in the study of Lyman and O'Brien (1986), the expected reduction of contribution by NST to total heat production would then be $0.456 \times 0.8 \times 0.8 = 0.29$ or 29%. Thus, on theoretical grounds, this heat deficit could be still compensated for by enhanced shivering thermogenesis resulting in no major difference in rewarming rate during arousal. It is obvious that a definitive answer cannot be provided as to the quantitative importance of NST by BAT during arousal in these reported species (Lyman and O'Brien 1986) until a similar analysis has been performed incorporating the quantitative aspects of their maximum thermogenesis and the proportional contribution of shivering and NST.

As to the status of NST by BAT within each hibernation episode, there is a significant reduction in the magnitude of GDP binding during hibernation in the golden hamsters, indicating that the NST by BAT is depressed in this state (Horwitz et al 1986). However, GDP binding is quickly reversed to the higher, euthermic level within 90 min of arousal even though the BAT temperature at this time has only increased from 8° to 18 °C (Horwitz et al. 1986). This indicates either an unmasking and/or an activation of existing binding sites for GDP rather than synthesis of new sites on the GDP-binding protein. This unmasking of binding sites may be mediated by NE since incubation of BAT isolated from the hibernating squirrel with NE at 7 °C can reverse the reduction of GDP binding found during hibernation (Horwitz et al. 1986). There is also evidence that in the golden hamster, lipids from the BAT is preferentially utilized during arousal since no measurable loss of lipids from the white adipose tissue is detected following arousal (Nedergaard and Cannon 1984). The lipids from the BAT may either be combusted in situ, or may be exported and combusted elsewhere (e.g., shivering muscle). Based on the quantity of lipids lost from the BAT, a major contribution (up to 90 – 100%) to total heat production could be derived from this substrate for both shivering and NST (Nedergaard and Cannon 1984).

4 Cellular, Subcellular, and Membrane Adaptations

4.1 Seasonal Variation of Metabolism and Enzyme Activities

4.1.1 Lipid Metabolism

The characteristic annual fat cycle of the hibernators (Mrosovsky 1983) indicates seasonal changes in neuroendocrine regulation of food intake, energy metabolism, and activities of enzymes involved in lipogenesis. The incorporation of ^{14}C-glucose into adipose tissue lipids increases some 88 – 108-fold between June and August in the juvenile Richardson's ground squirrels in preparation for hibernation under field conditions (Bintz and Strand 1983). In the Richardson's (Melnyk and Martin 1985) and arctic ground squirrels (*S. undulatus*; Feist et al. 1986) and the marmot, *M. flaviventris* (Florant et al. 1985; Florant and Greenwood 1986), fall fattening is associated with hyperinsulinemia, hypertriacylglycerolemia, peripheral insulin resistance, and increased lipoprotein lipase activity in the white adipose tissue. However, these metabolic changes appear to be dictated by an endogenous rhythm rather than consequences of hyperphagia and weight gain. For instance, in the marmot (Florant and Greenwood 1986) hyperinsulinemia and peripheral insulin resistance are observed in September even though the animals' body weights are comparable to those of June. Similarly, food restriction, although preventing an anticipated weight gain in September in the arctic ground squirrel (Feist at al. 1986), does not prevent the occurrence of hyperinsulinemia, hypertriacylglycerolemia, and apparent fattening (i.e., increased proportion of fat to a constant or decreasing body weight). Thus, physiological signals which trigger lipogenesis may be generated in the absence of weight gain and hyperphagia. One of such signals may be insulin. It has been shown in the Richardson's ground squirrel (Melnyk and Martin 1985) that insulin binding by hypothalamic tissues is greater in animals in the weight-gain phase than it is in the weight-loss phase. Also, maximum insulin binding in the hypothalamus is positively correlated with the insulinosecretory responsiveness of the pancreatic islets to glucose and with maximum insulin-stimulated lipogenesis (Melnyk and Martin 1985). These observations suggest that insulin may exert a central effect on food intake and appetite regulation and that circulating insulin could act on the CNS via the arcuate nucleus of the hypothalamus. In addition, elevated insulin is also known to stimulate the activity of the lipoprotein lipase in the white adipose tissue (Florant and Greenwood 1986) and thus the supply of lipogenic substances (free fatty acids) for this tissue.

An interesting question related to the annual weight cycle is the role of the BAT as an energy buffer at the time of weight gain. It has been shown in rats that hyperphagia does not necessarily lead to obesity because of increased NST by the BAT disposing the excessive caloric intake (Himms-Hagen 1984). As ground squirrels only become hyperphagic and fattened prior to hibernation but maintain a very steady weight at other times of the year, it might be predicted that BAT may become inactivated or atrophied during the weight-gain phase. In the arctic ground squirrel (Feist et al. 1986), the weight of BAT is increased fourfold, total mitochondria protein eightfold, mitochondria protein per unit of

BAT twofold, and total GDP binding ninefold. These changes indicate a BAT hypertrophy and increased thermogenic capacity during the weight-gain phase in preparation of hibernation quite contrary to that predicted. This study, however, provides the cellular evidence for our in vivo observations in the Richardson's ground squirrel on seasonal changes of NST and weight cycle, namely: (1) during the weight-gain phase, the NE-stimulated NST is comparable to that seen in the weight-stable phase, suggesting patency of sympathetically stimulated thermogenesis of the BAT (Wang and Lee, unpubl. observations); and (2) that in the hibernation phase, the NE-NST is highest, suggesting hypertrophy of the BAT (Abbotts and Wang 1980). Taken together, these observations indicate that during the weight-gain phase, BAT hypertrophy occurs but thermogenesis is apparently not activated to dissipate the excess caloric intake following hyperphagia. As activation of the BAT depends on sympathetic stimulation, it is logical to suspect that sympathetic activity to the BAT is temporarily depressed to allow maximum weight gain. Preliminary experiments on BAT NE turnover as a measure of sympathetic activity indicate that this is indeed the case (Wang and Lee, unpubl. observations). How this temporary reduction of sympathetic activity is brought about and why such a reduction in sympathetic activity does not lead to a negative trophic effect on BAT hypertrophy as would be predicted based on studies in the nonhibernators (Mory et al. 1982) are interesting questions and need to be answered.

In the marmot (Florant and Greenwood 1986), activity of the lipoprotein lipase from the white adipose tissue is ten times higher during fall fattening than it is in summer. However, only a 50% increase is seen in this enzyme activity in the arctic ground squirrel during its weight-gain phase (Feist et al. 1986). In the dormouse kept under natural photoperiod and temperature (Castex and Sutter 1981), the activity of the white adipose tissue glucose-6-phosphate dehydrogenase, a key enzyme in the pentose phosphate shunt which generates NADPH for lipogenesis, is six times greater in the fall than it is in the spring and summer. The insulin-stimulated glucose utilization by the adipocytes is also highest during the fall. Interesingly, this seasonal increase in insulin-stimulated lipogenesis is manifested in spite of a decrease in insulin receptor number and no change in binding affinity. It is thus apparent that the postreceptor increase in enzyme activity has more than sufficiently compensated for the decrease in insulin receptor number in insulin-stimulated lipogenesis from glucose (Castex and Sutter 1981). In another study on the dormouse but kept under 12L:12D photoperiod and 22 °C (Melnyk et al. 1983), the insulin-stimulated glucose transport, oxidation, and lipogenesis are highest during the weight-gain phase and lowest during the weight-loss phase. The loss of insulin sensitivity for lipogenesis is also due to a postreceptor defect, but in addition, a decrease in insulin receptor number. In the 13-lined ground squirrel (Whitten and Klain 1969), the activities of liver enzymes involved in lipogenesis are twice as great in June as they are in September when annual fattening occurs.

As the onset of hibernation typically follows peaking of the body weight, it is not surprising that lipogenesis becomes depressed during the hibernation season. This is evidenced by the greatly decreased glucose-6-phosphate dehydrogenase activity in the liver of the 13-lined ground squirrel (Whitten and Klain 1969) and the adipose tissue of the hedgehog (Olsson 1972), and the reduced incorporation of ^{14}C-glucose into total lipids in the adipose tissue of the hibernating golden-mantled ground squirrel (Tashima et al. 1970). With respect to energy utilization during hibernation, fat is the primary fuel. In the hibernating big brown bat, *Eptesicus fuscus* (Yacoe 1983a), fat metabolism supports nearly all the energy needs and glucose oxidation is severely inhibited; muscle mitochondria preferentially oxidize fatty acids over pyruvate and physiological concentrations of palmitoyl-carnitine inhibits pyruvate oxidation. In addition to the mitochondrial aspect, the inhibition of glucose oxidation during hibernation could also be due to depression of glycolysis due to inhibition of phosphofructokinase by the combination of low intracellular pH and low temperature prevailing during hibernation (Hand and Somero 1983). The preferential utilization of fat results in ketosis and increased plasma ketone bodies have been observed during hibernation in the arctic (Rauch and Behrisch 1981) and in the Belding's ground squirrels, *S. beldingi* (Krilowicz 1985). At a plasma ketone body level of 1.0 – 1.7 mM, uptake of glucose is reduced by 60 and 100%, respectively, in the heart and pectoralis. Oxidation of the ketone body supplies 43, 21, and 31%, respectively, of the energy requirement of the heart, pectoralis, and diaphragm during hibernation (Krilowicz 1985). The consequence of these collective changes observed in hibernation is glucose sparing. Since the hibernating animal is typically in a fasting state (for those species which do not feed between hibernation bouts), glucose sparing also conserves tissue proteins since they provide the major gluconeogenic precursors in addition to glycerol and lactate. The importance of tissue protein preservation is easily perceivable as not only motility of the animal relies on muscle proteins following spring emergence but heat production by shivering is also an essential function of muscles during periodic arousal. As will be seen from the discussion below that despite these glucose and protein sparing measures, significant loss of tissue proteins still occurs, nevertheless, during hibernation.

4.1.2 Glucose Metabolism

Although the primary fuel during hibernation is fat, the central nervous system continues to utilize glucose especially in areas receiving thermal afferants (e.g., the partrigeminal nucleus) as illustrated by the tracer study utilizing [^{14}C] 2-deoxyglucose (Kilduff et al. 1983). During periodic arousal, intense shivering also requires glucose oxidation. The ability to maintain normal glycemia during hibernation has been construed as a key adaptation in surviving prolonged depression in body temperature (Musacchia 1984). In species which feed between hibernation bouts, glucose can be easily replenished. However, in species which rely on depot fat to last the whole hibernation season, glucose sparing, and gluconeogenesis from amino acid precursors, glycerol, and lactate are the only recourses. In addition to the glucose sparing measures described above, it is reasonable to expect that gluconeogenic activities should be enhanced during the hibernation season. Significantly greater gluconeogenic capacities have been observed in liver (Whitten and Klain 1968; Petrovic et al. 1985) and kidney cortex slices (Burlington and Klain 1967; Green et al. 1985) of hibernating ground squirrels as compared to their summer active counterparts. In the arctic ground squirrel, an elevation in phosphoenolpyruvate carboxylase, a key enzyme involved in hepatic gluconeogenesis is observed during hibernation, presumably favoring gluconeogenesis during this state (Behrisch et al. 1981). In the arctic ground squirrel (Fedotcheva et al. 1984), a rotenone, antimycin-insensitive pathway in liver mitochondria for the oxidation of NADH exists only in the hibernating state. The functional significance of this pathway is thought to provide an alternate means for the continuing production of NAD, which is needed for gluconeogenesis during hibernation, when NADH from lipolysis cannot be sufficiently oxidized by the normal mitochondrial respiratory chain due to inhibition of the dehydrogenase by low body temperature (Fedotcheva et al. 1984).

In the arctic ground squirrel (Galster and Morrison 1975), glycerol released from triglycerides accounts for two-thirds of the replenishment of carbohydrate reserves depleted during hibernation and arousal, whereby amino acid precursors account for less than 20% of total gluconeogenesis. On the other hand, a decrease in tissue protein with the progression of hibernation is typical for many other species: for instance, the 39% reduction in skeletal muscle, 12% in heart, and 22% in liver protein content in the hibernating golden-mantled ground squirrel (Tashima et al. 1970), and the 47% decrease in total muscle protein in the hibernating big brown bat (Yacoe 1983b). In the latter, liver and pectoris muscle protein synthesis and degradation are essentially ceased during torpor but rates of protein degradation increased significantly during periods of arousal (Yacoe 1983c), indicating that tissue protein catabolism accounts for the major supply of gluconeogenic precursors in this bat.

In terms of glucose utilization, seasonal variants of liver pyruvate kinase have been identified in the arctic ground squirrel (Behrisch 1978); the hibernating form has a higher isoelectric point, higher temperature sensitivity in V_{max} (Q10), and a much lower temperature dependency in its binding affinity to allosteric substrates (phosphoenolpyruvate and ADP), activator (fructose diphosphate), and inhibitor (ATP) than those found in the nonhibernating form. Similar findings have also been reported in the flight muscle and liver pyruvate kinase in the hibernating little brown bat (Borgmann and Moon 1976). The significance of the temperature-insensitive binding affinities to ligands and the high Q10 in V_{max} of the hibernating enzymes is that as body temperature decreases during entry into hibernation, the glycolytic flux can be greatly suppressed, whereas as body temperature rises during arousal, glycolytic flux can be greatly increased.

In species which feed between hibernation bouts (e.g., Syrian hamster), the blood glucose level during hibernation is maintained at a similar level as found in euthermia mainly through glycogenolysis from the liver (Musacchia and Deavers 1981). Prehibernation cold exposure in Syrian hamsters results in inhibition of lipogenesis in the liver (Denyes and Carter 1961) but a four- to sixfold increase in the in vitro capacity of lipogenesis from acetate in the white adipose tissue (Baumber and Denyes 1963). Further, this increased lipogenic capacity is retained during hibernation unlike the inhibition of lipogenic capacity with the onset of hibernation in ground squirrels and hedgehogs described above. During arousal from hibernation, hamsters utilize predominantly carbohydrate as an energy source as reflected by the profound depletion of liver and muscle glycogen in the late stages of arousal (Lyman and Leduc 1953). This is in contrast to the use of mainly fat but some carbohydrate in ground squirrels arousing from hibernation.

As to the regulation of carbohydrate metabolism during hibernation, the plasma level of immunoreactive insulin is markedly reduced in the hibernating hedgehog (Hoo-Paris and Sutter 1980) and the dormouse (Castex et al. 1984) and the glucagon level follows a similar trend in the hedgehog (Hoo-Paris et al. 1982) and dormouse (Hoo-Paris et al. 1985). Based on in vivo clearance of exogenous hormones (Hoo-Paris et al. 1982) and in vitro studies on pancreatic fragments (Hoo-Paris et al. 1984, 1985) there is a drastic reduction in hormone secretion during hibernation. Exogenous insulin or insulin antibody does not affect blood glucose level in the hibernating hedgehog (Hoo-Paris and Sutter 1980) but exogenous glucagon elicits a slow, but definitive hyperglycemic effect under the same condition (Hoo-Paris et al. 1982). Furthermore, the stimulatory role of glucose on insulin secretion is absent, whereas that of the arginine on glucagon apparently persists in the hibernating hedgehog (Hoo-Paris et al. 1984). Taken together, these observations indicate a total shutdown of pancreatic beta cell function during hibernation in the hedgehog and the dormouse but a retention of quantitatively reduced alpha cell function under the same state. This suggests a differential thermosensitivity in stimulation-secretion coupling between the alpha and beta cells of the pancreatic islets. Whether this represents differences in membrane lipid-protein interaction on ligang-receptor binding and/or postreceptor signal transduction and response remains to be further investigated. Nevertheless, the suppression of pancreatic hormonal functions during hibernation is apparently solely due to depressed body temperature since above a body temperature of 15° – 25° C during arousal, both the secretion and the action of these hormones are restored. These results are somewhat different from those recently observed in the hibernating marmot (Florant et al. 1986), in which both the alpha and beta cell functions are retained during hibernation although at quantitatively reduced levels. Whether this represents a species difference is currently unknown.

4.2 Ionic Regulation

The survival of cells at low temperature (e.g., 5° C) is critically dependent on the differential distribution of key ionic species: high intracellular K^+ and low intracellular Na^+ and Ca^{2+}. Loss of intracellular K^+ and gain of intracellular Na^+ and Ca^{2+} result in cell swelling and loss of cell function. Since ionic movement involves both active transport and passive leakage and since cooling reduces active pumping more rapidly than it does the passive leakage of ions (Willis 1986), it is intuitively clear that in order to maintain ionic balance at low temperature, the activity of the pump must keep pace with the change in passive leakage.

Tissues (kidney cortex, liver, skeletal muscle, aorta) and cells (cells from kidney cortex, red blood cells) from the hibernators (13-lined ground squirrel, hamster) have been shown to have greater cold tolerance than those of the nonhibernators (guinea pig, rat) as indicated by the greater retention of intracellular K^+ after several days of storage at 5 °C (Willis 1979). Using the red blood cells as the model (Kimzey and Willis 1971), it has been demonstrated that the greater ability of the hibernators to retain intracellular K^+ is due to both a greater capacity of the $(Na^+ + K^+)$ pump and a much reduced passive K^+ leak at 5 °C than that of their nonhibernating counterparts. In a systematic survey, the $(Na^+ + K^+)$ pump activity ratio at 5°/37 °C measured as ouabain-sensitive K^+ influx, ranges between 1.9 – 3.5% in six of seven hibernators and 0.18 – 0.78% in eight of nine nonhibernators (Willis et al. 1980). These results suggest that retention of a greater Na^+ pump activity at low temperature appears to be a consistent trend in the hibernators, although this is not universal. Further studies (Ellory and Willis 1982; Willis 1986) on the kinetics of the Na^+ pump indicate that a temperature decrease from 37° to 5 °C increases the Na^+ pump affinity for external K^+ and internal Na^+ in both hibernators and nonhibernators. The number of ouabain binding sites per cell or the amount of ouabain bound to cell at 37° and 5 °C are also similar for the two groups. The turnover numbers for Na^+ pumps at 37 °C are about equal between the two groups; however, at 5 °C, the turnover numbers are three to fivefold higher in the hibernators. The reason for this difference is presently unknown but it is not related to the blockage of converting the E2 (K^+-binding) to E1 (Na^+-binding) form at low temperature, but may be related to alteration in partial fluxes between $K^+ : K^+$ and $Na^+ : Na^+$ exchange (Ellory and Willis 1982) or the distribution of pump cycle intermediates between the two species (Willis 1986).

With regards to the pass permeability for K^+, there are at least three pathways present in the red blood cell (Willis 1986). These are the Cl^--dependent $Na^+ + K^+$ cotransport which is inhibited by "loop" diuretics (e.g., furosemide or bumetanide), the intracellular Ca^{2+}-dependent K^+ efflux (Gardos channel) which is inhibited by quinine, and the residual K^+ leak which is linearly dependent on extracellular K^+ concentration. The Cl^--dependent $Na^+ + K^+$ cotransport is absent at 5 °C in the guinea pig red blood cell but present in the 13-lined ground squirrel (Hall and Willis 1984) so it is unlikely that this pathway has contributed to the greater loss of intracellular K^+ in the guinea pig. Further, the presence of this pathway is also highly variable among different species and therefore cannot be a universal mechanism for cellular cold tolerance. On the other hand, the ouabain- and furosemide-insensitive (i.e., both pump and cotransport blocked), Ca^{2+}-dependent K^+ efflux increases significantly in the presence of external Ca^{2+} at 5 °C in the guinea pig. This increase is not seen if quinine is present, or if Ca^{2+} is omitted in the assay medium (Hall and Willis 1984, Fig. 6). These observations indicate that the Gardos channel is activated at low temperature due to the influx and intracellular accumulation of Ca^{2+} beyond a critical level causing loss of intracellular K^+. In contrast, red cells of the 13-lined ground squirrel do not show this Ca^{2+} activated K^+ loss. It is possible that the Gardos channel may not exist in this species. However, this has been proven to be not the case since K^+ loss is greatly accelerated if the red cells are loaded with Ca^{2+} through the use of a cation ionophore, A23187 (Hall and Willis 1984). It is therefore likely that the reason why the Gardos channel is not activated at

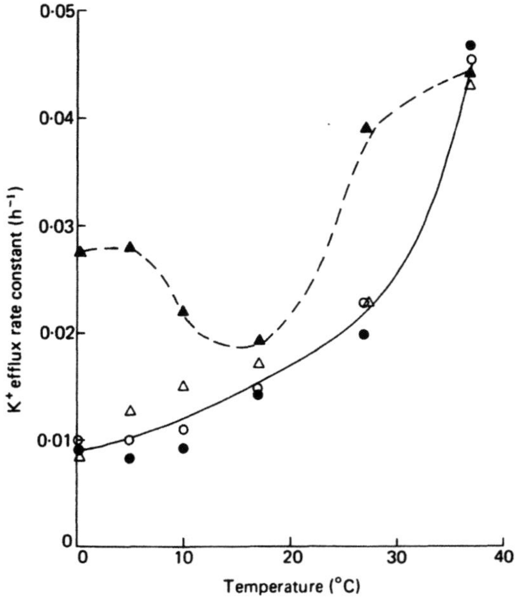

Fig. 6. The influence of Ca^{2+} on ouabain- and bumetanide-insensitive K^+ efflux from guinea pig and 13-lined ground squirrel erythrocytes at various temperatures. Efflux was determined in the presence (▲ 1.25 mM) and absence (△) of Ca^{2+} in the medium for guinea pig cells and in the presence (●) and absence (○) of Ca^{2+} for the ground squirrel cells. Note the increase in K^+ efflux below 18 °C in the guinea pig cells only in the presence of extracellular Ca^{2+} (Hall and Willis 1984)

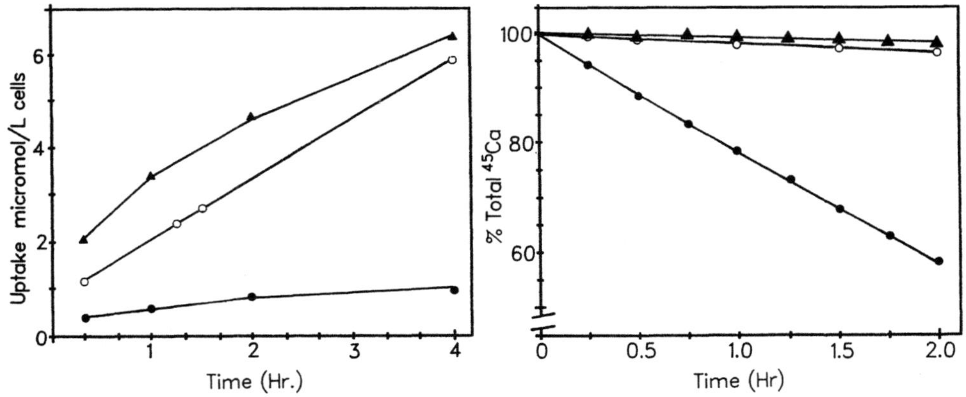

Fig. 7. The time course of Ca^{2+} uptake (*left*) and efflux (*right*) at 5 °C in fresh (●) and ATP-depleted (▲) Richardson's ground squirrel erythrocytes and in fresh (○) guinea pig erythrocytes. Methods of Ca uptake and efflux are identical to those descrbed in Hall et al. (1986) (Hall, Wolowyk, Wang, and Ellory, unpubl. data)

5 °C in the hibernator is either due to a low membrane permeability to external Ca^{2+} so that little Ca^{2+} enters the cells or that once entered, the excess Ca^{2+} is pumped outward by the plasma membrane Ca^{2+} pump so that the critical Ca^{2+} level for activation of the Gardos channel is not reached. Using ATP depletion as a means to deenergize the Ca^{2+} pump, it has been shown that the passive permeability to Ca^{2+} entry into the red blood cells at 5 °C is similar between the guinea pig and hedgehog (Ellory and Hall 1983), and between the guinea pig and the Richardson's ground squirrel (Hall et al. 1987). However, the active Ca^{2+}

pump for Ca^{2+} efflux in the guinea pig is inhibited at 5° C, whereas that of the hedgehog and the ground squirrel is not (Fig. 7). Thus, the ability to retain intracellular K^+ and consequently cell survival at low temperature in the hibernators may reside in their superior ability in regulating intracellular Ca^{2+} concentration as compared to their nonhibernating counterparts. Whether this is due to a difference in the pump protein itself, or due to differences in membrane lipid-protein interaction at low temperature remains to be studied.

4.3 Membrane Aspects

Traditionally, the temperature sensitivity of a membrane-related function (e.g., transport) the activity of a membrane enzyme, or the thermotropic behavior of a membrane component (lipid or protein molecules) has often been assessed using the Arrhenius plot (log maximum velocity vs the reciprocal of absolute temperature). A linear plot (i.e., single slope) across the physiological temperature range (e.g., 2° – 37 °C) indicates a constant activation energy (Ea) for the process in question, whereas a nonlinear plot (curve, curvilinear, or multisloped) indicates multivariate changes possibly involving enzyme-substrate affinity, conformation of the protein, and enthalpy and entropy of activation, etc., and Ea cannot be estimated with confidence (Silvius and McElhaney 1981). In cases where a sharp break in the Arrhenius plot occurs (i.e., two straight lines) the intercept of the two lines has been termed the characteristic temperature (Tc; Aloia et al. 1986). Historically, Tc has been viewed as a critical temperature below which, a thermotropic lipid phase transition occurs in bulk membrane lipids from the liquid-crystalline to the gel phase (Raison and Lyons 1971). Because liver mitochondrial respiration shows a Tc at 23 °C in summer active ground squirrels but below 4° C in winter hibernating animals, a hypothesis has thus been advanced that a prerequisite for hibernation is the alteration of membrane lipids such that the membrane fluidity is increased at low temperature so as to allow continuing function of membrane enzymes during hibernation (Raison and Lyons 1971). Numerous studies have since followed aiming to unravel the specific aspects of membrane lipid alterations that may explain the mechanisms of improved cellular and organismic survival under profound hypothermia. To date, this subject remains to be one of the most challenging tasks facing hibernation research. The earlier studies have been extensively reviewed by Charnock (1978), Aloia and Pengelley (1979), Willis et al. (1981), and Wang (1987a). Only a brief summary and some recent studies will be presented here.

The Arrhenius plots of $(Na^+ + K^+)$-ATPase from kidney, brain, and heart membranes are nonlinear in both hibernators and nonhibernators and in hibernators between the hibernating and nonhibernating phase (Charnock 1978). The SR Ca^{2+}-ATPase from the leg muscles of the 13-lined ground squirrel (Becker and Willis 1972) and that from the heart of the Richardson's ground squirrel (Belke et al. 1987) both show qualitatively similar, nonlinear Arrhenius plots regardless of the hibernating status. These observations suggest no specific changes in thermotropic behavior in these enzymes of their lipid environments in conjunction with hibernation. In the Syrian hamster (Houslay and Palmer 1978),

Arrhenius characteristics of eight liver plasma membrane-bound enzymes are different between the hibernating and nonhibernating state and the differences appear to be related to the location of the enzyme in the lipid bilayer. Of the three enzymes (fluoride-stimulated adenylate cyclase, basal adenylate cyclase, and cAMP phosphodiesterase) located on the cytoplasmic side of the membrane, the Tc remains constant at near 25 °C regardless of the hibernating state. Of the single enzyme (phosphodiesterase I) on the outer bilayer, the Tc decreases from 13° to 4 °C in hibernating hamsters. Of the remaining four enzymes [glucagon-stimulated adenylate cyclase, 5′-nucleotidase, $(Na^+ + K^+)$-ATPase and Mg^{2+}-ATPase] which span both bilayers, two Tc's (25° and 13 °C) are observed in the nonhibernating state, the lower Tc decreases to 4 °C during hibernation, whereas the higher Tc remains unchanged. Addition of a membrane lipid fluidizer (benzyl alcohol) decreases the Tc's by 7° – 8 °C in the glucagon- and fluoride-stimulated adenylate cyclase activities, suggesting the dependence of these enzymatic rates on membrane fluidity. Furthermore, fluorescence polarization (DPH) measurements also indicate an increase in membrane fluidity associated with hibernation in that the Tc's decrease from 25° and 13 °C in the active, to 25° and 4 °C, respectively, in the hibernating hamsters. Taken together, the authors interpret the differential changes in Tc's amongst different enzymes as due to a lipid phase separation occurring around 25 °C in the inner membrane and 13 °C in the outer membrane in active hamsters and the latter Tc decreases to 4 °C during hibernation. Thus, depending on the physical location of the particular enzyme in the lipid bilayer, its Tc or Tc's is either unchanged or lowered with the onset of hibernation (Houslay and Palmer 1978). This suggests membrane asymmetry in lipid composition and Tc's for phase transition. Whether this is a common mechanism to account for the differential thermotropic behaviors of membrane enzymes from each bilayer requires more documentation.

In the mitochondrial membrane, Arrhenius plots of succinate-stimulated oxygen consumption in summer active ground squirrels are typically nonlinear, whereas in the hibernating animals they are linear (Raison and Lyons 1971; Pehowich and Wang 1981, 1984; Augee et al. 1984; Aloia et al. 1986). The succinate-stimulated mitochondrial proton ejection, calcium uptake, and oxygen consumption in spring, summer, and winter active (20° and 4 °C-acclimated) Richardson's ground squirrels are nonlinear but become linear in hibernating and aroused animals (Pehowich and Wang 1984). Interestingly, winter active animals (either 20° or 4 °C-acclimated) which do not show weight gain (presumably in their nonhibernating phase) exhibit nonlinear Arrhenius plots but with decreased Tc's of 11° – 13 °C from the 21 °C seen in summer squirrels (Fig. 8). In those winter active animals which have shown significant weight gain (presumably in their hibernating phase), their Arrhenius plots are linear with Tc's near below 4 °C even though these animals have not yet experienced any decrease in body temperature (Fig. 8). A similar pattern in seasonal change of Tc has also been observed in the liver mitochondrial succinate-cytochrome c reductase activity and spin label motion of mitochondrial membrane lipids in the Richardson's ground squirrel (Augee et al. 1984) and in the mitochondrial membranes of kidney, brown fat, and liver of the golden-mantled ground squirrel (Aloia et al. 1986). The mitochondria from the heart, however, show a Tc < 4 °C all year-

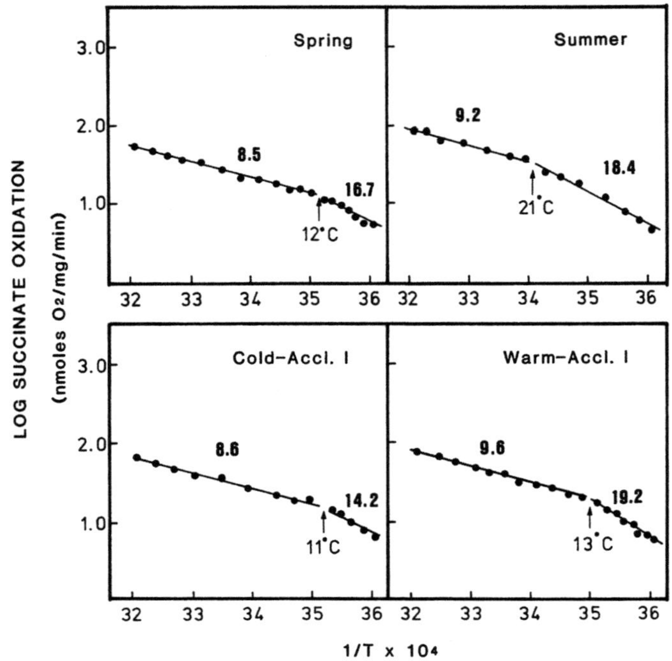
Spring
8.5
16.7
12°C
Summer
9.2
18.4
21°C
Cold-Accl. I
8.6
14.2
11°C
Warm-Accl. I
9.6
19.2
13°C
LOG SUCCINATE OXIDATION
(nmoles O2/mg/min)
1.0
2.0
3.0
32
33
34
35
36
1/T x 10⁴

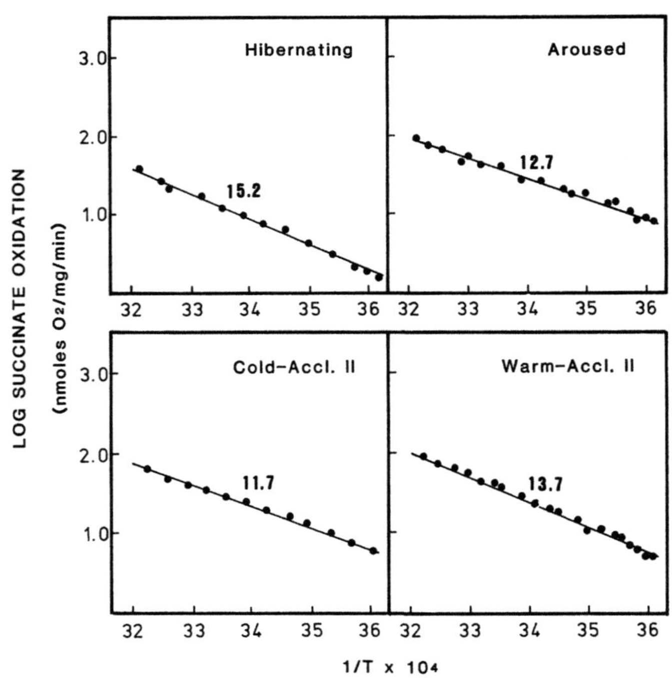
Hibernating
15.2
Aroused
12.7
Cold-Accl. II
11.7
Warm-Accl. II
13.7
LOG SUCCINATE OXIDATION
(nmoles O2/mg/min)
1.0
2.0
3.0
32
33
34
35
36
1/T x 10⁴

round in this latter species. Taken together, these observations suggest that the seasonal differences in the thermotropic behavior of some mitochondrial membrane functions are independent of the photoperiod and ambient temperature at which the animals are kept and the level of body temperature that prevails prior to sampling. Further, these changes are also independent of dietary conditions since standard laboratory chows are used for all animals. It is likely that these changes are related to the endogenous circannual rhythm for hibernation, although how they are initiated is unknown. Further, the membrane changes are already in place prior to hibernation and these changes remain throughout the hibernating phase irrespective of the changes in body temperature during periodic arousals.

To verify whether the observed seasonal changes in Arrhenius characteristics of membrane enzymes are a consequence of changes in membrane structure, the thermotropic behavior of the liver mitochondrial inner membrane has recently been studied in the Richardson's ground squirrel (Fig. 9; Pehowich, Macdonald, McElhaney, Cossins and Wang, unpubl.). Differential scanning calorimetric measurements on native membrane employing both heating and cooling scans at $10\,K\,min^{-1}$ indicate that the midpoint of transition is lowered from near $-2\,°C$ in the summer active to near $-5\,°C$ in the hibernating squirrels. Given the protocol used, this represents only a very slight decrease in melting point of membrane lipids in the hibernating state. Using 19F-palmitic acid NMR, a gradual increase in ordering of the acyl chain is seen with the decrease of temperature. Between 10° and 40 °C, the order parameter (Smol) is relatively constant and no difference is seen between the hibernating and nonhibernating state. This indicates a weak dependence of Smol on changing temperature, suggesting that the membrane lipids are already in a fluid state. Below 10 °C, the increase in Smol of the nonhibernating preparation is greater than that of the hibernating preparation but the difference is small. Even at $-3\,°C$, about half of the membrane lipids from summer animals and more than half from hibernating animals remain in the liquid-crystalline state. Using DPH fluorescence measurements which probe the fluidity change in the inner hydrocarbon core of the lipid bilayer, the polarization ratios at 4 °C are 0.344 and 0.337, respectively, for summeractive and hibernating squirrels. At 37 °C, these are 0.221 and 0.211, respectively, for the two states. None of these values are significantly different. Also, the Arrhenius plots of DPH fluorescence for both states are linear between 4° – 37 °C, indicating that no phase transitions occur in either summer or hibernating squirrels within this temperature range (Fig. 9). The significance of these results, collectively, indicate that no thermotropic lipid phase transition occurs in the bulk membrane lipids in summer squirrels until perhaps below 4 °C. Since

Fig. 8. Arrhenius plots of oxygen consumption as a measure of succinate dehydrogenase activity in the Richardson's ground squirrel under different physiological states. The *number beside each line* is the Arrhenius activation energy (Kcal mol^{-1}) and the *arrow* indicates the Tc. Note the nonlinear plots for spring, summer, and cold- and warm-acclimated animals which did not gain weight and the linear plots in hibernating and aroused animals as well as in cold-and warm-acclimated animals which gained weight but have not yet exhibited hibernation. Note also the seasonal difference in Tc associated with the nonlinear Arrhenius plots (Pehowich and Wang 1984)

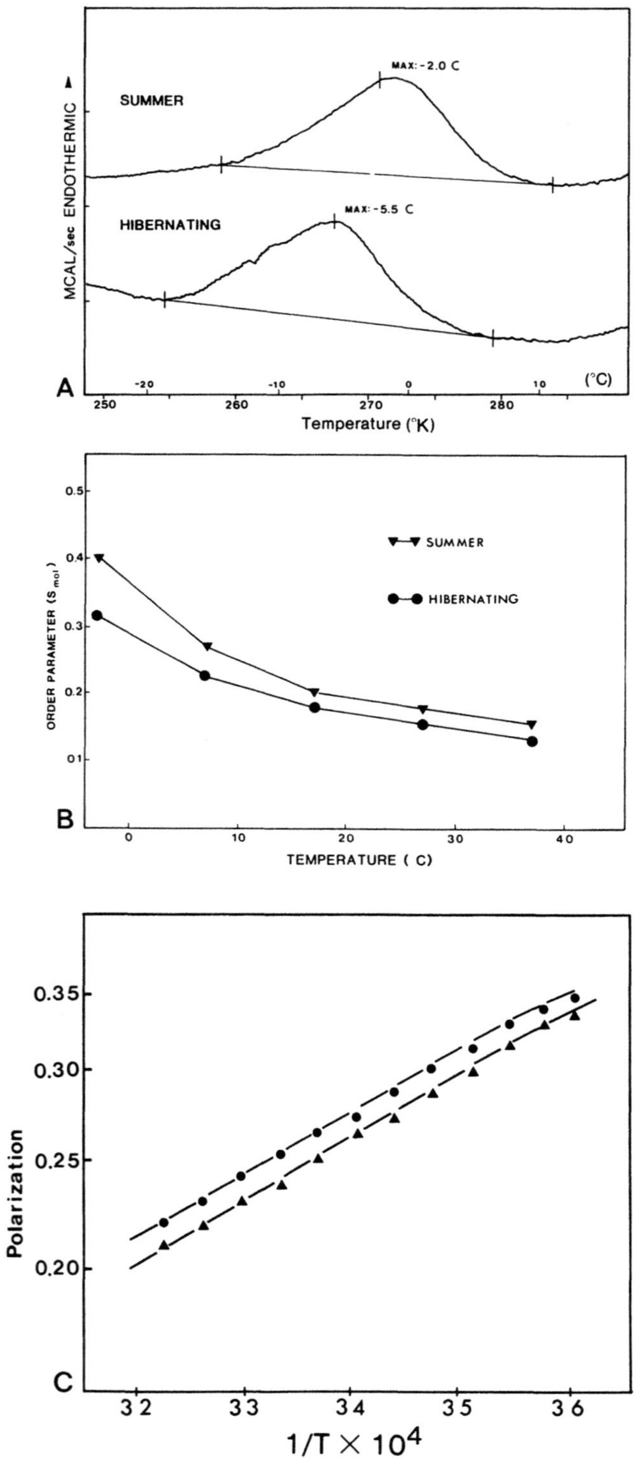
MAX: -2.0 C
SUMMER
MAX: -5.5 C
HIBERNATING
MCAL/sec ENDOTHERMIC
A
-20
-10
0
10
(°C)
250
260
270
280
Temperature (°K)
0.5
0.4
0.3
0.2
01
ORDER PARAMETER (S mol)
SUMMER
HIBERNATING
B
0
10
20
30
40
TEMPERATURE (C)
0.35
0.30
0.25
0.20
Polarization
C
3 2
3 3
3 4
3 5
3 6
1/T × 10[4]

enzymatic studies have consistently shown Tc near 21 °C in summer active animals and 11° – 13 °C in fall animals prior to < 4 °C in hibernation, it is clear that these observed breaks in Arrhenius plots of enzymatic functions cannot be due to a bulk membrane lipid phase transition.

Other studies employing similar biophysical techniques have also yielded similar conclusions. For instance, in the cardiac membrane, analysis by differential scanning calorimetry and electron spin resonance (ESR) indicate only a small increase in membrane fluidity during hibernation as compared to the active state in the Richardson's ground squirrel (Charnock et al. 1980b; Raison et al. 1981). A minor increase in fluidity as detected by fluorescence probe (DPH) has also been observed in the brain microsomal membrane of hibernating Syrian hamsters (Goldman and Albers 1979) and European hamsters (Montaudon et al. 1984). However, using the same technique, such an increase in membrane fluidity has not been observed in the brain synaptosomes and kidney microsomes of the hibernating Syrian hamster (Cossins and Wilkinson 1982) or in the kidney total lipids, kidney microsomes, and kidney microsomal lipids from the hibernating European hamsters (Montaudon et al. 1986). Overall, these studies indicate no major change in bulk membrane fluidity in association with hibernation in these membranes.

Based on compositional analysis of lipids from the brain and kidney of hamsters, liver and heart mitochondria of ground squirrels (see Aloia and Pengelley 1979; Willis et al. 1981), and liver mitochondria of European hamsters (Cremel et al. 1979), there does not appear to be any consistent change in bulk membrane lipid fatty acid composition or in the unsaturation index of fatty acids in relation to hibernation. These observations are consistent with the observations of only minor increases in membrane lipid fluidity during hibernation as assessed by DPH fluorescence polarization described above. It is evident based on these studies that "homeoviscous adaption", a mechanism whereby ectothermic animals significantly increase their membrane fluidity in response to cold acclimation or winter acclimatization (Cossins and Lee 1985), clearly does not occur in the transition from nonhibernating state. This is perhaps not surprising since homeoviscouse adaptation underlies the foundation for capacity adaptation which allows constant energy output in a changing environment, whereas natural hibernation is a form of resistance adaptation which allows tolerance of an extreme environment through reduced energy outputs.

Clearly, a dilemma must be resolved: how to explain the consistently observed seasonal changes in some of the mitochondrial membrane enzyme func-

◄

Fig. 9A – C. Biophysical assessments on thermotropic behavior of liver inner mitochondrial membrane in the Richardson's ground squirrel in euthermia and in hibernation. *A* Representative differential scanning calorimetry thermogram of endothermic (heating) phase transition at a scan rate of 10 K min^{-1} and a sensitivity of 0.2 mcal s^{-1} full scale. Note that the average transition temperature is below 0 °C for both states. *B* 19F-NMR (19F – palmitic acid) orientational order parameter (Smol) profile showing little difference between the two states until below 0 °C (n = 3 for each group). *C* Arrhenius plots of polarization of the fluoroprobe DPH (1,6-diphenyl-1,3,5-hexatriene) between summer active (●) and hibernating (▲) states (n = 9 for each group). (Pehowich, Macdonald, McElhaney, Cossins, and Wang, unpubl. data)

tions from a nonlinear Arrhenius characteristic during the nonhibernating state to the linear Arrhenius characteristic during the hibernating state when the composition and the thermotropic behavior of the bulk membrane lipids are apparently constant year-round? At least two possibilities may be considered: (1) that the Arrhenius changes are due to changes in tertiary and quaternary structure of the proteins affecting rotational and lateral motions involved in the enzymatic reaction mechanism; and (2) a lipid annulus immediately surrounding the protein molecules may specifically modulate the protein's thermotropic behavior depending on the composition of the annulus lipids. In the first case, there has been increasing evidence (e.g., Biglow et al. 1986) that at least in the SR membrane of nonhibernators, the nonlinear Arrhenius plot of the Ca^{2+}-ATPase is not due to the thermotropic behavior of the bulk lipids since Arrhenius plots of the mobility of spin-labeled lipid hydrocarbon chains located either near the polar head region or at the center of the bilayer are linear between 0° – 40 °C in both the intact SR and its extracted lipids. Interestingly, the Arrhenius plot of the rotational mobility of the spin-labeled Ca^{2+}-ATPase itself in the intact SR is also linear, substantiating the results from the lipid mobility analysis since a change in lipid mobility would have also been reflected in protein mobility. The conclusion of these studies is that no Arrhenius break in bulk lipid hydrocarbon mobility or any large-scale change in protein structure or association is responsible for the observed nonlinear Arrhenius plot of Ca^{2+}-ATPase activity in the intact SR. Although not ruling out entirely the possible contribution of thermally induced mobility change of the annulus lipids in limiting the catalytic function of the enzyme, the authors consider it unlikely a mechanism since such a change would have also been reflected in the overall protein mobility measurements. The authors favor the view that subtle structural changes in the protein's active site or protein-lipid interface or a change in protein-protein association in the presence of ligands in the enzymatic reaction may be the reason that the nonlinear Arrhenius plot is seen in the activity of the SR Ca^{2+}-ATPase.

In the second case it has been argued that, based on ESR analysis, there is excellent correlation between Arrhenius characteristics of enzymes and spin label motion (e.g., Aloia et al. 1986) but since the bulk lipid does not show phase transition near 21 °C in the summer squirrel, the transition must be due to, by inference, that of the annulus lipid surrounding the protein. Currently, there is no direct experimental evidence in support of this possibility. Two other considerations also seem to weaken this argument. First, an inherent flaw has been recognized in the invasic nature of the ESR technique which may disturb the architecture of the native membrane and consequently the thermotropic behaviors of both the membrane and the reporting probe. This has led to criticisms on the reliability and diagnostic limitation of this technique (e.g., Martin 1986). Second, since the rate of exchange between bulk and annulus lipids is in the range of 10 – 7 and 10 – 5 s (Stubbs 1983), there is most likely constant "mixing" of the two domain lipids to eliminate any compositionally distinctive annulus lipid microdomains that might exist near the enzyme protein. Thus, although lipid annulus modulating protein activity cannot be ruled out as a possibility in causing an Arrhenius break, more direct, experimental support is needed to strengthen this argument.

Another important issue seldom addressed in the dispute of linear and nonlinear Arrhenius plots in relation to hibernation is the question of the functional significance these changes imply. To date, there has been no experimental documentation, to my knowledge, that a nonlinear Arrhenius plot of a given function is a direct indication of lack of cold tolerance for cell or organism survival at depressed body temperature. Our recent study on the thermotropic behavior of mitochondrial ion fluxes has actually suggested the contrary (Pehowich and Wang 1987). In our study, the Arrhenius plots of mitochondrial proton ejection, Ca^{2+} uptake, and oxygen consumption are nonlinear in summer active Richardson's ground squirrels with Tc's near 23 °C, but linear in the hibernating squirrels. These seasonal differences in Arrhenius plots, however, do not reflect any membrane impairment in ion transport functions below the Tc since the stoichiometric relationship between H/Ca and Ca/O remain constant at all temperatures despite the drastic change in Ea in the summer squirrels below 23 °C (Pehowich and Wang 1987). It is apparent that future research endeavors should include functional assessments at the cellular, organ, or organismic level in deciphering whether the specific thermotropic behavior of a membrane enzyme or a membrane process does indeed have a functional significance in low temperature survival.

In summary, based on the temperature coefficient of membrane-bound enzyme activities, physical measurements of membrane fluidity, and compositional analysis of membrane lipids, there is presently insufficient information to formulate a firm conclusion as to the role of membrane lipids in facilitating survival at low body temperature during hibernation. It is apparent, however, that based on most if not all membranes examined to date, a bulk membrane lipid phase transition (homeoviscous adaptation) similar to that observed in ectotherms following temperature acclimation (see Cossins and Lee 1985) does not occur within the physiological temperature range as a special adaptation for hibernation. However, there remains the possibility that subtle changes in protein structure or local changes in lipid-protein interaction may account for the observed changes in Arrhenius kinetics of enzymes commensurate with hibernation. Future studies incorporating functional, physical, and compositional analysis on the same membrane preparation may further advance our understanding on the fundamental aspects of membrane protein-lipid interactions and maintenance of biochemical homeostasis at low body temperature.

5 Antimetabolic Peptides, the "Hibernation Induction Trigger", and the Opioids

5.1 The "Antabolone"

It has long been proposed that certain "antimetabolic" substances may exist in the hibernators to allow their achievement of the characteristically depressed metabolic state during hibernation. The difficulties, as may be predicted, are to identify the nature of these substances and to locate their occurrence in the animals. Historically, as early as 1932, extracts from the lymphatic tissue, brown adipose tissue, and brain of the hibernators have been injected into either hibernators or nonhibernators for the purpose of inducing artificial hibernation in the recipients. However, none of these attempts have been met with success (see Swan 1981). Between 1963 and 1969, Swan proposed the existence of an antimetabolic substance in the brain of estivating lungfish which he termed "antabolone". Antabolone is apparently a peptide; when injected into the rat, it induces a state of torpor associated with a 35% fall in oxygen consumption and a 3 °C decrease in body temperature. These effects are not seen if brain extracts from the active lungfish are similarly tested (Swan 1981). Similar results are also demonstrated by Reinhard and his colleagues, whom in collaboration with Swan have been pursuing a similar goal since 1976. Using lungfish brain extracts and white mice as models they have partially purified the active agent which is a small peptide 500 – 1500 daltons in molecular weight (for reference see Swan 1981). Encouraged by these initial findings, Swan and his colleagues further proposed that a similar antabolone may also exist in the brain of hibernating ground squirrels. Indeed, they have shown that peptides (mol. wt. = 1000 – 10,000) extracted from the crude brain homogenates of hibernating 13-lined ground squirrels can induce a 35% reduction in metabolism and a 3 °C decrease in body temperature when given intravenously to white rats. In contrast, brain extracts from summer active squirrels are without these effects (Swan 1981). The target organ of these peptides is unknown but the heart is apparently not involved. However, more recent studies on these peptides have yielded conflicting results: when administered ICV into rats, there is no difference in metabolic and thermal responses to peptides isolated from either hibernating or active squirrels nor is there a dose-response relationship after the injection of peptides from the hibernating squirrels (Swan 1981). These results raise questions as to the functional significance of these brain peptides in centrally modulating metabolism. It is possible that either the actions of these brain peptides are only peripheral or that a diffusive barrier prevents the access of these peptides to their CNS active sites following intraventricular injection. Regardless of these possibilities, antabolone has not yet been tested for the induction of hibernation and its effect in this respect remains uncertain.

5.2 The "Hibernation Induction Trigger"

The existence of a bloodborne "hibernation induction trigger" (HIT) in hibernating rodents was first reported by Dawe and Spurrier (1969). When serum or plasma taken during hibernation from several species of ground squirrels and the woodchuck is transfused into active 13-lined ground squirrels, hibernation is induced in the recipients during the summer, at a time when hibernation is normally absent (Dawe 1978; Oeltgen and Spurrier 1981). However, using a similar approach but with other species of ground squirrels (Galster 1978; Abbotts et al. 1979) or hamsters (Minor et al. 1978), each serving as both the donor and the recipient, summer hibernation cannot be induced. These variant results seem to suggest that a species specificity may exist in the recipient but not in the donor (Abbotts et al. 1979).

This possibility has recently been extensively investigated in my laboratory involving two series of experiments (Wang, Lee, Jourdan, Westly, Nürnberger, and Belke, unpubl.). In the first series, conducted in early August, 1985, 13-lined ground squirrels in their nonhibernating phase as judged by weekly weight records were used as the recipient and either hibernating or active Richardson's ground squirrels were used as the donor. In the second series, conducted in June, 1986, in addition to the same arrangements as above, juvenile Richardson's ground squirrels were also used as the recipient. It is known that these animals do not normally hibernate until after mid-September based on field observations near Edmonton (Wang 1979). All our experimental protocols followed those described in Dawe (1978) and Abbotts et al. (1979) in animal housekeeping and in the preparation and injection of donor plasma. Groups of seven to ten animals each received one of the following treatments: (1) cold (0° – 4 °C), freshly isolated hibernating plasma; (2) cold, cryogenically preserved hibernating plasma (– 80 °C) obtained less than 3 months earlier; (3) cold, fresh summer active plasma; (4) cold, cryogenically preserved summer active plasma obtained less than 3 months earlier; (5) cold saline; (6) warm (23 °C) saline; and (7) sham injected (i.e., unsuccessful i.v. injection of cold or warm saline). The volume of injection was 0.5 ml in all cases. The results were quite consistent for both series of experiments. To summarize, 13-lined ground squirrels entered hibernation quite readily in the summer after each and every of the above treatments; the rate of occurrence of hibernation ranged from 78% after sham injection to 86% each after warm saline, fresh summer active plasma, and fresh hibernating plasma, respectively. There was no difference in the time delay prior to onset of hibernation after each treatment; the average was as short as 3 days (range = 1 – 33) after cold saline to 18 days (range = 4 – 40) after warm saline with hibernating plasma averaging 5 – 7 days (range = 1 – 21) and summer active plasma 6 – 13 days (range = 1 – 34) after treatment. In contrast, none of the juvenile Richardson's ground squirrels entered hibernation after any of these treatments up to the end of the 8-week observation period. These results strongly indicate that the 13-lined ground squirrel is easily induced into summer hibernation by a variety of treatments even though it is not in its endogenously determined hibernating phase based on body weight changes. These results also caution against the use of the 13-lined ground sqiurrel as a standard test animal for the bioassay of HIT since despite the precaution of preselecting the experimental animals which have been in captivity with a history of a weight record favoring nonhibernation, they entered hibernation, nevertheless, after receiving seemingly quite nonspecific stimuli. The failure of inducing summer hibernation in the juvenile Richardson's ground squirrels receiving hibernating plasma from the same species echoes the results published earlier from our laboratory (Abbotts et al. 1979) and further argues against the transferability of any bloodborne substances that may exist in the hibernating animal for induction of hibernation in the recipient out of season.

Based on earlier work of HIT, attempts have been made to isolate and purify this bloodborne substance. To date, only a heterogeneous preparation has been obtained; the bioactive substance is tightly bound to the serum albumin fraction and cannot be separated despite the use of techniques such as isotachophoresis, isoelectric focusing, and affinity chromatography (Oeltgen and Spurrier 1981). Consequently, only a lyophilized preparation of serum albumin which supposedly contains the bioactive substance is available for the bioassay of induction of summer hibernation.

Intravenous injection of this lyophilized albumin fraction to the 13-lined ground squirrel induces summer hibernation in 2 days to 5 weeks thereafter (Oeltgen and Spurrier 1981). Recently, this lyophilized fraction also causes bradycardia, hypothermia, behavioral depression, and aphagia in the macaque monkeys when given into the cerebral ventricle (Oeltgen et al. 1982); these effects may be retarded or reversed by the opiate antagonist, naloxone or naltrexone, suggesting the possibility that the HIT may be an endogenous opioid (Oeltgen et al. 1982). In addition, HIT has also been found to reduce renal creatinine clearance in euthermic macaque monkeys after i.v. injection (Oeltgen et al. 1985), but whether this effect is also antagonized by naloxone is unknown. The possible involvement of opioids in hibernation will be further discussed below.

Comparing the results consequent to treatment of the "lyophilized serum albumin" and those from the original studies of Dawe and colleagues using hibernating serum, plasma, or their dialysates, certain questions may be raised as to whether both series of studies are dealing with the same type of molecules. Certain discrepancies appear to exist in selecting the starting materials for the biochemical isolation and purification of HIT. In the original publications as summarized by Dawe (1978), the HIT is deemed to be a small, thermally labile molecule less than 5000 daltons in molecular weight. This is further supported by the biological effectiveness of dialysate in inducing hibernation after the hibernating serum is dialyzed against 4–10 vol saline for 30 min at 4 °C with a molecular cutoff of 10000. The success of this maneuver inspite of the shortness in duration of dialysis in the cold indicates high mobility and easy diffusibility of the HIT into the dialysis medium. In contrast, the starting materials used by Oeltgen (Oeltgen and Spurrier 1981) for biochemical isolation are those larger than 5000 mol. wt. since any substances less than 5000 are disgarded in their dialysis pretreatment prior to purification. This means the small, easily diffusable bioactive substances used by Dawe and colleagues for induction of hibernation may not have been included in further processing in the studies of Oeltgen and colleagues. Second, the very tightly bound nature of the bioactive molecule to serum albumin as described in Oeltgen's studies is in strong contrast to the apparently very loose, easily diffusable nature of HIT demonstrated in Dawe's studies. Third, one of Dawe's reported experiments involved the injection of HIT to pregnant ground squirrels and resulted in induction of hibernation in the newborns, indicating that the bioactive substances are capable of crossing the placenta. This may be a difficulty to resolve if the bioactive substance is tightly bound to serum albumin during deep hibernation and cannot be easily separated or transported across the placenta. Because of these apparent differences in basic characteristics in the nature and mode of action both series of studies claim to be studying the same HIT molecule, however, there exists the possibility that two different (or more) kinds of molecules are actually being investigated. Whether these molecules share similar or different physiological or pharmacological effects in the recipient is an open question and can only be answered when the purified substances are chemically characterized and systematically tested.

5.3 The Opioids

The similarity in the depressive state of physiological systems during hibernation to that after morphine administration raised the suspicion that endogenous opioids may be involved in the regulation of hibernation (Beckman 1986). It has been shown that the brain levels of Met- and Leu-enkephalins (Kramarova et al. 1983) and that of Met-enkephalin immunoreactivity in specific hypothalamic areas are increased during hibernation (Nürnberger 1983). Administration of an opioid antagonist, such as naloxone or naltrexone, to hibernating animals either reduces the duration of the hibernation bout (Beckman 1986) or induces premature arousal from hibernation (Kromer 1980; Margulis et al. 1979). In the Columbian ground squirrel, the responsiveness of thermoregulatory responses to ICV morphine and Met-enkaphalinamide is reduced during the hibernating phase (Wang et al. 1987). This is consistent with the observed decrease in dihydromorphine binding to receptors in the cerebral cortex, septum, hippocampus, thalamus, and the amygdala during hibernation (Beckman 1986). Further, in the golden-mantled ground squirrel, physical dependence on morphine fails to develop during hibernation but develops in euthermia (Beckman et al. 1981); this difference is not due to low body temperature per se since physical dependence to morphine may also be demonstrated in hypothermic ground squirrels experiencing similar depression in body temperature and for similar duration to those seen in the hibernating ground squirrel (Jourdan, Beckman, and Wang, unpubl. observations). This suggests that the lack of physical dependence to morphine during hibernation is probably due to an altered CNS state in opioid activity, involving perhaps greater occupancy of opiate receptors by endogenous opioids during hibernation. Such an increase in endogenous opioid activity may also explain the reduced receptor binding described above. Finally, as described previously, the HIT-induced hypothermia, hypophagia, and bradycardia in monkeys can be reversed by naloxone (Oeltgen et al. 1982). Taken together, it is possible that the maintenance of hibernation may be associated with an increased brain opioid activity. Whether this increase is manifested in the peripheral circulation and appears as the HIT is unknown. It is also interesting to note that ICV infusion of enkephalin and dynorphin (delta and kappa agonists, respectively) results in shortening rather than lengthening of hibernation duration (Beckman 1986). This indicates that subtypes of endogenous CNS opioid systems may have different effects on the maintenance and termination of the hibernating state. It is apparent that much needs to be learned in this area in future research endeavors.

Acknowledgments. The research described herein from my laboratory has been generously supported by Operating and Equipment Grants from the Natural Sciences and Engineering Research Council of Canada since 1971. I thank all my graduate students, postdoctoral fellows, collaborators, and technical staff who have contributed significantly toward our understanding of mammalian hibernation.

References

Abbotts B, Wang LCH (1980) Seasonal thermogenic capacity in a hibernator, *Spermophilus richardsonii.* J Comp Physiol B 140:235 – 240
Abbotts B, Wang LCH, Glass JD (1979) Absence of evidence for a hibernation "trigger" in blood dialysate of Richardson's ground squirrel. Cryobiology 16:179 – 183
Aloia RC, Pengelley ET (1979) Lipid composition of cellular membranes of hibernating mammals. In: Florkin M, Scheer BT (eds) Chemical zoology, vol 11. Academic Press, London New York, pp 1 – 47
Aloia RC, Augee ML, Orr GR, Raison JK (1986) A critical role for membranes in hibernation. In: Heller HC, Musacchia XJ, Wang LCH (eds) Living in the cold. Elsevier, Amsterdam New York, pp 19 – 26
Augee ML, Pehowich DJ, Raison JK, Wang LCH (1984) Seasonal and temperature-related changes in mitochondrial membranes associated with torpor in the mammalian hibernator, *Spermophilus richardsonii.* Biochim Biophys Acta 776:27 – 36
Baumber J, Denyes A (1963) Acetate-1-^{14}C metabolism of white fat from hamsters in cold exposure and hibernation. Am J Physiol 205:905 – 908
Becker JH, Willis JS (1972) Absence of cold resistance in sarcoplasmic reticulum Ca-ATPase in a hibernating mammal. Comp Biochem Physiol 52B:533 – 535
Beckman AL (1978) Hypothalamic and midbrain function during hibernation. In: Veale WL, Lederis K (eds) Current studies of hypothalamic function, vol 2. Karger, Basel, pp 29 – 43
Beckman AL (1986) Functional aspects of brain opioid peptide systems in hibernation. In: Heller HC, Musacchia XJ, Wang LCHW (eds) Living in the cold. Elsevier, Amsterdam New York, pp 225 – 234
Beckman AL, Stanton TL (1982) Properties of the CNS during the state of hibernation. In: Beckman AL (ed) The neural basis of behavior. Spectrum, New York, pp 19 – 45
Beckman AL, Llados-Eckman C, Stanton TL, Adler MW (1981) Physical dependence on morphine fails to develop during the hibernating state. Science 212:1527 – 1529
Behrisch HW (1978) Metabolic economy at the biochemical level: the hibernator. In: Wang LCH, Hudson JW (eds) Strategies in cold: natural torpidity and thermogenesis. Academic Press, London New York, pp 461 – 479
Behrisch HW, Smullin DH, Morse GA (1981) Life at low and changing temperatures: molecular aspects. In: Musacchia XJ, Jansky L (eds) Survival in the cold. Elsevier/North Holland, Amsterdam, pp 191 – 205
Berger RJ (1984) Slow wave sleep, shallow torpor and hibernation: homologous states of diminished metabolism and body temperature. Biol Psychol 19:305 – 326
Belke DD, Pehowich DJ, Wang LCH (1987) Seasonal variation in calcium uptake by cardiac sarcoplasmic reticulum in a hibernator, the Richardson's ground squirrel. J Therm Biol 12:53 – 56
Biglow DJ, Squier TC, Thomas DD (1986) Temperature dependence of rotational dynamics of protein and lipid in sarcoplasmic reticulum membrane. Biochemistry 25:194 – 202
Bintz GL, Strand CE (1983) Radioglucose metabolism by Richardson's ground squirrels in the natural environment. Physiol Zool 56:639 – 647
Borgman AI, Moon TW (1976) Enzymes of the normothermic and hibernating bat, *Myotis lucifugus*: temperature as a modulator of pyruvate kinase. J Comp Physiol 107:185 – 199
Brück K, Wünnenberg W (1970) Meshed control of two effector systems: nonshivering and shivering thermogenesis. In: Hardy JD, Gagge AP, Stolwijk JAJ (eds) Physiological and behavioral temperature regulation. Thomas, Springfield, pp 562 – 580
Burlington RF, Klain GJ (1967) Gluconeogenesis during hibernation and arousal from hibernation. Comp Biochem Physiol 22:701 – 708
Burlington RF, Dean MS, Darvish A (1986) Cardiovascular responses to low temperature in isolated hearts from rats and 13-lined ground squirrels (*Spermophilus tridecemlineatus*). In: Heller HC, Musacchia XJ, Wang LCH (eds) Living in the cold. Elsevier, Amsterdam New York, pp 549 – 556
Canguilhem B, Miro JL, Kemph E, Schmitt P (1986) Does serotonin play a role in entrance into hibernation? Am J Physiol: Regul Integrative Comp Physiol 20:R755 – R761
Cannon B, Nedergaard J (1983) Biochemical aspects of acclimation to cold. J Therm Biol 8:85 – 90

Castex Ch, Sutter BChJ (1981) Insulin binding and glucose oxidation in edible dormouse (*Glis glis*) adipose tissue: seasonal variations. Gen Comp Endocrinol 45:273 – 278

Castex Ch, Tahri A, Hoo-Paris R, Sutter BChJ (1984) Insulin secretion in the hibernating edible dormouse (*Glis glis*): in vivo and vitro studies. Comp Biochem Physiol 79A:179 – 183

Charnock JS (1978) Membrane lipid phase-transitions: a possible biological response to hibernation? In: Wang LCH, Hudson JW (eds) Strategies in cold: natural torpidity and thermogenesis. Academic Press, London New York, pp 417 – 460

Charnock JS, Dryden WF, Marshall RJ, Lauzon P (1980a) Dissociation of inotropic and toxic effects of the semisynthetic cardiac glycoside actodigin during hibernation in the ground squirrel. Comp Biochem Physiol 67C:1 – 7

Charnock JS, Gibson RA, McMurchie EJ, Raison JK (1980b) Changes in the fluidity of myocardial membranes during hibernation: relationship to myocardial adenosine triphosphatase activity. Mol Pharmacol 18:476 – 482

Charnock JS, Dryden WF, Marshall RJ (1983) Seasonal variations in drug response and staircase phenomena in atrial muscle from a hibernating rodent (*Spermophilus richardsonii*). Br J Pharmacol 78:151 – 158

Collins VE, Cameron DM (1984) The effects of diet and photoperiod on hibernation in the woodland jumping mouse *Napaeozapus insignus*. Can J Zool 62:1938 – 1945

Cossins AR, Lee JAC (1985) The adaptation of membrane structure and lipid composition to cold. In: Gilles R (ed) Circulation, respiration and metabolism. Springer, Berlin Heidelberg New York Tokyo, pp 543 – 552

Cossins AR, Wilkinson HL (1982) The role of homeoviscous adaptation in mammalian hibernation. J Therm Biol 7:107 – 110

Cranford JA (1986) Field and laboratory annual cycles of activity and hibernation in the Uinta Basin ground squirrel (*Spermophilus armatus*). In: Heller HC, Musacchia XJ, Wang LCH (eds) Living in the cold. Academic Press, London New York, pp 411 – 418

Cremel G, Robel G, Canguilhem B, Rendon A, Waksman A (1979) Seasonal variation of the composition of membrane lipids in liver mitochondria of the hibernator *Cricetus cricetus*. Relation to intra-mitochondrial intermembranal protein movement. Comp Biochem Physiol 63A: 159 – 167

Davis DE (1976) Hibernation and circannual rhythms of food consumption in marmots and ground squirrels. Q Rev Biol 51:477 – 514

Dawe AR (1978) Hibernation trigger research updated. In: Wang LCH, Hudson JW (eds) Strategies in cold: natural torpidity and thermogenesis. Academic Press, London New York, pp 541 – 563

Dawe AR, Spurrier WA (1969) Hibernation induced in ground squirrels by blood transfusion. Science 163:298 – 299

Denyes A, Carter JD (1961) Utilization of acetate-1-C14 by heptic tissue from cold exposed and hibernating hamsters. Am J Physiol 200:1043 – 1046

Draskoczy PR, Lyman CP (1967) Turnover of catecholamines in active and hibernating ground squirrels. J Pharmacol Exp Ther 155:101 – 111

Duker G, Sjoquist PO, Svensson O, Wohlfart B, Johansson BW (1986) Hypothermic effects on cardiac action potentials: difference between a hibernator, hedgehog, and a nonhibernator, guinea pig. In: Heller HC, Musacchia XJ, Wang LCH (eds) Living in the cold. Elsevier, Amsterdam New York, pp 565 – 572

Ellory JC, Hall AC (1983) Ca^{2+} transport in hibernator and non-hibernator species' red cells at low temperature. J Physiol (London) 344:148P

Ellory JC, Willis JS (1982) Kinetics of the sodium pump in red cells of different temperature sensitivity. J Gen Physiol 79:1115 – 1130

Fedotcheva NI, Mironova GD, Iljasova EN, Kolaeva SG (1984) Appearance of the external pathway of NADH oxidation in mitochondria during hibernation. Cryo-Letters 5:27 – 32

Fedotcheva NI, Sharyshev AA, Mironova GD, Kondrashova MN (1985) Inhibition of succinate oxidation and potassium transport in mitochondria during hibernation. Comp Biochem Physiol 82B:191 – 196

Feist D, Florant G, Greenwood MRC, Feist C (1986) Regulation of energy stores in arctic ground squirrels: brown fat thermogenic capacity, lipoprotein lipase and pancreatic hormones during fat deposition. In: Heller HC, Musacchia XJ, Wang LCH (eds) Living in the cold. Elsevier, Amsterdam New York, pp 281 – 285

Fisher KC, Dawe AR, Lyman CP, Schönbaum E, South FE (eds) (1967) Mammalian hibernation, vol 3. Oliver and Boyd, Edinburgh

Florant G, Greenwood MRC (1986) Seasonal variations in pancreatic function in marmots: the role of pancreatic hormones and lipoprotein lipase in fat deposition. In: Heller HC, Musacchia XJ, Wang LCH (eds) Living in the cold. Elsevier, Amsterdam New York, pp 273–280

Florant GL, Weitzman ED, Jayant A, Cote LJ (1982) Plasma catecholamine levels during cold adaptation and hibernation in woodchucks (*Marmota monax*). J Therm Biol 7:143–146

Florant GL, Lawrence AK, Williams K, Bauman WA (1985) Seasonal changes in pancreatic B-cell function in euthermic yellow-bellied marmots. Am J Physiol 249:R159–R165

Florant GL, Hoo-Paris R, Castex Ch, Bauman WA, Sutter BChJ (1986) Pancreatic alpha and beta cell stimulation in euthermic and hibernating marmots, *Marmota flaviventris*: effects of glucose and arginine administration. J Comp Physiol B 156:309–314

Foster DO, Frydman ML (1979) Tissue distribution of cold-induced thermogenesis in conscious warm- or cold-acclimated rats reevaluated from changes in tissue blood flow: the dominant role of brown adipose tissue in the replacement of shivering by nonshivering thermogenesis. Can J Physiol Pharmacol 57:257–270

French AR (1985) Allometries of the durations of torpid and euthermic intervals during mammalian hibernation: a test of the theory of metabolic control of the timing of changes in body temperature. J Comp Physiol B 156:13–19

French AR (1986) Patterns of thermoregulation during hibernation. In: Heller HC, Musacchia XJ, Wang LCH (eds) Living in the cold. Elsevier, Amsterdam New York, pp 393–402

Galster W (1978) Failure to initiate hibernation with blood from the hibernating arctic ground squirrel, *Citellus undulatus*, and eastern woodchuck, *Marmota monax*. J Therm Biol 3:93

Galster W, Morrison PR (1975) Gluconeogenesis in arctic ground squirrels between periods of hibernation. Am J Physiol 228:325–330

Gehnrich SC, Aprille JR (1986) The suppression of mitochondrial metabolism during hibernation. Fed Proc 45:1923

Glass JD, Wang LCH (1979a) Thermoregulatory effects of intracerebroventricular injection of serotonin and a monoamine oxidase inhibitor in a hibernator, *Spermophilus richardsonii*. J Therm Biol 4:149–156

Glass JD, Wang LCH (1979b) Effects of central injection of biogenic amines during arousal from hibernation. Am J Physiol 236:R162–R167

Goldman SS, Albers RW (1979) Cold resistance of the brain during hibernation: changes in the microviscosity of the membrane and associated lipids. J Neurochem 32:1139–1142

Green CJ, Marley SPE, Fuller BJ (1985) Gluconeogenesis in stored kidneys from hibernating and non-hibernating ground squirrels. Cryo-Letters 6:353–360

Hall AC, Willis JS (1984) Differential effects of temperature on three components of passive permeability to potassium in rodent red cells. J Physiol (London) 348:629–643

Hall AC, Wolowyk MW, Wang LCH, Ellory JC (1987) The effects of temperature on Ca^{2+} transport in red cells from a hibernator (*Spermophilus richardsonii*). J Therm Biol 12:61–63

Hand SC, Somero GN (1983) Phosphofructokinase of the hibernator *Citellus beecheyi*: temperature and pH regulation of activity via influences on the tetramer-dimer equilibrium. Physiol Zool 56:380–388

Harlow HJ, Phillips JA, Ralph CL (1980) The effect of pinealectomy on hibernation in two species of seasonal hibernators, *Citellus lateralis* and *C. richardsonii*. J Exp Zool 213:301–303

Hayward JS (1968) The magnitude of noradrenaline-induced thermogenesis in the bat (*Myotis lucifugus*) and its relation to arousal from hibernation. Can J Physiol Pharmacol 46:713–718

Hayward JS, Lyman CP (1967) Non-shivering heat production during arousal from hibernation and evidence for the contribution of brown fat. In: Fisher KC, Dawe AR, Lyman CP, Schönbaum E, South FE (eds) Mammalian hibernation, vol 3. Oliver and Boyd, Edinburgh, pp 346–355

Heller HC (1979) Hibernation: neural aspects. Annu Rev Physiol 41:305–321

Heller HC, Walker JM, Florant GL, Glotzbach SF, Berger RJ (1978) Sleep and hibernation: electrophysiological and thermoregulatory homologies. In: Wang LCH, Hudson JW (eds) Strategies in cold: natural torpidity and thermogenesis. Academic Press, London New York, pp 225–265

Heller HC, Musacchia XJ, Wang LCH (eds) (1986) Living in the cold. Elsevier, Amsterdam New York

Hemingway A (1963) Shivering. Physiol Rev 43:397 – 422
Himms-Hagen J (1984) Brown adipose tissue thermogenesis, energy balance, and obesity. Can J Biochem Cell Biol 62:610 – 617
Hoffman RA (1964) Terrestrial animals in cold: hibernators. In: Handbook of physiology, sect 4. Adaptation to the environment. Am Physiol Soc, Washington DC, pp 379 – 403
Hoo-Paris R, Sutter BChJ (1980) Blood glucose control by insulin in the lethargic and arousing hedgehog (*Erinaceus europaeus*). Comp Biochem Physiol 66A:141 – 143
Hoo-Paris, Hamsany M, Sutter BChJ, Assan R, Biollol J (1982) Plasma glucose and glucagon concentrations in the hibernating hedgehog. Gen Comp Endocrinol 46:246 – 254
Hoo-Paris R, Aina E, Castex Ch, Sutter B (1984) In vitro B cell response to glucose in the hibernating hedgehog: comparison with the homeothermic hedgehog and the rat. Comp Biochem Physiol 78A:559 – 563
Hoo-Paris R, Castex Ch, Hamsany M, Thari A, Sutter B (1985) Glucagon secretion in the hibernating edible dormouse (*Glis glis*). Comp Biochem Physiol 81A:277 – 281
Horwitz BA, Kott KS, Hamilton JS, Moore BJ (1986) Regulation of brown fat thermogenesis in hibernators. In: Heller HC, Musacchia XJ, Wang LCH (eds) Living in the cold. Elsevier, Amsterdam New York, pp 101 – 108
Houslay MD, Palmer RW (1978) Changes in the form of Arrhenius plots of the activity of glucagon-stimulated adenylate cyclase and other hamster liver plasma-membrane enzymes occurring on hibernation. Biochem J 174:909 – 919
Hudson JW (1973) Torpidity in mammals. In: Whittow GC (ed) Comperative physiology of thermoregulation, vol 3. Academic Press, London New York, pp 97 – 165
Hudson JW, Wang LCH (1979) Hibernation: endocrinologic aspects. Annu Rev Physiol 41:287 – 303
Jacobs HK, South FE (1976) The effect of temperature on the electrical potentials of hibernating hamster ventricular strips. Am J Physiol 230:403 – 409
Jansky L (1973) Non-shivering thermogenesis and its thermoregulatory significance. Biol Rev 48:85 – 132
Jansky L (1978) Time sequence of physiological changes during hibernation: the significance of serotonergic pathways. In: Wang LCH, Hudson JW (eds) Strategies in cold: natural torpidity and thermogenesis. Academic Press, London New York, pp 299 – 326
Jansky L, Novotona R (1976) The role of central aminergic transmission in thermoregulation and hibernation. In: Jansky L, Musacchia XJ (eds) Regulation of depressed metabolism and thermogenesis. Thomas, Springfield, pp 64 – 80
Jones SB, Romano FD (1984) Functional characteristics and responses to adrenergic stimulation of isolated heart preparations from hypothermic and hibernation subjects. Cryobiology 21:615 – 626
Jouvet M (1972) The role of monoamines and acetylcholine-containing neurons in the regulation of sleep waking cycle. Ergeb Physiol Biol Chem Exp Pharmakol 64:166 – 307
Kamm KE, Zatzman ML, Jones AW (1979) Maintenance of ion concentration gradients in the cold in aorta from rat and ground squirrel. Am J Physiol 237:C17 – C22
Kayser C (1961) The physiology of natural hibernation. Oxford Pergamon Press, New York
Kilduff TS, Sharp FR, Heller HC (1983) Relative 2-deoxyglucose uptake of the paratrigeminal nucleus increases during hibernation. Brain Res 262:117 – 123
Kimzey SL, Willis JS (1971) Temperature adaptation of active sodium-potassium transport and of passive permeability in erythocytes of ground squirrels. J Gen Physiol 58:634 – 649
Kondo N, Shibata S (1984) Calcium source for excitation-contraction coupling in myocardium of nonhibernating and hibernating chipmunks. Science 225:641 – 643
Kramarova LI, Kolaeva SH, Yukhananov RYu, Rozhanets VV (1983) Content of DSIP, enkephalins and ACTH in some tissues of active and hibernating ground squirrels (*Citellus suslicus*). Comp Biochem Physiol 74C:31 – 33
Krilowicz Bl (1985) Keton body metabolism in a ground squirrel during hibernation and fasting. Am J Physiol 249:R462 – R470
Kromer W (1980) Naltrexone influence on hibernation. Experientia 36:581 – 582
Krueger JM, Shoham S (1986) Homologies between sleep and hibernation triggers. In: Heller HC, Musacchia XJ, Wang LCH (eds) Living in the cold. Elsevier, Amsterdam New York, pp 235 – 243
Kuhnen G, Petersen P, Wunnerberg W (1983) Hibernation in golden hamsters (*Mesocricetus auratus*, W.) exposed to 5% CO_2. Experientia 39:1346 – 1347

Langer GA (1980) The role of calcium in the control of myocardial contractility: an update. J Mol Cell Cardiol 12:231 – 239
Lyman CP (1982) Who is who among the hibernators. In: Lyman CP, Willis JS, Malan A, Wang LCH (eds) Hibernation and torpor in mammals and birds. Academic Press, London New York, pp 12 – 36
Lyman CP (1984) Pharmacological aspects of mammalian hibernation. Pharmacol Ther 25:371 – 393
Lyman CP, Blinks DC (1959) The effect of temperature on the isolated hearts of closely related hibernators and non-hibernators. J Cell Comp Physiol 54:53 – 63
Lyman CP, Chatfield PO (1955) Physiology of hibernation in mammals. Physiol Rev 35:403 – 425
Lyman CP, Dawe AR (eds) (1960) Mammalian hibernation. Bull Mus Comp Zool Harvard Univ 124
Lyman CP, Leduc EH (1953) Changes in blood sugar and tissue glycogen in the hamster during arousal from hibernation. J Cell Comp Physiol 41:471 – 492
Lyman CP, O'Brien (1986) Is brown fat necessary? In: Heller HC, Musacchia XJ, Wang LCH (eds) Living in the cold. Elsevier, Amsterdam New York, pp 109 – 116
Lyman CP, Willis JS, Malan A, Wang LCH (eds) (1982) Hibernation and torpor in mammals and birds. Academic Press, London New York
Malan A (1982) Respiration and acid-base state in hibernation. In: Lyman CP, Willis JS, Malan A, Wang LCH (eds) Hibernation and torpor in mammals and birds. Academic Press, London New York, pp 237 – 282
Malan A (1986) pH as a control factor in hibernation. In: Heller HC, Musacchia XJ, Wang LCH (eds) Living in the cold. Elsevier, Amsterdam New York, pp 61 – 70
Malan A, Rodeau JL, Daull F (1985) Intracellular pH in hibernation and respiratory acidosis in the European hamster. J Comp Physiol B 156:251 – 258
Margulis DL, Goldman B, Finck A (1979) Hibernation: an opoid-dependent state? Brain Res Bull 4:721 – 724
Marshall JM (1957) Effects of low temperatures on transmembrane potentials of single fibers of the rabbit atrium. Circ Res 5:664 – 669
Marshall JM, Willis JS (1962) The effects of temperature on the transmembrane potentials in isolated atria of the ground squirrel, *Citellus tridecemlineatus*. J Physiol (London) 164:64 – 76
Martin S (1986) Arrhenius plots and the involvement of thermotropic phase transitions of the thylakoid membrane in chilling impairment of photosynthesis in thermophilic higher plants. Plant Cell Environ 9:323 – 331
Melnyk RB, Martin JM (1985) Insulin and central regulation of spontaneous fattening and weight loss. Am J Physiol 249:R203 – R208
Melnyk RB, Mrosovsky N, Martin JM (1983) Spontaneous obesity and weight loss: insulin binding and lipogenesis in the dormouse. Am J Physiol 245:R403 – R407
Miller UM, Miller WL, South FE (1986) Vascular smooth muscle responsiveness in a hibernator: effects of season and temperature. Am J Physiol 250:R77 – R82
Milsom WK, McArthur MD, Webb CL (1986) Control of breathing in hibernating ground squirrels. In: Heller HC, Musacchia XJ, Wang LCH (eds) Living in the cold. Elsevier, Amsterdam New York, pp 469 – 475
Minor JD, Bishop DA, Badger CR (1978) The golden hamster and the blood-borne hibernation trigger. Cryobiology 15:557 – 562
Montaudon D, Robert J, Canguilhem B (1984) Fluorescence polarization study of lipids and membranes prepared from brain hemispheres of a hibernating mammal. Biochem Biophys Res Commun 119:396 – 400
Montaudon D, Robert J, Canquilhem B (1986) Fluorescence anisotropy of kidney lipids and membranes of a hibernating mammal. Cryobiology 23:177 – 183
Mory G, Ricquier D, Nechad M, Hemon P (1982) Impairment of trophic response of brown fat to cold in guanethidine-treated rats. Am J Physiol 242:C159 – C165
Mrosovsky N (1978) Circannual cycles in hibernators. In: Wang LCH, Hudson JW (eds) Strategies in cold: natural torpidity and thermogenesis. Academic Press, London New York, pp 21 – 65
Mrosovsky N (1983) Cyclical obesity in hibernators: the search for the adjustable regulator. In: Hirsh J, Itallie TB van (eds) Recent advances in obesity research, vol 4. Libey, London, pp 45 – 56
Mrosovsky N (1986) Thermal effects on the periodicity, phasing and persistence of circannual cycles. In: Heller HC, Musacchia XJ, Wang LCH (eds) Living in the cold. Elsevier, Amsterdam New York, pp 403 – 410

Musacchia XJ (1984) Comparative physiological and biochemical aspects of hypothermia as a model for hibernation. Cryobiology 21:583 – 592
Musacchia XJ, Deavers DR (1981) The regulation of carbohydrate metabolism in hibernators. In: Musacchia XJ, Jansky L (eds) Survival in the cold. Elsevier/North Holland, Amsterdam, pp 55 – 75
Musacchia XJ, Jansky L (eds) (1981) Survival in the cold. Elsevier/North Holland, Amsterdam
Myers RD (1980) Hypothalamic control of thermoregulation: neurochemical mechanisms. In: Morgane P, Panksepp J (eds) Handbook of hypothalamus. Dekker, New York Basel, pp 83 – 210
Nedergaard J, Cannon B (1984) Preferential utilization of brown adipose tissue lipids during arousal from hibernation in hamsters. Am J Physiol 247:R506 – R512
Nicholls DG, Locke RM (1984) Thermogenic mechanisms in brown fat. Physiol Rev 64:1 – 64
Novotona R, Jansky L, Drahota Z (1975) Effect of hibernation on turnover of serotonin in the brain stem of golden hamster (*Mesocricetus auratus*). Gen Pharmacol 6:23 – 26
Nürnberger F (1983) Der Hypothalamus des Igels (*Erinaceus europaeus* L.) unter besonderer Berücksichtigung des Winterschlafes. Cytoarchitektonische und immuncytochemische Studien. Ph D Diss, Univ Marburg, Federal Republic of Germany
Oeltgen PR, Spurrier WA (1981) Characterization of a hibernation trigger. In: Musacchia XJ, Jansky L (eds) Survival in the cold. Elsevier/North Holland, Amsterdam, pp 139 – 157
Oeltgen PR, Walsh JW, Hamann SR, Randall DC, Spurrier WA, Myers RD (1982) Hibernation "trigger": opioid-like inhibitory action on brain function of the monkey. Pharmacol Biochem Behav 17:1271 – 1274
Oeltgen PR, Blouin RA, Spurrier WA, Myers RD (1985) Hibernation "trigger" alters renal function in the primate. Physiol Behav 34:79 – 81
Olsson SOR (1972) Dehydrogenases (LDH, MDH, G-6-PDH, and a-GPDH) in the heart, liver, white and brown fat. Acta Physiol Scand Suppl No 380:62 – 95
Pehowich DJ, Wang LCH (1981) Temperature dependence of mitochondrial Ca^{2+} transport in a hibernating and nonhibernating ground squirrel. Acta Univ Carolinae Biol 1979:291 – 293
Pehowich DJ, Wang LCH (1984) Seasonal changes in mitochondrial succinate dehydrogenase activity in a hibernator, *Spermophilus richardsonii.* J Comp Physiol B 154:495 – 501
Pehowich DJ, Wang LCH (1987) Stoichiometry of H^+ efflux to respiration-dependent Ca^{2+} uptake and oxygen consumption in liver mitochondria from a hibernator. Physiol Zool 60: 114 – 120
Pengelley ET, Asmundson SJ (1974) Circannual rhythmicity in hibernating mammals. In: Pengelley ET (ed) Circannual clocks: annual biological rhythms. Academic Press, London New York, pp 95 – 160
Petrovic VM, Rajcic O, Janic-Sibalic V (1985) Accelerated gluconeogenic processes in the ground squirrel (*Citellus citellus*) during the arousal from hibernation. Comp Biochem Physiol 80A:477 – 480
Phillips JA, Harlow HJ (1982) Long-term effects of pinealectomy on the annual cycle of golden-mantled ground squirrel, *Spermophilus lateralis.* J Comp Physiol 146B:501 – 505
Pivorum EB (1986) Hypothalamic thermosensitivity in hibernating chipmunks *Tamias striatus.* Physiol Zool 59:194 – 200
Popova NK (1986) Thermoregulation effect of serotonin in hibernation. In: Heller HC, Musacchia XJ, Wang LCH (eds) Living in the cold. Elsevier, Amsterdam New York, pp 193 – 200
Popova NK, Voitenko NN (1981) Brain serotonin metabolism in hibernation. Pharmacol Biochem Behav 14:773 – 777
Raison JK, Lyons JM (1971) Hibernation: alteration of mitochondrial membranes as a requisite for metabolism at low temperature. Proc Natl Acad Sci USA 68:2092 – 2094
Raison JK, McMurchie EJ, Charnock JS, Gibson RA (1981) Differences in the thermal behaviour of myocardial membranes relative to hibernation. Comp Biochem Physiol 69B:169 – 174
Raths P, Kulzer E (1976) Physiology of hibernation and related lethargic states in mammals and birds. Bonn Zool Monogr 9:1 – 93
Rauch JC, Behrisch HW (1981) Ketone bodies: a source of energy during hibernation. Can J Zool 59:754 – 760
Reeves RB (1977) The interaction of body temperature and acid-base balance in ectothermic vertebrates. Annu Rev Physiol 39:559 – 586

Riedesel ML, Griego JE, Espinosa DC (1986) Respiratory patterns in hibernating *Spermophilus lateralis*. In: Heller HC, Musacchia XJ, Wang LCH (eds) Living in the cold. Elsevier, Amsterdam New York, pp 461–468

Sauerbier I, Lemmer B (1977) Seasonal variations in the turnover of noradrenaline of active and hibernating hedgehogs (*Erinaceus europaeus*). Comp Biochem Physiol 57C:61–63

Silvius JR, McElhaney RN (1981) Non-linear Arrhenius plots and the analysis of reaction and motional rates in biological membranes. J Theor Biol 88:135–152

South FE, Jacobs H (1973) Contraction kinetics of ventricular muscle from hibernating and non-hibernating mammals. Am J Physiol 225:444–449

South FE, Hannon JP, Willis JS, Pengelley ET, Alpert NR (eds) (1972) Hibernation and hypothermia: perspectives and challenges. Elsevier, Amsterdam New York

South FE, Miller VM, Hartner WC (1978) Neuronal models of temperature regulation in euthermic and hibernating mammals: an alternate model for hibernation. In: Wang LCH, Hudson JW (eds) Strategies in cold: natural torpidity and thermogenesis. Academic Press, London New York, pp 187–224

Spafford DC, Pengelley ET (1971) The influence of the neurohumor serotonin on hibernation in the golden-mantled ground squirrel, *Citellus lateralis*. Comp Biochem Physiol 38A:239–250

Stanton TL, Craft CM, Reiter RJ (1986) Evidence for the involvement of pineal melatonin in the control of the hibernation cycle in *Spermophilus lateralis*. In: Heller HC, Musacchia XJ, Wang LCH (eds) Living in the cold. Elsevier, Amsterdam New York, pp 309–316

Steffen JM, Riedesel ML (1982) Pulmonary ventilation and cardiac activity in hibernating and arousing golden-mantled ground squirrels (*Spermophilus lateralis*). Cryobiology 19:83–91

Stubbs CD (1983) Membrane fluidity: structure and dynamics of membrane lipids. In: Campbell PN, Marshall RD (eds) Essays in biochemistry. Academic Press, London New York, pp 1–39

Sturmwasser F (1960) Some physiological principles governing hibernation in *Citellus beecheyi*. In: Lyman CP, Dawe AR (eds) Mammalian hibernation. Bull Mus Comp Zool Harvard Univ 124:285–320

Suomalainen P (ed) (1964) Mammalian hibernation, vol 2. Ann Acad Sci Fenn Ser A IV Biologica 71

Swan H (1981) Neuroendocrine aspects of hibernation. In: Musacchia XJ, Jansky L (eds) Survival in the cold. Elsevier/North Holland, Amsterdam, pp 121–138

Tashima LS, Adelstein SJ, Lyman CP (1970) Radioglucose utilization by active, hibernating, and arousing ground squirrels. Am J Physiol 218:303–309

Torke KG, Twente JW (1977) Behavior of *Spermophilus lateralis* between periods of hibernation. J Mammal 58:385–390

Twente JW, Twente J, Brack V Jr (1985) The duration of the period of hibernation of three species of vespertilionid bats II. Laboratory studies. Can J Zool 63:2955–2961

Walker JM, Berger RJ (1980) Sleep as an adaptation for energy conservation functionally related to hibernation and shallow torpor. In: McConnell PS, Boer GJ, Romijn HJ, Corner MA (eds) Progress in brain research, vol 53: Adaptative capabilities of the nervous system. Elsevier/North Holland, Amsterdam, pp 255–278

Wang LCH (1978) Energetics and field aspects of mammalian torpor: the Richardson's ground squirrel. In: Wang LCH, Hudson JW (eds) Strategies in cold: natural torpidity and thermogenesis. Academic Press, London New York, pp 109–145

Wang LCH (1979) Time patterns and metabolic rates of natural torpor in the Richardson's ground squirrel. Can J Zool 57:149–155

Wang LCH (1982) Hibernation and the endocrine. In: Lyman CP, Willis JS, Malan A, Wang LCH (eds) Hibernation and torpor in mammals and birds. Academic Press, London New York, pp 206–236

Wang LCH (1985) Life at low temperature: mammalian hibernation. Cryo-Letters 6:257–274

Wang LCH (1987a) Mammalian hibernation. In: Morris GJ, Grout B (eds) The effects of low temperatures on biological systems. Arnold, London, pp 349–386

Wang LCH (1987b) Some neuroendocrine aspects of mammalian hibernation. In: Boissin J (ed) Endocrine regulations as adaptative mechanisms to the environment. C N R S Publ, Paris (in press)

Wang LCH, Abbotts B (1981) Maximum thermogenesis in hibernators: magnitudes and seasonal variations. In: Musacchia XJ, Jansky L (eds) Survival in the cold. Elsevier/North Holland, Amsterdam, pp 77–97

Wang LCH, Hudson JW (eds) Strategies in cold: natural torpidity and thermogenesis. Academic Press, London New York
Wang LCH, Lee TF, Jourdan ML (1987) Seasonal difference in thermoregulatory responses to opiates in a mammalian hibernator. Pharmacol Biochem Behav 26:565 – 571
Whitten BK, Klain GJ (1968) Protein metabolism in hepatic tissues of hibernating and arousing ground squirrels. Am J Physiol 214:1360 – 1362
Whitten BK, Klain GJ (1969) NADP-specific dehydrogenases and hepatic lipogenesis in the hibernator. Comp Biochem Physiol 29:1099 – 1104
Willis JS (1979) Hibernation: cellular aspects. Annu Rev Physiol 41:275 – 286
Willis JS (1982) The mystery of the periodic arousal. In: Lyman CP, Willis JS, Malan A, Wang LCH (eds) Hibernation and torpor in mammals and birds. Academic Press, London New York, pp 92 – 103
Willis JS (1986) Membrane transport at low temperature in hibernators and nonhibernators. In: Heller HC, Musacchia XJ, Wang LCH (eds) Living in the cold. Elsevier, Amsterdam New York, pp 27 – 34
Willis JS, Ellory JC, Wolowyk MW (1980) Temperature sensitivity of the sodium pump in red cells from various hibernators and nonhibernator species. J Comp Physiol 138:43 – 47
Willis JS, Ellory JC, Cossins AR (1981) Membranes of mammalien hibernators at low temperatures. In: Morris GJ, Clarke A (eds) Effects of low temperatures on biological membranes. Academic Press, London New York, pp 121 – 142
Wünnenberg W, Baltruschat D (1982) Temperature regulation of golden hamsters during acute hypercapnia. J Therm Biol 7:83 – 86
Wünnenberg W, Merker G, Speulda E (1978) Thermosensitivity of preoptic neurons and hypothalamic integrative function in hibernators and nonhibernators. In: Wang LCH, Hudson JW (eds) Strategies in cold: natural torpidity and thermogenesis. Academic Press, London New York, pp 267 – 297
Wünnenberg W, Kuhnen G, Laschefski-Sievers R (1986) CNS regulation of body temperature in hibernators and non-hibernators. In: Heller HC, Musacchia XJ, WAng LCH (eds) Living in the cold. Elsevier, Amsterdam New York, pp 185 – 192
Yacoe ME (1983a) Adjustments of metabolic pathways in the pectoralis muscles of the bat, *Eptesicus fuscus*, related to carbohydrate sparing during hibernation. Physiol Zool 56:648 – 658
Yacoe ME (1983b) Maintenance of the pectoris muscle during hibernation in the big brown bat, *Eptesicus fuscus*. J Comp Physiol 152B:97 – 104
Yacoe ME (1983c) Protein metabolism in the pectoralis muscle and liver of hibernating bats, *Eptesicus fuscus*. J Comp Physiol 152B:137 – 144
Zhou ZQ, Dryden WF, Wang LCH (1987) Seasonal and temperature dependent differences in the staircase phenomenon of heart tissues from Richardson's ground squirrel. J Therm Biol 12:167 – 169
Zimmerman ML (1982) Carbohydrate and torpor duration in hibernating golden-mantled ground squirrel (*Citellus lateralis*). J Comp Physiol 147B:129 – 135
Zucker I, Dark J (1986) Neuroendocrine substrates of circannual rhytms in ground squirrels. In: Heller HC, Musacchia XJ, Wang LCH (eds) Living in the cold. Elsevier, Amsterdam New York, pp 351 – 358

Chapter 2

Water Vapor Absorption by Terrestrial Organisms

M. J. O'Donnell[1] and *J. Machin*[2]

Contents

1 Introduction

1.1 Biological Significance of Water Vapor Absorption

Water vapor absorption (WVA) refers to the energy-dependent processes utilized by terrestrial organisms to obtain water directly from the atmosphere. It provides a source of water that is separate and distinct from that obtained by imbibition of liquids, ingestion of hydrated solids, or metabolism of food. This phenomenon was first noted in mealworm larvae by Mellanby (1932), and has subsequently

[1] Department of Biology, McMaster University, 1280 Main Street West, Hamilton, Ontario, Canada L8S 4K1
[2] Department of Zoology, University of Toronto, 25 Harbord Street, Toronto, Ontario, Canada M5S 1A1

Advances in Comparative and Environmental Physiology. Vol. 2

been documented in over 60 species of arthropods and one desert plant. The phenomenon is of significance to a number of biological subdisciplines, including biophysics, comparative physiology, evolutionary biology, as well as medical and economic entomology.

Biophysicists are often intrigued by WVA because it is a form of biological transport requiring not only that water undergo a change of phase as it is transported, but that it also be moved against enormous differences in its chemical potential. These differences can be more than 50 times larger than those established by electrolyte concentration in the salt glands of birds and reptiles or the loop of Henle of the mammalian kidney. Furthermore, WVA is of some historical interest to biophysicists, because at one time it was thought that this process might represent a form of primary active water transport. This belief probably arose because early experiments could not detect any obvious links between primary solute transport and secondary passive water movements. At one time it was believed that this primary water pump might consist of the lipid layer covering the entire external surface of absorbing arthropods (Beament 1964, 1965). This view has been dispelled by repeated discoveries of localized uptake sites (Noble-Nesbitt 1970a, b; Rudolph and Knülle 1974, 1982; Machin 1975; Dunbar and Winston 1975; O'Donnell 1977; Rudolph 1982b, 1983). Because epithelia are often closely associated with the sites of absorption, WVA has come increasingly to be interpreted and modeled in terms familiar to those involved in epithelial transport (Gerenscer 1983). Indeed, the majority of current models of WVA consider the water vapor to be absorbed as a secondary and entirely passive consequence of primary solute transport by epithelia. Absorption by the desert cockroach is exceptional in that epithelia are not an integral part of the absorption mechanism.

Studies of WVA are of relevance to evolutionary biologists and comparative physiologists because it is a transport process which has evolved independently a number of times. Dramatically different mechanisms are utilized, involving rectal or oral sites, modified salivary glands or excretory organs, or structures which appear to have arisen de novo solely for the absorption of water vapor.

WVA is of importance to medical entomologists because it represents a significant component of water balance in disease-carrying ticks which survive long periods off the host. Other acarines capable of absorption, such as house dust mites, are of medical significance as causative agents of allergies.

Lastly, the temperature and humidity regimes over which WVA occurs in various arthropods are parameters of great importance to economic entomologists. Conditions suitable for WVA may occur in stored products, thereby permitting infestations of mites and some types of beetle larvae.

WVA has recently been reviewed as a component of overall water balance (Machin et al. 1982; Wharton 1985; Needham and Teel 1986), or from an evolutionary and ecophysiological standpoint (Knülle 1984). Experimental techniques used in the study of WVA have been reviewed by Machin (1984). The present review emphasizes mechanistic and comparative aspects of the process. We are concerned here with principles of functional design which emerge from comparisons of a number of water vapor absorption systems. We believe that such a review is timely because the number of species known to be capable of WVA is rapidly increasing. In addition, a number of quite different localized sites for WVA have been discovered and mechanisms involving electroosmosis, hygroscopic solutions of organic of inorganic solutes, and hydrophilic properties of specialized areas of cuticle have been proposed.

1.2 Thermodynamic and Energetic Aspects

1.2.1 Water – Some Physical Principles

The partial pressure exerted by water molecules is defined as the vapor pressure (vp_1). The saturation vapor pressure (vp_0) corresponds to the equilibrium condition when a sample of air is enclosed over a plane surface of pure water, and varies only with the temperature.

If the air is not saturated, its vapor pressure can be expressed as a fraction of its value at saturation. This value is referred to as the water activity (a_w):

$$a_w = vp_1/vp_0 \ . \tag{1}$$

The percent relative humidity (% RH) is then given by:

$$\%\ RH = (100)\, a_w \ . \tag{2}$$

Water activity can also be calculated for solutions. For pure water in the liquid state, a_w is assigned a value of unity. For an ideal solution, a_w can be shown to be equal to the mole fraction of solvent (X_1) as follows:

$$X_1 = n_1/(n_1 + n_2) \ , \tag{3}$$

where n_1 = moles of solvent, n_2 = moles of solute.

The relationship between vapor pressure and the mole fraction of solvent is given by Raoult's law;

$$vp_1 = X_1 vp_0 \ . \tag{4}$$

From Eqs. (1) and (4) it is apparent that:

$$a_w = X_1 = n_1/[n_1 + n_2] \ . \tag{5}$$

For an aqueous solution the concentration of pure water is 1000/V = 1000/18.0145 = 55.5 mol kg^{-1}, and n_2 is the osmotic pressure in osmol kg^{-1}. Substitution of these values into Eq. (5) gives:

$$a_w = [55.5\ \text{mol kg}^{-1}]/[55.5 + \text{osmol kg}^{-1}] \ . \tag{6}$$

By rearrangement,

$$\text{osmolality (osmol kg}^{-1}) = [55.5 - 55.5\,(a_w)]/a_w \ .$$

To convert osmolality to an equivalent hydrostatic pressure, in kPa, the following relationship is used:

$$\text{kPa} = (RT)\ \text{osmol kg}^{-1} \ . \tag{7}$$

For example, at 25 °C, 1 osmol kg^{-1} = (8.31471 · kPa deg^{-1}) (298) = 2477.8 kPa (= 24.45 atm).

Osmotic pressure is often measured in terms of the freezing point depression (Δt). The relationship is:

$$\Delta t = 1.86\,°C\ osmol^{-1}\,kg^{-1}\ . \tag{8}$$

It is important to note that the water activity of the blood of terrestrial animals, even those able to take up water vapor from lower ambient water activities, is in the region of 0.995, equivalent to a relative humidity of 99.5%. At all water activities (humidities) below this value, water is lost to the environment. However, atmospheric absorbers are distinguished by their ability to maintain a fluid compartment that is distinct from the blood and is of sufficiently reduced activity for water vapor to pass thermodynamically "down hill" into the absorbing fluid. The minimum activity of the absorbing fluid represents the pump threshold of the absorption mechanism.

Some idea of the magnitude of the forces involved in absorption can be gained by converting the lowest absorption threshold known, 43% RH, into osmotic and hydrostatic terms. This humidity is equivalent to the water activity in a solution whose osmolality is 74 osmol kg^{-1}, which in hydrostatic terms represents a pressure of 1.8×10^5 kPa (1800 atm).

1.2.2 Critical Equilibrium Activity and Pump Threshold

The pump threshold is distinct from a parameter related to whole animal water balance, the critical equilibrium activity (Wharton and Devine 1968). At the critical equilibrium activity (CEA) the rate of water influx from the environment onto the site of uptake equals the rate of efflux from the nonabsorbing surfaces of the body. The CEA is the lowest activity, therefore, at which net gain of water occurs, and is usually calculated by extrapolating plots of uptake rate as a function of a_v to zero uptake. The critical equilibrium humidity (CEH) is the equivalent parameter expressed in terms of relative humidity (% RH; Knülle and Wharton 1964). Although these values are of considerable ecological significance, they are of less utility for analysis of the mechanisms of absorption. This is because the minimum humidity necessary for whole animal water balance depends on both gains and losses, whereas the minimum humidity for active uptake is a characteristic of the absorption mechanism alone. The term "pump threshold" has been suggested to denote the true physiological threshold of the pump site (Machin 1979a). In species with highly waterproofed cuticles, loss rates are a negligible fraction of uptake rates, and the CEA is the same as the pump threshold. However, if loss rates are significant, the CEA value is an overestimate of the pump threshold. A graphic procedure (Fig. 1) is used to calculate the pump threshold from gain and loss rates above and below the CEA (Machin 1979a).

1.2.3 Absorption Capacity

Figure 1 also shows that the relationship between uptake flux and atmospheric activity can be of two forms, either linear where the uptake mechanism appears to be nonsaturable, or curved so that there is evidence for at least partial saturation.

Comparison of WVA in organisms of different species or sizes is difficult because of wide variations in absorption rate, threshold, and the conducting properties of the absorption surfaces. Coutchié and Machin (1984) introduced the parameter "absorption capacity" (Fig. 1, shaded areas) as a means of integrating the various components of absorption to permit such comparisons. Absorption capacity is calculated from the function describing maximum uptake flux in saturated air ($a_w = 1.0$) multiplied by the activity range over which absorption occurs (1.0, threshold activity). In other words, it is the product of the driving force established by the uptake mechanism and its conductive capacity. Conductance increases with the area of the absorbing structures, and decreases with the diffusion distance between these structures and the atmosphere. It also increases with frequency if the mechanism happens to be oscillatory. A second parameter introduced by Coutchié and Machin (1984) "absorption conductance" permits the conductive properties alone to be compared between species. This is obtained by dividing maximum flux by activity

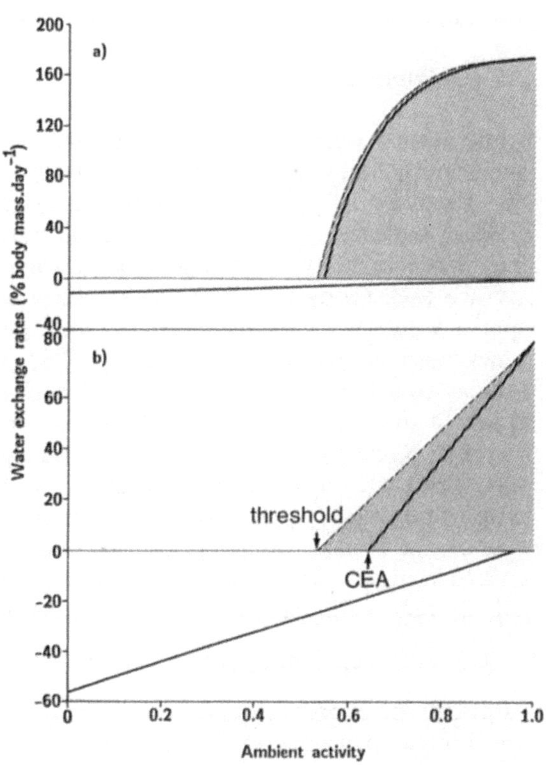

Fig. 1a, b. Diagram illustrating different patterns of net water uptake (positive values) and net loss (negative values) shown by the *bold lines*. Losses above CEA are obtained either by extrapolation from sub-CEA values or by using animals with disabled vapor uptake mechanisms. Total vapor uptake (*dotted line*) is obtained by correcting net gains for net losses at corresponding activities. The *shaded area* represents "uptake capacity". ***a*** A species (*Liposcelis rufus*) showing low integumental permeability and nonlinear or saturation uptake kinetics. It can seen that the difference between CEA and vapor uptaked mechanism is negligible. ***b*** A species (*Xenopsylla cheopis*) showing linear, nonsaturable uptake kinetics and a comparatively permeable integument, leading to a significant discrepancy between CEA and threshold. (Redrawn from Machin 1979b)

range over which vapor uptake is possible (1.0, threshold activity). An example of the use of these parameters is given below (Sect 3.2.1).

1.2.4 Energetics of Absorption

Although water vapor uptake involves transport against steep gradients, theoretical estimates and experimental measurements both indicate that the energetic cost is relatively small. Edney (1977) used the relationship between free energy change and the ratio of vapor pressures for air and hemolymph;

$$\Delta G = RT \ln vp_1/vp_0 \tag{9}$$

to show that absorption by *Arenivaga* at 6% body weight per day involves an expenditure of energy which is less than 1.5% of the animal's basal metabolic rate. Maddrell (1971) analyzed uptake by *Thermobia* and showed that even if the uptake mechanism works at an efficiency of only 10%, only 18 μg glucose, less than 0.07% of the body weight, would need to be catalyzed to provide energy for 1 h of absorption. As noted by Maddrell (1971), the economy of WVA is perhaps best appreciated by comparing its cost to that of insect flight, which consumes fuel at the rate of 0.6 to 20% of the body weight per hour, depending on the species.

1.3 Impact of Vapor Uptake on Water Balance

1.3.1 Allometric Considerations

Although it is common practice to express exchange rates in terms of a percentage of initial body mass to standardize body size, this practice may in fact hinder the discovery of important allometric relationships. For example, Rudolph (1982a) collected data for psocopterans ranging over almost three orders of magnitude in body mass. Potentially, when water loss rates are scaled with surface area (body mass$^{0.67}$) other factors being equal, the smallest of these species would be expected to lose the same proportion of its water reserves four times more rapidly than the largest (Fig. 2a). Taking vapor absorbers as a whole, ranging over five orders of body mass (Machin 1979a) relative water loss may be expected to range by a factor of over 50 from the smallest to the largest individual. Recalculated data from Rudolph's psocopterans shows absolute water loss to be scaled nearly proportionately with body mass$^{1.16}$. This means that the proportion of body mass lost per unit time (flux/body mass) is nearly independent of size. Thus, it would appear that in psocopterans, disadvantageous surface area to volume ratios in smaller species are compensated for in some way, probably by reduced integumental conductances, scaled as follows:

$$\text{Conductance} \propto \text{body mass}^{1.16}/\text{body mass}^{0.67} = \text{body mass}^{0.49}.$$

Although the opportunities for comparison with nonvapor-absorbing arthropods are limited, psocopterans appear to be unusual in compensating for the dis-

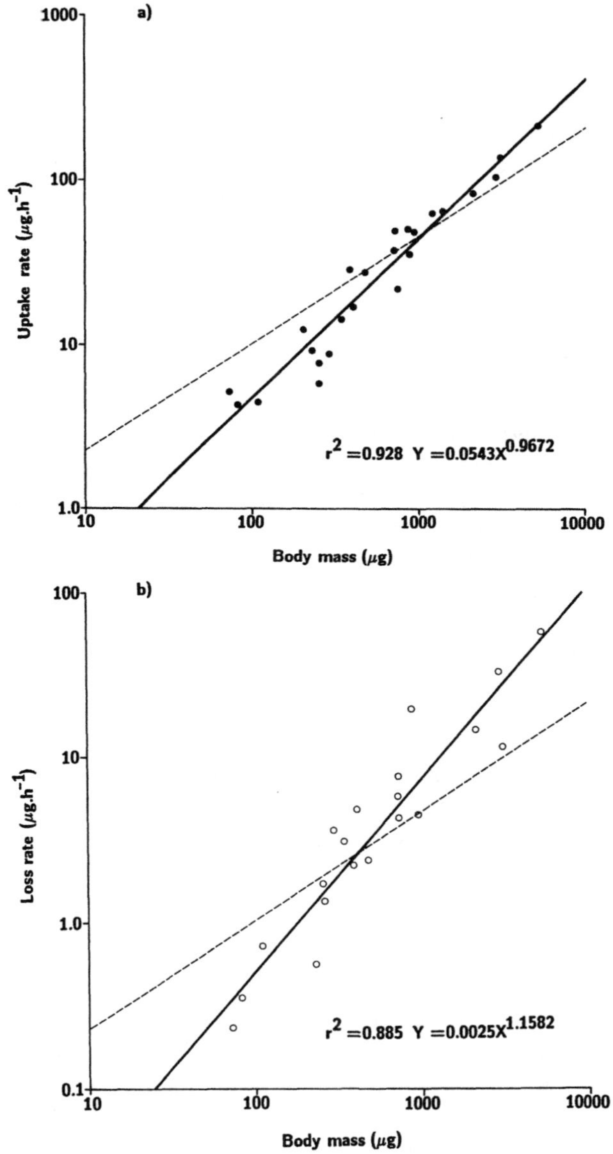

Fig. 2a, b. Whole animal net uptake (*a*) and loss (*b*) rates of psocopteran species, compared with their initial body mass, recalculated from data in Rudolph (1982a). Equations and r^2 values for the relationships are also given. If rates were scaled with surface area, in proportion to body $mass^{0.67}$, they would parallel the *dashed line*

advantages of small size. Weight loss rates in a group of desert tenebrionid beetle species or of differently sized individuals of the same species of desert beetle (Edney 1977) and a scorpion (recalculated data from Hadley 1970) were found to be scaled with body mass exponents from 0.31 to 0.66, rather than 1.0. In these examples integumental conductances either remain constant or increase in smaller individuals, thus failing to compensate for disadvantages of small size.

Water vapor absorption in Rudolph's psocopterans (Fig. 2b) as well as in a collection of insect and acarine species (Machin 1979a) is also scaled almost proportionately with body mass, suggesting that a simple relationship might exist between water loss and the ability to compensate for it by vapor uptake. Uptake of water vapor as a means of replacing evaporative water loss has number of practical advantages. It generally takes place more rapidly than evaporative loss, therefore occupying comparatively short periods, it does not require extra solutes to be excreted, and it does not involve the complications of food handling and elimination. Since the absorption threshold is not correlated with body size, it must be supposed that the scaling of uptake with body mass is achieved by disproportionately enlarged absorbing structures in the heavier species.

1.3.2 Environmental Considerations

Many but not all species capable of WVA inhabit environments in which sources of liquid water are widely separated or infrequently available. Desert species such as the Tenebrionid beetles of Namibia and the burrowing cockroach *Arenivaga investigata* of the southwestern United States exploit humid microenvironments beneath the surface of the sand.

However, even in mesic environments, small terrestrial arthropods, especially mites, are in constant danger of desiccation because their surface area is so large relative to their volume. For example, the house dust mite *Dermatophagoides farinae* has a surface area which is over 50-fold larger, per unit volume, than that of a large insect such as a locust. Because there are limits to the ability of organisms to limit cuticular and respiratory water losses, WVA may be of critical importance for water balance in the terrestrial environment.

WVA is also important to water balance in ticks, which may live for several years between blood meals and pass up to 98% of their life cycle off the host (Needham and Teel 1986).

Lastly, since WVA provides an alternative source of water, some vapor-absorbing species exploit relatively dry foodstuffs, such as dried hair or feathers for biting lice (Rudolph 1983), or stored grain products for flour mites (Wharton 1985) and mealworm beetle larvae (Machin 1975).

However, it is important to point out the studies by Rudolph (1982a, b, 1983) and Knülle (1984) which demonstrate that the adaptive value of WVA to some species is not always readily apparent. Probably all Psocoptera are capable of WVA, even though the majority inhabit environments where liquid water is readily available. Conversely, some species of Mallophaga have lost the ability to absorb water vapor, and others, feeding on the same keratinous diet, have retained it.

2 General Features of Absorption Systems: Sites, Structures, and Mechanisms

2.1 Sites and Structures

As noted above, sites of uptake in arthropods are limited to the rectal or buccal region. Vapor absorption may have exploited existing water-transporting capabilities since the gut in arthropods is the principal organ of energy-related water transport. In all cases initial vapor absorption in the buccal region takes place superficially but several different types of structures and mechanisms are involved (Fig. 3). In the ticks, it appears likely that a solution secreted onto the mouthparts during periods of absorption (Fig. 3a) is produced by the salivary glands. In mites fluid is produced by supracoxal glands and flows through the podocephalic canal, which is an open gutter carrying secretion from the openings of the salivary and supracoxal glands above the base of the foreleg across the supracoxal plate and the base of the palps and into the prebuccal cavity. In the booklice and biting lice, modified hypopharyngeal structures called lingual sclerites are protruded from the mouth during absorption (Fig. 3b, c). Secretions of the dorsal labial glands flow across grooves in the sclerites in response to suction produced by a complex pump located at the posterior of the prebuccal cavity. The desert cockroach utilizes a quite different variant of oral uptake. The hypopharynx is quite elastic, rather than sclerotized, and is inflated with hemolymph so as to protrude from the mouth (Plate 1a; Fig. 3d). Two lateral diverticulae, termed hypopharyngeal bladders, are the sites for condensation.

In order to exploit rectal reabsorption mechanisms humid air must first be drawn internally through the anus. There appear to be at least three different types of rectal mechanisms (Fig. 4). The rectal uptake site of the thysanurans is probably the simplest anatomically (Fig. 4a, b). The posterior region of the rectum consists of a single layer of epithelial cells thrown into folds so as to form saccules. In flea larvae, the rectal sac consists of two distinct cell types in a gutter-like arrangement (Fig. 4c, d). Cells of the dorsal gutter have highly folded apical plasma membranes, whereas cells of the ventral gutter show deep infoldings of the basal (hemolymph-facing) plasma membrane. The most anatomically complex absorption site is the cryptonephridial system found in tenebrionid larvae (Plate 1b; Fig. 4e, f). In the excretory system of these insects, the distal lengths of the Malpighian tubules are held closely to the surface of the rectum by an enveloping membrane. Much of the rectal surface is utilized for condensation, and the associated rectal epidermis appears to be a passive compartment between the rectal lumen and the site of reduced water activity within the Malpighian tubules of the cryptonephridial complex.

The succulent-leaved Atacama desert shrub, *Nolana mollis*, utilizes salt glands for WVA. The glands produce salts that cover the leaves and are effective in condensing moisture.

2.2 Mechanisms: Theoretical Overview

There is a limited number of mechanisms by which an organism could conceivably reduce water activity sufficiently for absorption of atmospheric water vapor.

2.2.1 Active Depression of Dew Point

Insects such as the fog basking beetles of the Namibian desert (Seely and Hamilton 1976) may exploit environmental temperature differences to condense

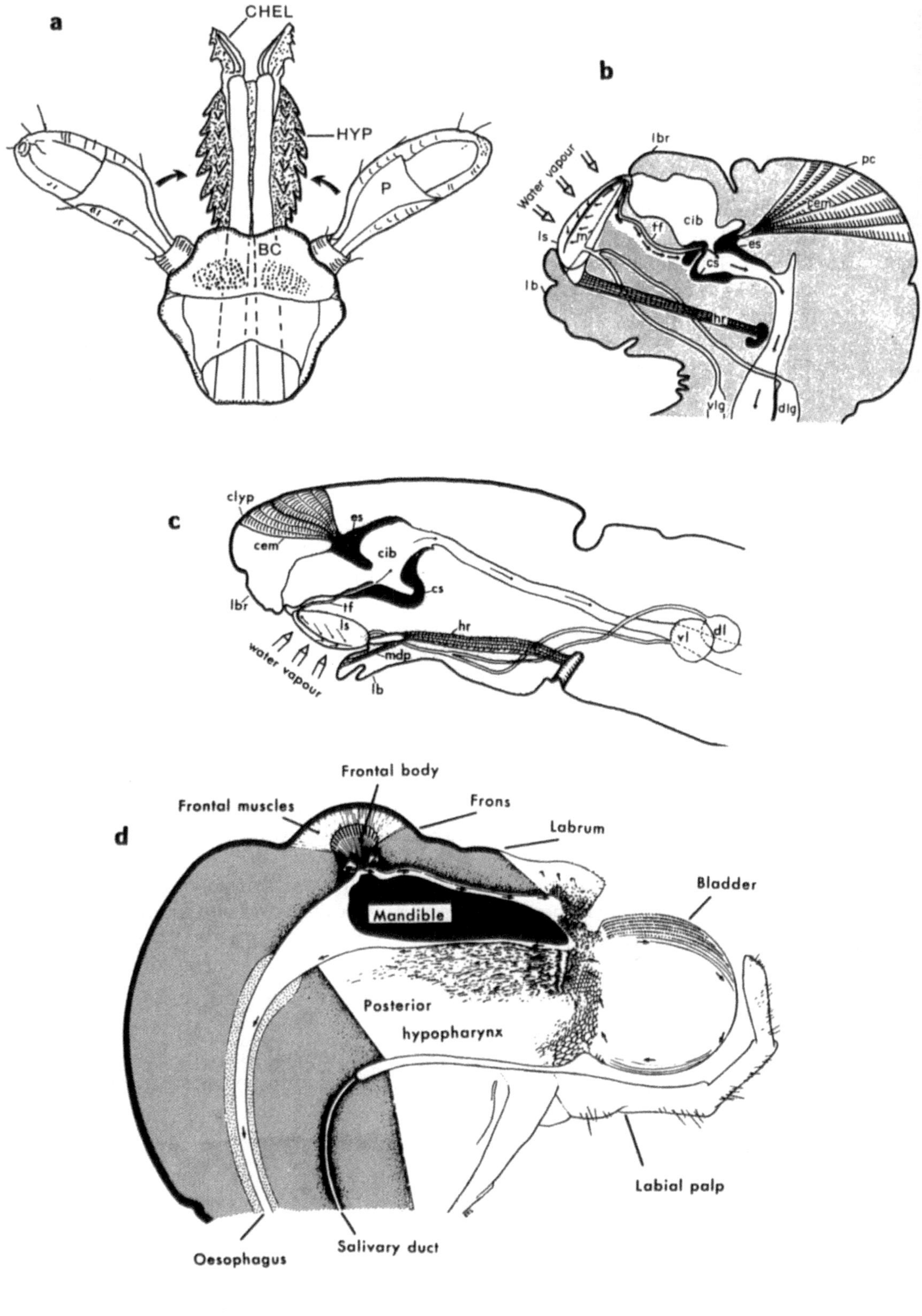
a
CHEL
HYP
P
BC
b
lbr
pc
cem
Water vapour
cib
tf
es
cs
ls
m
lb
hr
vlg
dlg
c
clyp
es
cem
cib
cs
lbr
tf
ls
hr
vl
dl
mdp
lb
water vapour
d
Frontal body
Frontal muscles
Frons
Labrum
Bladder
Mandible
Posterior
hypopharynx
Labial palp
Oesophagus
Salivary duct

water from the atmosphere. But no insects or other animals have been shown to actively reduce the temperature of part or all of the body surface below the dew point of the surrounding air.

2.2.2 Colligative Properties – Organic Versus Inorganic Solutes

An absorption mechanism based on the colligative properties of a solution of inorganic salts will be physically limited by the solubility of the particular salts. However, in practice a sizable difference exists between the absorption threshold and the activity of the saturated solution of the salt involved. It seems reasonable to assume that this discrepancy represents a margin of safety, ensuring that salts in the absorbing fluid are not precipitated. This is a potential danger with vapor absorption in fluctuating humidities. For example in *Tenebrio* larvae, where the composition of the absorbing fluid has been shown to contain about 95% KCl (Ramsay 1964) minimum threshold activities were 0.894, whereas saturated KCl has an activity of 0.85. The detection of salts in tissue compartments suspected to comprise the absorption system, therefore, must be consistent with the empirical absorption threshold of the absorption site, making an allowance for the margin of safety described above.

If we assume that there is a correlation between CEA and the solubility of the salts employed to lower activity, errors in CEA not withstanding, Fig. 5 provides some indication of the probable importance of the different solutes involved in vapor uptake. CEAs most commonly found are higher than that of saturated NaCl but lower than KCl, suggesting that NaCl is the salt of preference. A small group to the right of this peak with CEAs greater than that of saturated KCl, suggests some animals, such as *Tenebrio* larvae, may be more dependent on this salt alone. However, it can be seen that a sizable proportion of species have CEAs to

Fig. 3a–d. Survey of buccal structures involved in WVA. *a* Drawing showing positions of tick mouthparts during absorption. The palps (*P*) are splayed, as indicated. During non-absorbing periods they are moved close to the hypostome (*HYP*), as indicated by the *arrows*. *CHEL*, chelicerae; *BC*, basis capituli. (After Evans et al. 1961). *b* Sagittal section through a psocopteran head showing structures involved in WVA. The probable route of condensed water from the site of condensation, i.e., the ventral surface of the hypopharynx, to the gut is indicated by the *arrows*. *cem*, clypeo-epipharyngeal muscle; *cib* cibarium, *clyp* clypeus; *cs*, cibarial sclerite; *dlg*, dorsal labial gland; *es*, epipharyngeal sclerite; *hr*, hypopharyngeal retractor; *lb*, labium; *lbr*, labrum; *ls*, lingual sclerite; *m*, membranous portion of the ventral hypopharynx; *pc*, postclypeus; *tf*, tubular filament; *vlg*, ventral labial gland (Rudolph 1982b). *c* Sagittal section through the head of an absorbing biting louse showing structures involved in WVA. The mouthparts are presented in the typical absorbing position. *Arrows* indicate the probable route of condensed water movement from the site of condensation, the lingual sclerites, to the gut. Abbreviations differing from those of *b* are as follows: *dl*, dorsal (acidophile) labial gland; *mdp*, musc. dorsalis praementi; *vl*, ventral (basophile) labial gland (Rudolph 1983). *d* Drawing of the head of the desert cockroach with portions removed so as to expose the distal end of the epipharynx and the adjacent hypopharynx. Parts of the head in sagittal section are *stippled*. The *arrows* indicate fluid movements from the site of ultrafiltration in the frontal bodies to the bladders and then to the esophagus. Atmospheric water condenses onto the bladders and is entrained in fluid movements toward the esophagous (O'Donnell 1981b)

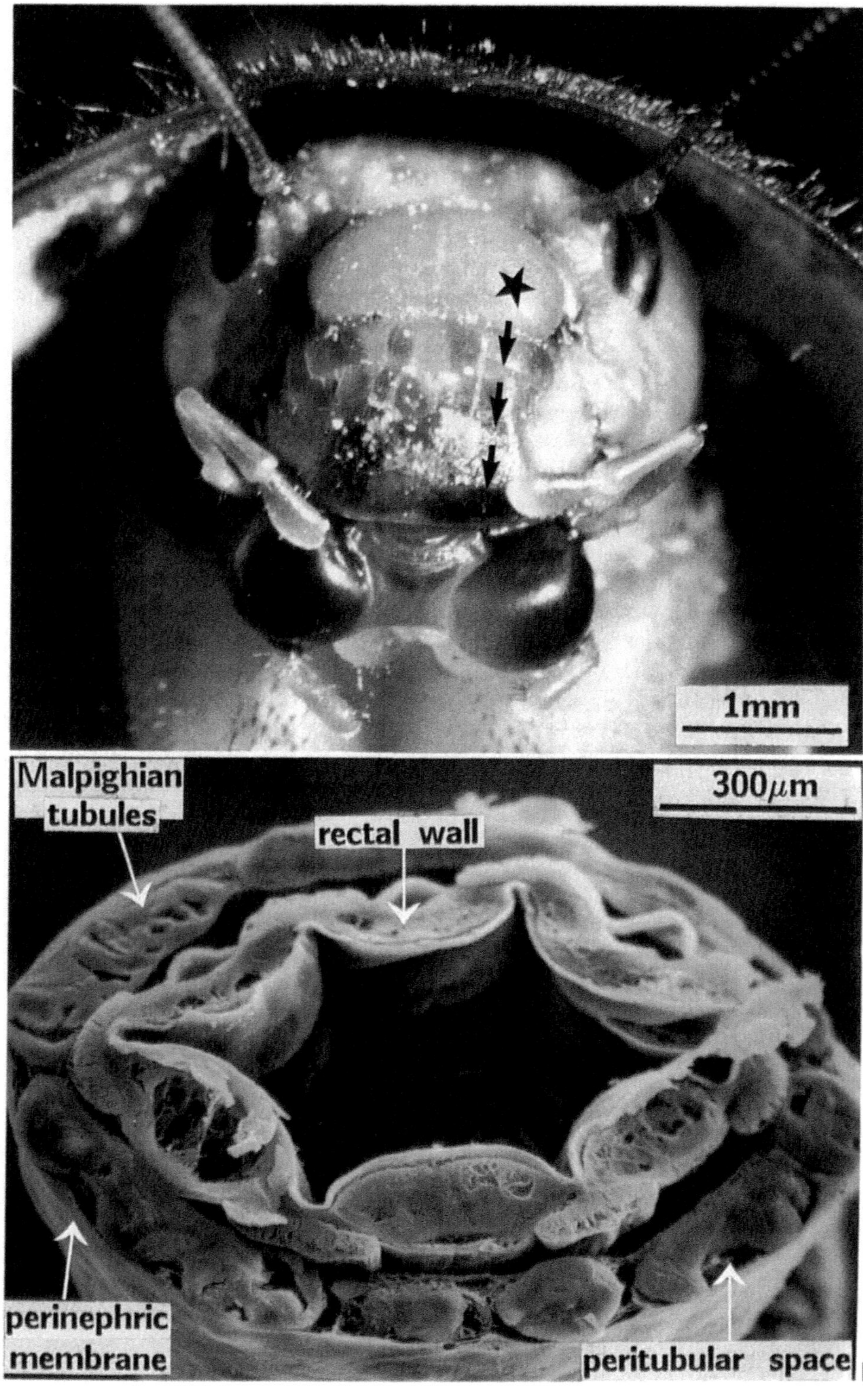
1mm
a
Malpighian tubules
300μm
rectal wall
perinephric membrane
peritubular space
b

the left of the mixed saturated Na/KCl activity, 0.70. Since there are no other salts which could be concentrated by known biological mechanisms, or indeed tolerated in such high concentrations, the use of inorganic solutes are considered likely to be employed in this group.

In contrast to inorganic salts, hydrophilic organic solutes do not show a clear limit of solubility. It has been suggested a number of times that insects might make use of concentrated organic solutions. These could be produced either directly, much as freezing-resistant insects produce high concentrations of glycerol (Zachariassen 1979), or by depolymerization of a macromolecular precursor. Insect epidermal cells are known to produce lytic enzymes which convert glycoproteins of the endocuticle into smaller molecules such as amino acids and glucosamines (Locke 1974). However, Edney (1977) has pointed out that release and solution of amino acids would probably generate no more than a 1–2 osmolar solution, much less than the 50–60 osmolar gradients found in insects absorbing at thresholds of a_v 0.5 or below.

2.2.3 Surface Phenomena and Capillarity

The fine tubular structures observed in electron micrographs of insect cuticle lead a number of authors to suggest that reduction of vapor pressure over the surface of fluid contained in these capillarylike spaces might be involved in absorption (Beament 1964, 1965). The trough-shaped spaces between fine hairlike extensions of the cuticle might act similarly. The Kelvin equation (Defay and Prigogine 1966) relates the equilibrium vapor pressure over a curved surface to its radius of curvature (r):

$$a_v = p_1/p_0 = e^{-2\gamma V/rRT} , \tag{10}$$

where γ = surface tension (dynes cm^{-1}) and V is the partial molar volume of water ($cm^3\ mol^{-1}$).

For a trough, the surface is curved in only one direction and equation reduces to $a_v = e^{-\gamma V/rRT}$.

These equations assume that the surface of the pore or trough is wettable, i.e., that the contact angle is zero.

Plate 1a, b. Photographs of (*a*) the head of the desert cockroach *Arenivaga investigata* during vapor absorption and (*b*) a scanning electron micrograph of the cryptonephridial complex of the mealworm *Tenebrio molitor*, showing the concentric arrangement of the rectum and Malpighian tubule sheath. These examples from either end of the arthropod gut, functioning by entirely different mechanisms, symbolize the wide range of specialized water vapor transporting structures available for study. In *Arenivaga* the bladders, the side of vapor absorption, are the two dark structures protruded from the mouth. Fluid produced internally by the frontal bodies (one indicated by the *star*) follows the path to the bladders indicated by the *arrows*

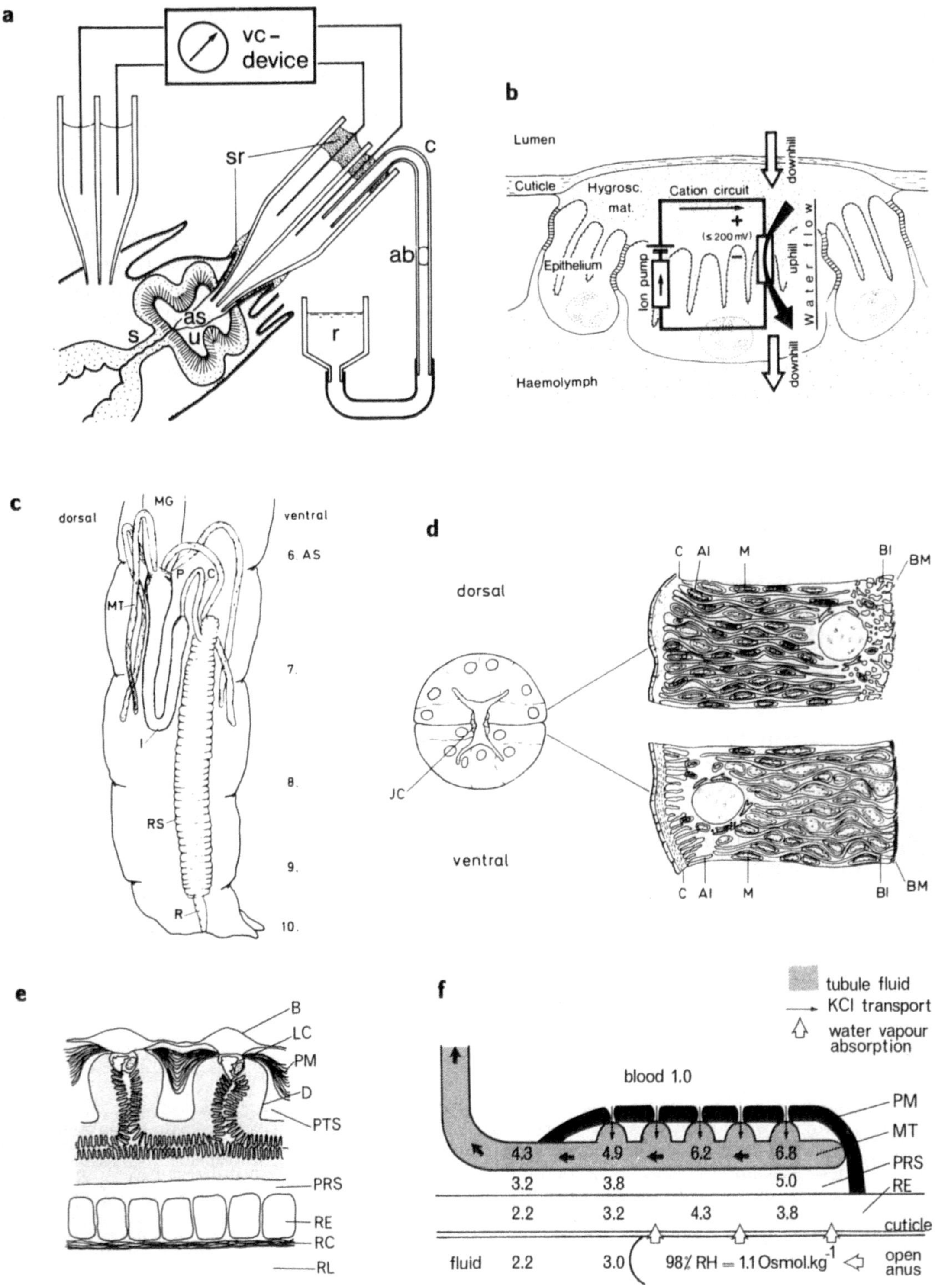
a
vc -
device
sr
sr
c
ab
as
s
u
r
b
Lumen
Cuticle
Hygrosc.
mat.
Cation circuit
+
(≤ 200 mV)
−
Epithelium
Ion pump
uphill
downhill
downhill
Water flow
Haemolymph
c
dorsal
ventral
MG
6. AS
P
C
MT
7.
I
8.
RS
9.
R
10.
d
dorsal
ventral
C
AI
M
BI
BM
JC
C
AI
M
BI
BM
e
B
LC
PM
D
PTS
PRS
RE
RC
RL
f
tubule fluid
KCl transport
water vapour
absorption
blood 1.0
PM
MT
PRS
RE
cuticle
4.3
4.9
6.2
6.8
3.2
3.8
5.0
2.2
3.2
4.3
3.8
fluid
2.2
3.0
98% RH = 1.1 Osmol.kg^{-1}
open
anus

2.2.4 *Hydrophilic Macromolecules*

It is well known that proteins, if moved away from the isoelectric point, expose hydratable polar groups with a resultant increase in net attraction for water molecules. Changes in temperature, ionic strength, or voltage can also bring about discontinuous changes in the volume and water content of polyelectrolytes such as proteins (Tanaka 1981).

Polyelectrolytes shrink as ionic strength is increased because increasing salt concentration reduces the mutual repulsive forces between like charges on a protein or gel structure. As a result, more noncovalent cross-links are possible, and shrinkage results. The swelling of a polyelectrolyte, which increases with the square of the gel's net charge, therefore varies inversely with the ionic strength of the surrounding medium. Figure 6 shows that over a particular range of ionic strengths, infinitesimal changes in the concentrations of inorganic salts can produce enormous changes in the water affinity, and therefore the volume, of polyelectrolytes. Such changes are fully reversible. Cyclical changes in ionic strength are proposed as the means by which water is periodically released from the hydrophilic mouthparts during WVA by the desert cockroach (Sect. 3.3).

2.2.5 *Electroosmosis*

A current (I) passing through aqueous channels will induce an electroosmotic volume flow (J_v) as given by:

$$J_v = BI,$$

where B is the electroosmotic permeability. Maximal efficiency of electroosmotic water transport will be achieved when the frictional force between the water and the pore wall is much less than that between the counterions and water. Under

Fig. 4a–f. Survey of rectal structures involved in WVA. ***a*** Experimental setup for measuring electroosmotic absorption in the rectum of ***Thermobia***. *vc-device*, voltage clamp; *sr*, silicon rubber seal; *as*, lumen of anal sac; *u*, subcuticular space; *s*, sphincter; *c*, calibrated cannula; *ab*, air bubble; *r*, reservoir. Outer diameter of the anal sac: 300 to 400 μm; setup not to scale (Küppers et al. 1986). ***b*** The proposed model for electroosmotic uptake of water vapor (Küppers et al. 1986). ***c*** The different regions of the hindgut in larvae of the oriental rat flea. *AS*, abdominal segment; *C*, colon; *I*, ileum; *MG*, midgut; *MT*, Malpighian tubule; *P*, pylorus; *R*, rectum; *RS*, rectal sac (Bernotat-Danielowski and Knülle 1986). ***d*** A schematic representation of the rat flea rectal sac in transverse section (*left*), and the morphology of its two different cell types (*right*). *AI*, apical infoldings; *BI*, basal infoldings; *BM*, basement lamina; *C*, cuticle; *JC*, junctional cell; *M*, mitochondria (Bernotat-Danielowski and Knülle 1986). ***e*** Sagittal section of *Tenebrio* larvae rectal complex showing structural interrelationships between the Malpighian tubules, the perinephric membrane, and the rectum. *B*, basement membrane blister; *LC*, leptophragma cell; *PM*, perinephric membrane; *PTS*, peritubular space; *PRS*, perirectal space; *RE*, rectal epidermis; *RC*, rectal cuticle; *RL*, rectal lumen (Machin 1983). ***f*** Diagram summarizing osmotic gradients and presumed water and ion flows in *Tenebrio* larvae rectal complex during WVA. *PM*, perinephric membrane; *MT*, Malpighian tubule; *PRS*, perirectal space; *RE*, rectal epidermis (Machin 1983)

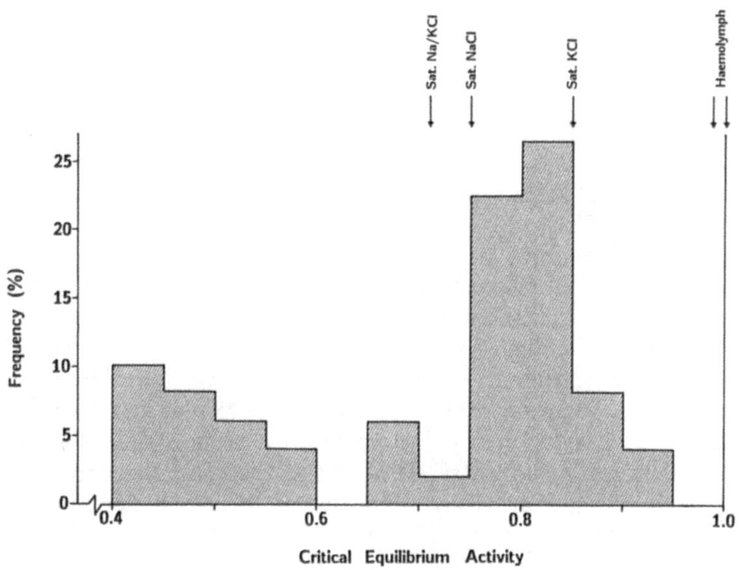

Fig. 5. Frequency of CEA from all animal taxa of vapor absorbers, taken from the literature (n = 48 species). The range of hemolymph activities as well as those of saturated potassium and sodium chlorides (Winston and Bates 1960) are also indicated

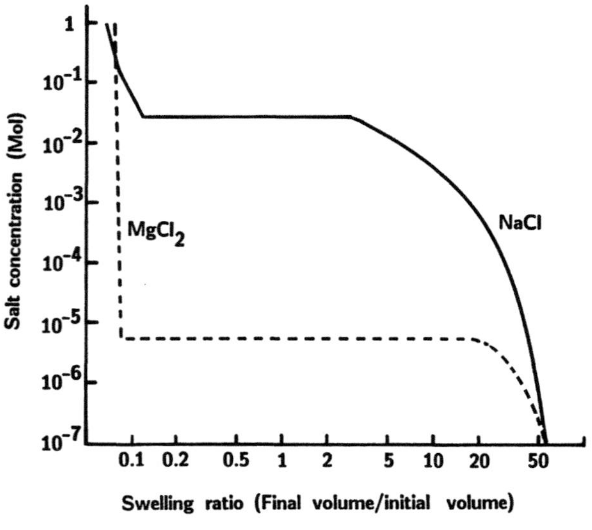

Fig. 6. Hypothetical logarithmic plot of the swelling ratio of a polyelectrolyte as a function of salt concentration. Although this plot is based on data for synthetic polyacrylamide gels (Tanaka 1981), biological polyelectrolytes such as proteins behave similarly

these conditions, $B = 1/(FC_i)$ (House 1974), where F is the Faraday and C_i is the concentration of ions.

2.2.6 Pressure Changes

The possible use of pressure changes in the rectum of *Thermobia* as a basis for WVA has been examined and rejected on theoretical grounds (Maddrell 1971).

Air might be taken in by dilating the hindgut, and then compressed by closing the anus and constricting the muscles of the hindgut. As the volume of air was reduced, its relative humidity would rise, facilitating the absorption of water vapor. However, calculations indicate that even if the volume were reduced twofold, the rectum would have to fill and empty at an unrealistically high frequency, more than 80 times s^{-1}. These calculations are supported by direct manometric measurements of the volume of enclosed specimens of *Thermobia*. During a period when uptake occurred and the anal valves pulsated, there were no measurable cyclical volume changes. Nonetheless, the function of the pulsations of the anal valves remains problematic. Noble-Nesbitt (1977) suggests that they are likely to be involved in tidal replenishment of the air in the tidal sacs. However, Maddrell cites calculations by the late T. Weis-Fogh indicating that water vapor can enter by diffusion at the necessary rate, providing that the absorbing surfaces are not deeper within the animal than about 1 mm.

Wharton (1985) has offered the intriguing suggestion that pressure changes in the respiratory system, normally thought to be a prime avenue for water loss in terrestrial arthropods, might under special circumstances result in a net gain of water. During one phase of a type of respiratory cycle known as intermittent respiration, the spiracles are partially occluded. Solution of O_2 in the body fluids results in a drop of pressure within the tracheal system, so that a consequent bulk entry of air may occur. Since this movement of air is opposite to the direction of evaporation from the respiratory surfaces, water loss is reduced. Enough water vapor may be carried with this bulk flow so that a net gain of water could result. Because the partial pressure of N_2 in the tracheal system increases above that in the atmosphere, N_2 will normally be expected to diffuse out of the animal, and oppose the bulk entry of air due to solution of O_2. However, if N_2 could be lost through the cuticle, the quantity of air and water which could enter by bulk flow would increase. However, this net gain could only occur if the partial pressure of water vapor exceeded that in equilibrium with the hemolymph, which has a water activity in excess of 0.99. It seems, therefore, that net gain of water could only occur if ambient humidity was virtually saturated, i.e., in excess of 99% RH.

2.3 Mechanisms: Experimental Criteria for Assessment of Absorption

A number of techniques and analytical procedures have been shown to be particularly useful in characterizing absorption mechanisms. The discussions below also point out certain pitfalls inherent in some of these techniques.

2.3.1 Gravimetric Analysis of Weight Change in Controlled Humidities

Much of our understanding of WVA has been gained by periodic or continuous weighing of animals in known humidities. For humidity control, widespread use is still made of saturated salt solutions (Winston and Bates 1960). More sophisticated methods permitting precise control of any combination of temperature and humidity are available, and are reviewed elsewhere (Machin 1984). Periodic weighing of animals as a means of estimating the instantaneous rate of weight gain suffers from a serious flaw. Interference with the animal's behavior may prevent absorption for some time after the animal is returned to above threshold humidities, and rates of weight gain by absorption would be underestimated. An important advance, therefore, has been the introduction of continuously recording electronic microbalances with sensitivities in the microgram or even submicrogram range.

Analysis of the pattern of weight gain provides important clues to the role of WVA in water balance. It appears that there are two distinct patterns of vapor uptake. In species requiring daily, even hourly replenishment of water loss to maintain water balance, weight traces take on a "sawtooth" pattern of periods of gain alternating with periods of loss (see Rudolph 1982b; Machin 1984). Among such species the critical activity (CEA) (Wharton and Devine 1968) represents the minimum environmental humidity at which this type of regulation can take place. In a much smaller group of atmospheric absorbers, typified by xeric beetle larvae such as *Tenebrio*, whose water loss rates are fractions of a percent of body water reserves per day, atmospheric absorption is probably less important in routine water balance. Normal environmental conditions are probably such that long periods without water must be survived to exploit relatively rare periods of plenty. It is during the latter periods that the ability to absorb water vapor is particularly important.

Moreover, if humidities can be changed rapidly, analysis of nonsteady state weight changes permits important inferences to be made concerning the size and number of the absorbing compartments within an animal. This type of analysis, which has been successful in *Tenebrio* larvae in identifying a fluid compartment which is of reduced water activity and is distinct from the hemolymph (Machin 1976), could potentially be exploited in many of the larger species, particularly those with rectal sites of absorption.

2.3.2 Freezing Point Depression – Thermal Analysis of Frozen Sections as a Test for High Solute Concentrations

Estimation of the osmolarity of fluids or tissue compartments associated with WVA can be calculated from measurements of freezing point depression (Δt). Measurements of Δt in nanoliter quantities of fluid collected by micropipettes suffer from the disadvantage that spatial osmotic gradients may be disrupted by the micropipetting procedure, or that time-dependent alterations of such gradients may occur once the normal physiology of the animal is disrupted by collection of the animals and the dissection necessary to reveal the absorption sites or compartments.

Thermal analysis of frozen sections provides a simple yet effective means of circumventing these problems. Animals can be flash-frozen and cryo-sectioned, and the sections photographed at progressively warmer temperatures maintained by a temperature-controlled stage of a microscope which is itself maintained in the cryostat at -40 °C. Melting of ice crystals is clearly apparent in the photographs, and the direction of sectioning can be altered to permit detection of osmotic gradients within the tissue compartments which comprise the absorption system. Although the method is less precise than nanoliter osmometry, the difference in resolution is of little consequence because the depression of freezing points associated with WVA are very large. In the cryptonephridia of *Tenebrio* larvae, for example, the maximum values are as much as −8° to −12 °C (Machin 1979b). Useful information is also provided by the observed absence of significantly elevated osmolarities detected by Δt. In the desert cockroach, for example, a pair of globular organs beneath the frons of the head is responsible for the production of a fluid during WVA. However, Δt's within these structures are elevated only a few percent above the Δt of hemolymph, whereas values of more than −20 °C would be expected if they were in fact the source or reservoirs for a fluid which lowered a_w colligatively (O'Donnell 1982a).

We believe that the thermal analysis of frozen sections will be of considerable importance in future studies of absorption mechanisms in organisms such as ticks, biting lice, and booklice. In each of these groups, salivary or other glands implicated in absorption are connected by ducts to external surfaces of the mouthparts. It should be possible, therefore, to gain information on the osmolalities of fluids within the fluids of the glands and the ducts by thermal analysis of frozen sections.

2.3.3 Microanalytical Techniques for Organic and Inorganic Solutes

A number of highly sensitive micromethods have been developed within the last 10−15 years for detection and analysis of minute quantities of organic and inorganic solutes in biological tissues or fluids. Bombardment of a sample with a high density neutron flux and subsequent spectrometry of gamma emitting isotopes constitute the technique known as instrumental neutron activation analysis (INAA). This technique permits multielemental analysis on samples smaller than 1 mg dry wt. (Devine 1982a; O'Donnell 1982a) and higher density neutron fluxes should allow even smaller samples to be analyzed. Laser-generated X-ray microradiography can also be used for multielement analysis (Epstein et al. 1979).

An important caveat, however, is that these techniques measure mass, not concentration, of an element within the tissue sample. Moreover, samples are exposed to a vacuum within the experimental chamber during laser-generated X-ray microanalysis. Meaningful data on ionic concentration can only be obtained if the water content of the sample is determined prior to elemental analysis, and this measurement is difficult, of course, for microgram samples of material. A possible solution to this problem is the use of frozen sections (Gupta et al. 1976) in conjunction with an electron microprobe.

Much less attention has been paid to the analysis of organic solutes, in fluids or cellular compartments associated with WVA systems. A limited number of type of organic solutes are suitable for colligative lowering of a_w, because such compounds must necessarily be highly hydrophilic. Sugars, polyhydric alcohols such as glycerol, and primary amines or amino acids have all been suggested as possible candidates. The analysis of primary amines in fmol quantities has been greatly facilitated by the fluorometric technique based on orthopthalaldehyde (Roth 1971), whose use for microanalysis was pioneered by Stephens et al. (1978) for determination of uptake of free amino acids by marine organisms.

Sensitive fluorometric reagents are also available for the detection of reducing sugars and polyhydric alcohols. DANSYL hydrazine (1-naphtholene sulfonyl 5-dimethylamino hydrazine) will complex with reducing sugars with an efficiency of over 90% (Avigad 1977). Because DANSYL hydrazine is itself fluorescent, reaction products must be separated by thin layer chromatography (TLC). As little as 30 pmol glucose can be detected by this method (O'Donnell 1982a). DANSYL hydrazine can also be used to detect glycoproteins (Eckhardt et al. 1976) or polyhydric alcohols (O'Donnell 1982a) after the related aldehydes are produced by periodate oxidation. Visual detection limits for mannitol and glycerol on high performance TLC plates are 1 and 10 nmol, respectively. Because high concentrations, typically in excess of 1 M, are required to reduce water activity sufficiently for absorption, these techniques will probably be useful for analysis of polyhydric alcohols in microgram quantities of tissue or in nanoliter quantities of fluids obtained from the surfaces of absorption sites in arthropods such as ticks, booklice, and biting lice.

2.3.4 Optical Properties and Temperature of Absorption Surfaces

In those absorbing species of arthropods in which the external surfaces of the mouthparts form part of the absorption system, the appearance and the temperature of those surfaces can provide useful information concerning the pump threshold and the functional role of the fluids on the mouthparts.

Measurements of surface temperatures in the desert cockroach have shown that two bladderlike diverticula of the hypopharynx are as much as 0.2 °C warmer than the surrounding mouthparts. Moreover, this differential varies with ambient humidity and, therefore, with the rate of condensation onto the bladder surface. These data indicate that the temperature difference is the result of the heat released as water changes phase during condensation onto the bladders.

In species where fluid is applied to the external mouthparts, it is possible to section the connecting ducts or to temporarily interfere with apposition of the ducts to the absorbing surface. A functional role of the fluid in WVA is indicated if the temperature of the mouthparts approaches that of the nonabsorbing surfaces after interruption of the supply of fluid. This type of experiment has been performed for the desert cockroach *Arenivaga* (O'Donnell 1977).

In species with externally visible absorption sites, observation of the absorbing surfaces during rapid reduction in ambient humidity is a simple yet highly revealing experiment. If the absorbing surface changes form a shiny wet surface (i.e., condensing) to a dry mat appearance (i.e., nonabsorbing), the humidity at that point corresponds to the pump threshold. Such observations can provide confirmatory measurements of gravimetrically determined pump thresholds. However, such observations do not necessarily indicate that the fluid is hygroscopic (see Sect. 3.3).

2.3.5 In Situ and in Vitro Preparations

In vivo studies permit measurement of water uptake by gravimetric or other methods but do not provide information on the physiological processes necessary for uptake. Such information can only be gained when one can sample the fluids involved. In some cases, vapor-absorbing structures are discrete organs or are located in specific parts of the body such as the head. In vitro preparations of water-transporting tissues such as the midgut, rectum, Malpighian tubules, and salivary glands have been developed and modified, and it appears likely that the tissues involved in WVA in some species may be amenable as well to isolation and study in vitro. Almost nothing is known of the cellular physiology of such cells, and yet, as noted above and in the discussions below, mechanisms for the regulation of cell volume and ionic content are likely to be highly efficient in these tissues. The cell membranes of tissue compartments involved in WVA may show specializations necessary to limit diffusive backflux of actively transported solutes.

In *Tenebrio* larvae, an in vitro preparation of the cryptonephridial complex has been developed. The isolated complex will transport water from rectal lumen to Malpighian tubules against osmotic gradients of nearly 7 osmol maintained by sucrose (Tupy and Machin 1985). Because both the hemolymph-facing surface of the complex and the rectal lumen can be continuously perfused, the effects of variations in ionic composition, oxygen tension, or the addition of pharmacological agents on fluid production can be examined. Moreover, simultaneous measurements of electrical potential and ionic activity can be made (Sect. 3.2.1). These studies offer the potential for elucidation of the cellular mechanisms responsible for production of high concentrations of KCl which are the basis for the reduction of water activity necessary for both fecal dehydration and WVA.

In Thysanurans an in situ preparation has been developed for recording the transepithelial potential and current simultaneously with measurements of water flow in the anal sac (Küppers et al. 1986). Double-barreled electrolyte-filled electrodes and Ag/AgCl electrodes of 50 µm tip diameter are placed in the rectal lumen and the hemolymph and a voltage clamp circuit is used to measure short-circuit current. The entire animal is maintained within an experimental chamber which permits replacement of air in the chamber by CO_2 or N_2.

2.3.6 Tracer Methods

Measurements of the efflux and influx of tritiated water have been successfully used to describe whole animal exchange kinetics in acarines (Wharton and Devine 1968) and booklice (Devine 1982b). Tracer methods have some advantages in studying very small terrestrial organisms, and in a few cases attempts have been made to separate active and passive fluxes. These can be judged successfully if net sorption can be demonstrated over a limited range of ambient water activities, correlated with independently measured CEAs, and if different exchange kinetics for sorption and transpiration can be demonstrated. The house dust mite *Dermatophagoides farinae* has been shown by Arlian and Wharton

(1974) to exhibit two water compartments below the CEA: a small one characterized by a short time constant for tracer exchange, and a larger one with a longer time constant. Above the CEA, exchange is limited to the fast compartment. Knülle and Devine (1972) have successfully exploited the lack of spiracles in the tick *Dermacentor variabilis* and demonstrated that the rate constant associated with active uptake is sensitive to high CO_2 or low oxygen levels.

3 Functional Classification of Absorption Mechanisms

3.1 Absorption Mechanisms Involving Epithelial Transport – Oral Uptake Sites and Superficial Fluids

3.1.1 Booklice (Psocoptera) and Biting Lice (Mallophaga)

As noted above (Sect. 2.1), there are morphological similarities in the WVA systems of the Mallophaga and Psocoptera but significant differences in threshold exist. Early observations by Knülle and Spadafora (1969) showed the psocopteran *Liposcelis* to be capable of WVA, and the more recent papers of Rudolph (1982a, b) provide details of the mechanism and occurrence of the process in an additional 22 species. WVA appears to be characteristic of the group, and the ability does not appear to be related to environmental conditions. Fully winged, flying insects, which presumably gain frequent access to liquid water, are nonetheless capable of WVA.

Observation of animals coincident which continuous weight measurement in controlled humidities has demonstrated that in nonabsorbing animals the preoral cavity is closed apically by the close fitting margins of the labium and labrum, and laterally by the mandibles and maxillae. However, during absorption, the labium is lowered so as to expose the salivarium (Fig. 3b). The hypopharynx divides the preoral cavity into a dorsal cibarium, the start of the alimentary canal, and a ventral salivarium, into which the ducts of the salivary glands and the dorsal glands enter. The lateral surface of the hypopharynx is formed by a pair of elongate, oval structures called lingual sclerites. When WVA begins, these are positioned vertically between the distal margins of the labium and labrum. During absorption, the surface of the sclerites are coated with a thin, irridescent liquid film which dries quickly if ambient humidity is reduced to values below the CEH. Although direct evidence is lacking, it appears likely that the film is hygroscopic, and it is probably produced by the associated labial glands. The fluid is conveyed from the glands to the ventral region of the lingual sclerites by a duct. A shallow groove, running longitudinally over almost the entire length of the sclerites, may collect fluid and transfer it to a tubular filament. The apparent end of the tubular filament opens into a cuplike cibarial sclerite which is closely associated with a knoblike epipharyngeal sclerite. These two sclerites form the piston and cylinder of a suction pump. Large clypeo-epipharyngeal muscles operate the pump so that the fluid layer and condensate are removed from the exposed surfaces of the lingual sclerites and conveyed to the cibarium and hence the gut.

The WVA system of the Mallophaga (Fig. 3c) is exceptionally efficient; thresholds are among the lowest, and yet uptake rates are the highest known. The anatomy of the uptake system is virtually identical with that of the Psocoptera. However, the frequencies of contraction of the clypeo-epipharyngeal muscles, ranging from about 250 min^{-1} at 53% RH to 460 min^{-1} at 91% RH (Rudolph 1983), are higher than in the Psocoptera, about 100 – 125 min^{-1} between 84% and 94% RH. These higher frequencies are presumably related to the much higher rates of condensation in the Mallophaga. Furthermore, although WVA is apparently characteristic of the entire order in the Psocoptera, not all mallophagans absorb. The mouthparts which comprise the absorption system in the absorbing species are reduced or modified in the nonabsorbing species. Surprisingly, some of the nonabsorbing species share similar environments and nutritional modes with the absorbing species (Rudolph 1983). The source of water for recovery of water loss in the nonabsorbing species is unknown.

3.1.2 Mites and Ticks (Acari)

Oral sites of absorption are also found in the mites and ticks. Adult ticks are obligate blood feeders which spend 98% or more of their life off the host animal. They may endure for several years without blood meals, and WVA is thus an important component of water balance at these times (Needham and Teel 1986). It has been known for some time that the CEA of a number of species increases with age in the field (Lees 1946; Needham and Teel 1986). Even in females of a single species, *Amblyomma americanum,* reared under standard conditions, the CEA ranges from 0.80 to 0.90 (Jaworski et al. 1984). There is thus a great deal of variation in the CEA, not only between species and stages, but between ticks of a given stage within a species. This variation may be due either to changes in water gain by the "pump", or to changes in cuticular water loss (Needham and Teel 1986). For example, during prolonged starvation, lipids which are normally applied to the cuticle to repair the results of abrasion or degradation may be utilized instead for energy, and cuticular transpiration may increase as a result.

Uptake of water vapor occurs through the mouth in both ixodid (Rudolph and Knülle 1974) and argasid ticks (Rudolph and Knülle 1982). Way blocking of the mouthparts, but not of other body surfaces, prevents uptake. Accumulations of crystalline material on the mouthparts of dehydrated ticks may be the remnants of a hygroscopic solution. A clear fluid is secreted onto the mouthparts during absorption, and the salivary glands may well be the source of this possibly hygroscopic fluid (Rudolph and Knülle 1974). This view is supported by histological studies which indicate that the ultrastructure of the type I agranular alveoli resembles that of salt glands. Pumping movements in the region of the pharynx suggest that a suction pump may remove fluid from the external mouthparts and transfer it to the gut (Needham and Teel 1986).

Changes in the position of the mouthparts are correlated with uptake. The palps are retracted and splayed so that they no longer contact the chelicerae and hypostome (Fig. 3a), and the chelicerae move back and forth independently and at irregular intervals (M. Sigal and G. Needham, pers. commun.). The function of these changes is not known, but they may be related to the control of fluid flow over the mouthparts.

The fluid secreted onto the mouthparts contains significant quantities of sodium and potassium (Rudolph and Knülle 1974), but analysis of its composition at the time of secretion has been limited by the miniscule volumes involved. Although salivation can be greatly stimulated by injection of pilocarpine into the ticks, the secretion obtained in the latter case is more likely to be a component of the cement which ticks use for host attachment (Rudolph and Knülle 1974).

There is no definitive evidence as yet that the fluid secreted directly from the salivary glands is sufficiently hygroscopic to be responsible for net uptake of water. Moreover, unfed ticks may secrete two types of saliva (Needham and Teel 1986). One type may be secreted during severe desiccation as a form of storage excretion for inorganic ions and other hemolymph constituents. The other type may be secreted during WVA as a hygroscopic solution with an equilibrium humidity at or below the CEA. Although most attention has been directed toward type 1 salivary acini, it has been suggested that granules secreted by the b cells of type II acini from rehydrating unfed *A. americanum* females could be the source of either hygroscopic organic molecules, or a surfactant which allows salivary fluid secreted by other cells to spread out on the mouthparts. The importance of a surfactant, to prevent any hygroscopic secretion from forming a droplet, has been pointed out (Wharton 1985). However, it is worth noting that surfactants would be unnecessary if the absorbing surfaces are formed of wettable materials. Such appears to be the case in the desert cockroach, for example. Isolated parts of the absorption system of this species can be repeatedly washed, dried, and rewetted with distilled water in the absence of surfactants. Moreover, swelling of the hairs under conditions of low ionic strength shows that the cuticular surfaces are themselves wettable. Perhaps the absence of epicuticular lipids, normally responsible for the wetting of insect cuticle, obviates the need for surfactants.

Hygroscopic fluids have also been proposed as the basis for absorption by the mites as well (reviewed by Wharton 1985). Secretions of the supracoxal glands may play a role in absorption. Fluid produced by the glands is carried by the podocephalic canal, which is an open gutter carrying secretions from the openings of the glands above the forelegs, across the palps and into the prebuccal cavity. Salt plugs found in the openings of the ducts may be the remnants of a hygroscopic secretion, or alternatively, they may represent sites of storage excretion for ions which will be reingested during subsequent rehydration.

3.2 Absorption Mechanisms Involving Epithelia – Rectal Uptake and Internal Sites of Reduced a_w

3.2.1 The Rectal Complex of Tenebrio molitor Beetle Larvae (Coleoptera)

The cryptonephridial complex of *Tenebrio* larvae is the best understood epithelium-based vapor uptake system at the present time. Since it is a relatively large, behaviorably amenable experimental animal, comparative studies with different sized larvae and with related species, in situ measurements of the distribution of osmotic pressure combined with computer simulations of the coupling between water and ion flows, and a functional in vitro preparation have all contributed to this understanding.

Functional Morphology and Ultrastructure. The cryptonephridial complex was shown by Ramsay (1964) to be capable of generating extremely high osmotic pressures which were apparently employed in dehydrating the feces. A cryptonephridial complex is a morphological modification in which the normally free distal ends of the Malpighian tubules combine to form a cylindrical sheath surrounding the rectum. In *Tenebrio*, Machin (1979b) established that the highest osmotic pressures, up to 6.8 osmol kg^{-1}, occurred in the six tubules of the sheath and therefore concluded that they were generated by ion transport by the tubules. The same activity threshold for vapor uptake and fecal dehydration suggests that both processes are different parts of essentially the same reabsorptive process.

The tubules in the complex are folded in such a way that they completely cover the entire posterior two-thirds of the rectal surface. It is this part of the rectum that becomes water vapor filled during vapor uptake (MacLeod 1941; Machin 1979b). Surprisingly, the rectal wall does not appear to take an active part in reabsorption since the elaborate ultrastructural organization associated with active ion reabsorption seen in other insects is lacking in *Tenebrio* larvae.

Malpighian tubules in contact with the posterior two-thirds of the rectum are structurally distinct from other regions of the same tubules and from the tubules of other insects. Whereas the tubule cells are similar in ultrastructure to those of other insects, apical brush borders are so expanded that they reduce the tubule lumen to a series of narrow interconnecting spaces between the microvilli. *Tenebrio* microvilli are longer, slightly narrower, and more tightly packed and with a higher proportion occupied by mitochondria than their counterparts, e.g., in *Rhodnius*. *Tenebrio* tubules are studded on the side further from the rectum with triple compartmented diverticulae or "boursouflures" (Fig. 4e,f), the central compartment of which appears to communicate with the blood via a single specialized "leptophragma" cell and a blisterlike region in the complex multicellular perinephric membrane surrounding the tubule sheath. Other regions of blister walls are bordered by a peritubular space created by and separated from the blood by thicker regions of perinephric membrane. Elaborate structures on the outer tubule wall distinguish it from that adjacent to the perirectal space and rectal wall.

Dynamics of Water-Ion Coupling in the Rectal Complex. In general terms it is reasonably clear how the cryptonephridial complex works to absorb water vapor. Although countercurrent flow between tubule and rectal fluid may enhance fecal dehydration, rectal flow is reversed during atmospheric absorption becoming cocurrent with flow in the tubules. The possibilitiy that U-shaped loops in the tubules or the multicompartmented diverticulae might act as countercurrent multipliers can be discounted on the grounds that their orientation is perpendicular to the main longitudinal axis of concentration rather than parallel to it.

In nonfeeding mealworms water and ion reabsorption driven by tubular ion transport from both rectum and the blood eventually reduces rectal water content until air spaces begin to form between the remaining solid excreta. These spaces form a conducting pathway for water vapor entering through the open anus. Coutchié and Machin (1984) have argued that water vapor must enter the rectum by some form of convective or ventilatory process since calculations based on the dimensions of the rectal canal show that observed uptake rates could not be sustained by diffusion alone.

Even though reabsorption of ions from the rectum may cease during vapor uptake, it is evident that those transported from the blood continue to generate the radial osmotic gradients which passively convey water from the rectal lumen through the intervening compartments and barriers to the tubules (Tupy and Machin 1985). Volume concentration relations of protein-containing perirectal fluid located between the rectum and the tubules show that the fluid makes an ideal coupling medium between the two (Machin 1979b) rather than acting as a pump as suggested by Ramsay (1964). Tracer measurements which establish the low water permeability of the perinephric membrane show how the concentrated fluids in the tubules are protected from dilution from the blood overcoming the rather low permeability of the combined rectal wall and cuticle to produce the greatest water flux from the rectum. Figure 4f summarizes our current knowledge of the concentration and flows within the intact vapor-absorbing rectal complex.

Modified standing gradient models (after Diamond and Bossert 1967) have been applied to the tubular lumen to better understand the coupling between water and ion fluxes as water entering from the rectum displaces fluid anteriorly along the tubules (Brown and Machin 1985). Since tenebrionids show nonsaturable uptake kinetics, tubular solute concentrations must remain relatively unchanged in different uptake rates. Some evidence has been obtained with in vitro preparations (Tupy and Machin 1985) that this is somehow achieved by the complex matching of ion fluxes to water flows. Although the mechanism of this process is far from understood, it may be related to the fact that tubular osmotic pressures are observed to decline as fluid is passed anteriorly along the tubules (Ramsay 1964; Machin 1979b). The suggestion from computer simulation that this longitudinal gradient reflects changes in the physiological parameters which affect ion or water leakage to and from the blood so as to reduce the number of water molecules absorbed per ion transported, must be important. We speculate that the anterior gradient is part of a mechanism which matches overall ion fluxes to water flows and reduces the risk of tubular dehydration and KCl precipitation in fluctuating humidities.

Tubule Cell Physiology. One uncertainty involving the rectal complex is the precise route followed by ions in their passage from the blood to the tubule lumen. Electron micrographs (Grimstone et al. 1968) suggest that the leptophragma cell at the apex of each "boursouflure" is uniquely placed to conduct ions directly into the tubule lumen. Furthermore, leptophragma correspond to the blisterlike regions of the perinephric membrane, areas which might be more permeable to ions. On the other hand, the leptophragma make up less than 1% of the tubule wall area and lack elaborately folded membranes and dense mitochondria normally associated with transporting epithelia cells. The remaining tubule cells, however, do have these features. If these cells were responsible for generating the driving forces necessary for WVA, ions from the blood would first have to find their way into the peritubular spaces.

Cryoscopic analysis of sections of the rectal complex frozen during vapor absorption (Machin 1979b) showed a clear discontinuity between the melting points of tubule tissue and the blood. This indicates that the perinephric membrane represents a significant barrier protecting concentrated fluids in the tubules from dilution by the blood. Unfortunately, the lumen of the tubules cannot be distinguished from the cells in frozen sections, leaving some uncertainty as to the magnitude of intracellular osmotic pressure and the existence of an apical water barrier. We speculate that the unusually expanded brush border of *Tenebrio* cells may act as a diffusion barrier to water, so that dilution from the blood is controlled by a two-step process.

Tubule cells in the *Tenebrio* rectal complex show a remarkable capacity for volume regulation, perhaps a necessary requirement for all cells in organs adapted to withstand large changes in tissue osmotic pressure (Sect. 4.3.2). It is well established (Ryerse 1977; Machin 1979b) that rectal osmotic pressures are allowed to decline to blood levels at the onset of molting. Machin (1981) has shown that the loss of water from *Tenebrio* tubule cells which change from 0.6 to 5.4 osmol was only one-third the expected amount. This volume regulation as well as the maintenance of high intracellular osmotic pressure may well involve organic osmolytes in order to keep intracellular potassium within viable limits (Somero 1985).

Comparisons of ion and water fluxes sustained by tubule by the tubule walls with those of comparable epithelia indicate fluxes in *Tenebrio* to be very low (Machin 1983). It is possible that the very high osmotic pressures generated by the rectal complex of *Tenebrio* can only be achieved at the expense of flow rates.

Other Tenebrionids. Allometric studies of vapor uptake in two tenebrionid beetle larvae (Coutchié and Machin 1984) show that "uptake capacities" in *Onymacris marginipennis* exceed those of *Tenebrio molitor* by a constant amount over the complete size range. Since "uptake conductances" were essentially the same, differences between the two could be attributed only to threshold values. Mean threshold activities for *Onymacris* were 0.841 rather than 0.897 for *Tenebrio*, giving rise to higher uptake rates in the former at the same humidity. *Onymacris* thresholds are so close to the minimum for saturated KCl ($a_w = 0.85$ at 25 °C) that the use of the more soluble NaCl is suggested. This would seem to make some sense as *Onymacris* comes from the Namibian desert (Southwest Africa), which is known to be affected by sodium-rich oceanic fogs (Seely and Hamilton 1976).

The suggestion based on analyses of uptake rates in a substantial number of different species that vapor-absorbing structures might be scaled with body mass does not apply to tenebrionid larvae. In the only direct study of scaling of absorbing structures within the same species to date, the rectal absorption area was found to be proportional to body mass$^{0.546}$ in *Tenebrio molitor* and body mass$^{0.590}$ in *Onymacris marginipennis* (Coutchié and Machin 1984). Taking the scaling of rectal diffusion distance into account predicted conductances, which increase with body mass$^{0.370}$ in *Tenebrio* and body mass$^{0.389}$ in *Onymacris*, represent a reasonable agreement with observed exponents of 0.355 and 0.418 for the two species, respectively. Failure of absorption to increase proportionately with body mass in tenebrionid larvae may reflect structural limitations for internal absorbing structures. These limitations may not apply to absorbing structures of different anatomical origin especially if they are located on the animal's external surface.

3.2.2 Rectal and Anal Sacs in Flea Larvae (Siphonaptera) and Thysanurans

In the firebrat *Thermobia* and the silverfish *Lepisma* absorption is accomplished by a special organ referred to as the posterior rectum or anal sac (Fig. 4a). The sac is formed from a highly folded single-layered epithelium. The subcuticular space between the rectal cuticle and epithelial layer is distended by an accumulation of hydrophilic material (Fig. 4b) which is similar to glycerol, but is not identical with it (Küppers and Thurm 1980), and may also contain mucopolysaccharide (Noirot and Noirot-Timothee 1971). A number of authors have suggested that the first step in WVA by lepismatids is condensation of atmospheric water vapor onto this material. The electroosmotic model of water transport developed by Küppers et al. (1986) suggests that an electrogenic cation pump, comparable to that of the insect midgut (Harvey et al. 1983), is ultimately responsible for WVA in lepismatids. It is suggested that K^+ is actively transported across the apical membrane of the anal sac epithelium, generating a transepithelial potential difference of as much as 200 mV, lumen-positive (Fig. 4b). In response to this voltage, cations passively diffuse back through cation-selective channels. Water is coupled electroosmotically to the inward current, so that water which condenses from the atmosphere onto the hygroscopic material of the subcuticular space is removed to the cytoplasm of the epithelial cells. Subsequent movement of water from cell to hemolymph is assumed to occur passively.

We consider this hypothesis to be intriguing but untenable in its present form for several reasons. Firstly, the electrical measurements involved filling the anal sac with solutions which contained 500 mM KCl or more. These concentrations are clearly unphysiological in the context of water vapor absorption, and may stimulate ion-coupled water transport that is distinct from the mechanism used for WVA. The suggestion that electrolyte-free solutions would possibly wash out ions required for the system (Küppers et al. 1986) seems unlikely in view of the location of the proposed pump at some distance beneath the rectal cuticle. Moreover, water absorption from electrolyte-free solutions is readily demonstrated in *Tenebrio* (Tupy and Machin 1985) and the nonabsorbing *Schistocerca* (Phillips 1964). It is also possible that the solutions used are hyperosmotic, but not hypertonic. In other words, ions may diffuse down their concentration gradient into the epithelial cells and hemolymph, with water following passively. It would be of interest to determine if water transport occurs when the anal sac is filled with a solution containing a low concentration of electrolytes (10 – 20 mM), sufficiently conductive for the electrical measurements, and a higher concentration (4 – 5 M) of a nonelectrolyte such as sucrose to provide an osmotic gradient.

Secondly, the use of KCl in this system is problematic because the proposed coupling of electrogenically transported ions to water occurs in the subcuticular space at water activities which are below the saturation point of KCl. In other words, why does KCl not precipitate in the posterior rectum at an activity of 0.40, the pump threshold (Machin 1979a) since it becomes saturated at an a_w of 0.84?

Thirdly, the evidence for a K^+ current is circumstantial, and is based primarily on the assumption that electron dense particles apparent in electron micrographs represent the potasssium-transporting portasomes present in other epithelia for which ion substitution experiments and radiotracer data clearly demonstrate potassium transport. Insect rectal cells are also known to transport chloride from lumen to hemolymph (Hanrahan and Phillips 1984) and any diffusive backflux in the latter case would, in fact, lead to loss of water if electro-osmotic coupling occurred.

Lastly, and most importantly, the pump threshold estimated from interpolation of uptake rates of 24 nl h^{-1} at a_w 0.75 and – 5 nl h^{-1} at a_w 0.65 is in excess of 0.66, which is difficult to align with the more generally accepted value of 0.40 calculated by Machin (1979a) and assumed by Küppers et al. (1986).

Despite these reservations, the development of methods to measure water uptake concurrent with measurements of transepithelial potential represents a major technical advance. This elegant methodology shows great promise in further studies of WVA by lepismatids.

We speculate below (Sect. 4.4) that the basis for absorption in *Lepisma* may involve cyclical modulation of the hydrophilic properties of the material within the subcuticular space, and that in this characteristic, the WVA system of lepismatids may be comparable to that of the desert cockroach.

In larvae of the oriental rat flea *Xenopsylla cheopis* WVA occurs through the anus. During absorption rhythmical opening and closing of the anal valves (49 cycles min^{-1} at 93% RH and 25 °C) is accompanied by wavelike constrictions which pass from the distal to the proximal end of the rectal sac, the enlarged anterior portion of the rectum (Fig. 4c). These movements probably serve to ventilate the rectal sac, ensuring that the absorbing epithelia are adequately supplied with ambient air (Rudolph and Knülle 1982).

The histology of the rectal sac differs distinctly from that of the lepismatids and tenebrionids. There is a clear dorsoventral asymmetry of the cells (Bernotat-Danielowski and Knülle 1986; Fig. 4d). Those of the dorsal surface have highly folded apical plasma membranes on the lumen sides, whereas the cells of the ventral surface have deep infoldings of the basal plasma membrane on the hemolymph side. Subcuticular spaces containing material which appears granular, fibrillar, or sometimes without structure are found in the rectal sac. These spaces are reminiscent of those seen in the epithelia of the anal sac of the Lepismatidae, and the material may play some role in the condensation of water vapor. The asymmetric arrangement of the apical and basal infoldings suggests dissimilar functions of the dorsal and ventral cells of the rectal sac. It has been suggested that an absorbing and secretory function may be considered (Bernotat-Danielowski and Knülle 1986).

3.3 Absorption Mechanisms Independent of Epithelial Transport – Hydrophilic Cuticle in the Desert Cockroach

Morphology and Ultrastructure. The desert cockroach *Arenivaga investigata* is a burrowing species found in the Mojave desert of southern California. The animal is well suited to resist natural desiccation. Hemolymph osmotic pressure is strongly regulated during dehydration, and the cuticle is more resistant to water loss than nonxeric cockroach species (Edney 1968; Edney and McFarlane 1974). Studies by Edney and co-workers also demonstrated that nymphs and adult females absorb water vapor from humidities above 82.5% RH, and that these humidity levels can be attained year-round at depths of less than 30 cm below the surface (Edney 1966; Edney et al. 1974). Typical rates of uptake are 4% body, day^{-1} at 96% RH, amounting to about 20 mg day^{-1} for an adult female (O'Donnell 1980).

During absorption, and at no other times, two bladderlike diverticula of the hypopharynx are protruded from the prebuccal cavity by inflation of the hypopharynx with hemolymph. Observation of animals during continuous weight recording under controlled temperatures and humidities shows a perfect correlation between weight gain under suprathreshold humidities and protrusion of the hypopharyngeal bladders. The bladders are covered with a tightly packed mat of cuticular hairs less than 180 nm in diameter over much of their length. A helicoid groove extends along the 80 – 100 μm length of the hairs. Surface temperatures of the bladders (see Sect. 1.3.4) under conditions of controlled temperature and humidity indicate a humidity-dependent temperature differential between the bladders and the surrounding mouthparts that is consistent with the bladders acting as sites for condensation.

During absorption, fluid which is produced by a pair of structures termed frontal bodies is applied to the bladder surface (Plate 1a; Fig. 3d). In contrast to the glandular structures associated with oral uptake sites in ticks, mites, psocids, and mallophagans, frontal bodies are nonglandular derivatives of the integument, and the fluid they produce is nonhygroscopic. Each spheroidal frontal body is derived from an invagination of the cuticle of the preoral cavity and communicates with the cavity through a porous cuticular plate. Cyclic contractions of a heavy mass of frontal muscles distort the frontal bodies and apparently produce internal pressure changes which result in exudation of a hemolymph ultrafiltrate through the cuticular plate (O'Donnell 1981a). This plate opens into a groove in the dorsal surface of the preoral cavity; the distal end of the groove contacts the bladder surface. Breaking this contact by lifting the labrum away from the bladder or cutting the groove prevents condensation onto the associated bladder.

Muscle contraction frequency increases from 40 – 50 cycles min^{-1} at 88% RH, to 130 cycles min^{-1} at 99% RH. In other words, contraction frequency is correlated to condensation rate. The frequencies also increase transiently if nanoliter quantities of distilled water are applied by pipette to the bladder surface. These observations suggest that the animal senses some aspect of the rate of condensation, and that a sophisticated feedback mechanism permits maximal uptake rates by ensuring that the amount of water condensed per frontal body cycle is relatively constant (Machin et al. 1982).

The path of fluid and condensate movements has been followed by the use of tracer solutions applied to protruded bladders. Cinematography of fluorescent tracer solutions indicates that frontal body fluid and condensate move posteriorly and laterally away from the point where the epipharyngeal groove contacts the bladder surface (O'Donnell 1981 b). Movements of fluid on the mouthparts which are not directly visible have been monitored by fast-freezing animals at timed intervals after application of solutions containing radiolabeled solutes. Liquid scintillation counting of dissected portions of lyophilized animals indicate that the tracers continue to move posteriorly on the mouthparts and then enter into the alimentary canal, accumulating in the esophagous and crop.

Evidence Against Solute-Dependent Water Vapor Absorption. The most dramatic difference between the WVA scheme proposed for *Arenivaga* and the schemes for other arthropods concerns the nature of the fluid exuded onto the bladder surface during absorption. Evidence from a variety of experimental techniques indicates that the frontal body fluid is not hygroscopic. Freezing point depression measurements in frontal body tissues indicate osmolarities of 645 mOsmol kg^{-1}, only 30% above that of hemolymph (O'Donnell 1981 a). In comparison, a hygroscopic solution capable of absorption at the pump threshold of 81% RH would require osmolalities in excess of 11,700 mOsmol kg^{-1}.

This view is supported by analysis of frontal body fluid and hypopharyngeal bladder cuticle by neutron activation and atomic absorption spectrophotometry. By assuming that all the salts in bladder cuticle samples are contained in the fluid trapped between the cuticular hairs, an upper limit of 275 mEq, of total inorganic salts, or approximately 275 mOsmol, is placed upon the solute concentration of the fluid. It appears, therefore, that the concentrations of inorganic salts in the absorption system are not elevated above normal physiological levels. Moreover, the use of fluorometric reagents indicate that the sum of the concentrations of hydrophilic organic molecules such as peptides, amino acids, polyhydric alcohols, and reducing sugars on the bladder surface is less than 2 mM (O'Donnell 1982a).

Lastly, the frontal bodies do not appear to be active sites for synthesis of hydrophilic proteins or polysaccharides. This conclusion is based upon the nonglandular nature of the tissue apparent in ultrastructural studies, and on the absence of significant accumulation of radiolabeled precursors injected into the hemolymph (O'Donnell 1982a).

The accumulated evidence against high concentrations of hydrophilic solutes on the surface of absorbing bladders suggests that the frontal bodies, and the fluid which they produce, play a subordinate or ancillary role in the absorption mechanism. Analysis of the water content/humidity relationship of isolated samples of bladder cuticle suggests that the reduction of water activity at the absorption site may be, in fact, a property of the cuticular hairs which cover the bladder surface.

Hydrophilic Cuticle. The water affinity of the bladder cuticle is much higher than that of cuticle from other regions of the body of *Arenivaga* and other species. In other words, the water content of bladder cuticle (g H_2O g^{-1} dry weight) at a given humidity is greater than that of nonspecialized cuticle. These differences remain in cuticle samples which have been washed in distilled water so as to remove dissolved salts. The hydrophilic properties of the bladder cuticle, therefore, are not due to the presence of salts in the fluid applied to the bladders in vivo. The addition of physiologically appropriate quantities of salts to the bladder cuticle actually reduces water affinity. This finding is consistent with the swelling of the hairs under conditions of low ionic strength, as observed in scanning electron micrographs.

How Might Condensed Water be Released from Hydrophilic Cuticle? Absorption of water vapor by the bladders of *Arenivaga* appears to represent a unique form of water transport; it is neither solute-coupled nor based upon ultrafiltration (O'Donnell 1982b, 1986). The mechanism appears to involve cyclical swelling of the cuticular hairs in response to periodic alteration of the ionic strength. It is well known that polyelectrolytes, such as chitin and protein, or synthetic polyacrylamide gels, swell as the ionic strength of the surrounding fluid is reduced. Increasing salt concentration reduces the mutual repulsive forces between like charges on a protein or gel structure. As a result, a greater number of noncovalent cross-links, possibly of the van der Waals type, are possible, and shrinkage occurs (Katchalsky 1954). Importantly, over particular ranges of temperature, pH, or ionic strength, the swelling or shrinkage of a gel is discontinuous. Within these ranges an infinitesimal change in the external conditions can produce dramatic changes in the volume of the gel (Fig. 6).

Cyclical modulation of the water affinity of the bladder hairs is of cardinal importance in the proposed mechanism of absorption of atmospheric water vapor onto the bladder surface. This modulation is brought about by the periodic alteration of ionic strength around the hairs by the exudation of frontal body fluid onto the bladder surface. A crucial difference between this model of absorption and others is that the energy input into the system is necessary to promote the release of condensed water. The following scheme is suggested. When the bladders are first protruded, the cockroach salivates copiously, and much of this saliva evaporates. Gradually, the rate of evaporation declines and the water activity of the bladder cuticle approaches equilibrium with atmospheric water activity.

Exudation of a small quantity of frontal body fluid perturbs this equilibrium. Contraction of the frontal muscles results in a short pulse of fluid having a relatively high ionic strength being applied to the bladders and moving posteriorly and laterally over the bladder surface as a zone or wave of increased ionic strength. The water affinity of the bladder hairs is lowered, and they release water and shrink. Some of the water released from the hairs is swallowed.

At this point in the cycle, it is suggested that the water affinity of the hairs is reduced, but highly unstable. There may be a gradient of ionic strength on the bladders, highest closest to the point where frontal body fluid is first applied, so that water affinity of some regions of the bladder surface may be altered by the release of water from adjacent hairs. Furthermore, minute temperature fluctuations and the randomness of molecular motions will result in some condensation onto the bladder surface. An infinitesimal amount of condensation will decrease the ionic strength of the fluid on the bladder hairs, and their water affinity will increase, causing further net condensation. This positive feedback upon the water affinity of the hairs will continue rapidly, until the volume change of the hairs is constrained by polymer elasticity or by the next addition of frontal body fluid.

3.4 Absorption by a Desert Plant

WVA has also been demonstrated in *Nolana mollis*, a succulent-leaved shrub found in the extreme coastal deserts of Chile (Mooney et al. 1980). The leaves are covered with salt glands which occur in depressions on the leaf surface and are connected to each other by a series of troughs or canals. The gland-produced salts, mostly NaCl, cover the leaves and are effective in condensing moisture from unsaturated atmospheres. In one experiment, 9.5 μl H_2O cm^{-2} leaf surface accumulated overnight when the highest humidity was 82% RH. It seems likely that water which accumulates on the leaf surface drops off, wetting the ground surface, and is taken up by the roots. The major nonresistive pathway directly from the leaf surface to the mesophyll would be through the stomata, but it seems improbable that condensed water follows this route because the stomata are positioned above the leaf surface. In conjunction with modifications which limit water loss during photosynthesis, WVA provides the plant with a means of meeting its water requirements through the expenditure of respiratory energy.

Although the histology of the salt glands of *Nolana* has not been described, salt glands of nonabsorbing plants (reviewed by Hill and Hill 1976) are enclosed within a cuticular envelope which extends continuously around the gland cell complex except for localized gaps which open into the surrounding leaf mesophyll. The exudate is produced by ionic pumping at the plasma membrane, and a Cl^--stimulated ATPase is probably involved. The osmotic pressure of salt gland exudates is accounted for mostly by NaCl, as in *Nolana*. NaCl concentrations of 0.3 to 1.0 M have been measured. Transporting tissues of halophytic plants containing structures called salt hairs show even greater salt concentrating abilities. In the extreme halophyte *Atriplex*, salt hairs contain a total of 5 – 6 M Na^+ and K^+, and 9.2 M Cl^- (Mozafar and Goodin 1970).

4 Concluding Remarks

4.1 Independent Evolution of Water Vapor Uptake Mechanisms

WVA is by no means a physiological anomaly whose occurrence is restricted to a few closely related species of arthropods. Mechanisms for the absorption of water vapor have evolved at least eight times, and the process can be characteristic of entire orders such as the ticks and the psocids. In some cases these mechanisms are variants of preexisting water and ionotransport capabilities, but in others the structures involved appear to have evolved and are utilized solely for atmospheric water uptake.

In ixodid ticks, for example, the salivary glands function in an iono- and osmoregulatory capacity during blood feeding. A saliva containing excess ions and water is injected back into the host (Tatchell 1967; Kaufman and Phillips 1973). The ability to produce a highly concentrated saliva at times other than feeding may have evolved later so as to provide a mechanism for storage excretion of ions and possibly for the production of a hygroscopic solution utilized for WVA. A similar process may have led to the ability of the excretory organs of mites, the coxal glands, to produce highly concentrated fluids either for storage excretion or WVA. As noted above (Sect. 3.1.2), definitive evidence for hygroscopic solutions in acarines is lacking.

In tenebrionid larvae it seems that WVA is ancillary to the more primitive and widespread function of fecal dehydration. Osmotic gradients produced by the concentration of KCl in the cryptonephridia can provide reduced water activities sufficient to promote reabsorption of water from fecal material in the rectal lumen, or condensation of atmospheric water vapor onto the rectal cuticle. In flea larvae, ultrastructural studies (Bernotat-Danielowski and Knülle 1986) suggest that the two types of cells of the rectal sac may function in secretion and absorption. Perhaps here as well, a primitive excretory or osmoregulatory function has been subsequently modified or augmented so as to permit WVA.

Structures and mechanisms utilized specifically for water vapor absorption appear to have arisen de novo in the psocopterans and mallophagans, lepismatids, and the desert cockroach. Knülle (1984) was the first to point out that the hypopharyngeal structures used by termites for imbibition may provide a clue to the hypopharyngeal WVA systems. Water is apparently taken up by capillary attraction between densely packed cuticular hairs on the anterior hypopharynx of termites. The hypopharynx is protruded during imbibition, and the hairs are dramatically similar to the cuticular hairs of the desert cockroach. It is not known if frontal bodies or homologous structures are found in termites, although a pulsating muscle movement is apparent in the frontal area of the head capsule when the hypopharynx is protruded (Sieber and Kokwaro 1982). Perhaps an intermediate stage in the evolution of the WVA capabilities of the desert cockroach involved the drinking of water from successively finer porous surfaces, so that eventually water activity in the porous material was comparable to that in the atmosphere under humid conditions.

In the Psocoptera and Mallophaga, the source of the fluid on the lingual sclerites is believed to be the dorsal labial glands. In spinning species of psocids, these glands are responsible for the production of silk. They are retained, however, in many nonspinning forms. It seems possible that the glands might have become modified so as to produce some form of hydrophilic macromolecule which does not form silk but is hygroscopic. This type of secretion might still be quite viscous, and the function of the large clypeo-epipharyngeal muscles may be related to the need to maintain a continuously flowing, thinly drawn out layer of this material on the lingual sclerites.

The threshold for WVA by the cigarette beetle *Lasioderma serricorne* is 43% RH (Knülle and Spadafora 1970), well below the equilibrium humidity over KCl or other common biological salts. It seems unlikely, therefore, that the mechanism of WVA in this species is related to that of the tenebrionid beetles. In larvae of the clothes moth *Tinea pellionella* uptake occurs above humidities of 93% RH (Chauvin and Vannier 1980). Although the site of uptake is unknown, it is conceivable that a rectal uptake site based upon reduced water activities produced within a cryptonephridial complex is possible. Cryptonephridial arrangements are characteristic of the Lepidoptera, although they are thought to be involved in ion regulation rather than water regulation (Ramsay 1976).

4.2. Sensory Implications of Absorption

4.2.1 Maintenance of Body Water Content

In general, insects capable of WVA are able to control periods of absorption so as to maintain body water levels within a narrow range. Among Psocoptera, for example, body water is maintained to within ± 1% under equilibrium conditions in some species, and to within ±3% in the majority. The Mallophaga typically maintain body water content to ±2% (Rudolph 1983). The same level of control is achieved by the ticks, which alternate loss periods of 8 – 14 days with uptake periods of 2 – 6 days (Rudolph and Knülle 1979). Equilibrium body water content is adjusted very finely in the firebrat, with oscillations of only 0.1% of total weight under equilibrium conditions. In the desert cockroach, animals starved in 96% RH lose weight at a fairly constant rate of 0.2% day^{-1} for 20 days, suggesting that when losses of dry weight are accounted for, this species also maintains body water content quite well (Edney 1968).

Results of another type of experiment also suggest that species capable of WVA sense some aspect of the rate and/or degree of water gain. In psocids, the desert cockroach, and thysanurans, the same water content is achieved after periods of desiccation, irrespective of the degree to which ambient humidity exceeds threshold or the extent of previous desiccation. In contrast, many tick species (Baloshov 1960; Hafez et al. 1970; Rudolph and Knülle 1978; Shih et al. 1973), the grain mite (Knülle 1965), and larvae of the oriental rat flea (Knülle 1967) maintain body water content at higher levels in higher suprathreshold humidities. Fasting mealworms also lack adequate means of regulating atmospheric absorption during prolonged exposure to high humidities (Machin 1975). For example, after dehydration by 15% of initial body weight, larvae exposed to humidities above threshold gain water and overcompensate by 28% of the predesiccation weight (Dunbar and Winston 1975). However, in this species the apparent lack of control may be of adaptive significance. Machin (1975) has shown that initial absorption in fasting animals permits accelerated growth when they are subsequently fed.

4.2.2 Effect of Nutritional Status

Among the psocids, the length of the period for maintaining water balance by adsorption of atmospheric water vapor differs considerably among species. Those which normally feed on organic matter and molds remain in water balance for about a week when starved under high humidities. In contrast, members of the suborder which feeds on algae, lichens, and fungi on tree trunks and foliage maintain body water equilibrium for only 1 – 2 days in the absence of food. The end of the equilibrium phase is usually brought about by the failure of the uptake mechanism to start at normal intervals. Uptake mechanisms can be repeatedly reactivated by feeding the animal for a few hours each time the ability to absorb water vapor is lost. The ability to absorb is also restored by feeding after starvation in the firebrat (Noble-Nesbitt 1977) and desert cockroach (O'Donnell 1980), but several weeks of starvation is required in these species before absorption fails (Okasha 1972; O'Donnell 1980). In the Mallophaga and Psocoptera the continuous digestion of food material present in the gut may be essential for an adequate energy supply, not only to the uptake system but to the whole organism (Rudolph 1983). Food is available ad libitum for the mallophagans in their natural environment, i.e., on the host's body, and they feed at frequent intervals. Under natural conditions, therefore, it appears that unrestricted WVA and consequent maintenance of water balance is assured.

Mealworms provide another example of insects living in an environment where water is scarce but food is not. In the laboratory periods of water shortage can be represented in cultures when only dry grain products are supplied and when the ambient humidity in winter drops to < 25% RH. Even though abundant food is present its consumption would result in net water loss. Mealworms under these conditions do not feed, they survive primarily by fat oxidation, where metabolic water provides enough water for survival. They have stopped growth since few are found in a premolt condition, and are evidently water stressed since they are very shrunken, blood volume being much reduced. The rectum is highly concentrated and ready to absorb water vapor, as experimental transfer to high humidity readily shows.

It is well known that mealworm cultures with added fruit and vegetables of high water content rapidly develop, with the proportion of molting mealworms greatly increased. Blood volume will be greatly increased and few will gain weight on exposure to high humidity. The transition from the former water-depleted state to the hydrated condition, which permits further growth without compromising water balance, is rapidly brought about by uninterrupted atmospheric absorption lasting several days. Again vapor uptake is superior to other methods of gaining water because it is two to three times more rapid than obtaining water from the food (Machin 1976) and takes place without the associated processing, elimination, and excretion. In mealworms water vapor absorption appears less important in the routine maintenance of water balance as it is in animals which show "sawtooth" patterns of weight gain (Sect. 2.3.1), but does provide the means for rapid recovery from a water-depleted state.

4.2.3 Humidity Detection

Most of the analyses of humidity detection have been in species with oral uptake sites, partly because it is often possible in these animals to directly observe some correlate of the uptake mechanism under different conditions of temperature and humidity. The most detailed study of humidity detection has been provided from field (Edney et al. 1974) and laboratory (Edney et al. 1978) studies of the desert cockroach. Field observations suggested that the desert cockroach utilizes small, microclimatically favorable niches for the maintenance of heat and water balance. Animals accumulate in appropriate thermal and temperature environments. For example, they remain below the surface of the sand dunes until the surface temperature drops below 40 °C in the evenings. Even on the hottest summer days, when the surface temperatures can exceed 70 °C, temperatures of 40 °C or less are available no more than 15 cm below the surface. Although surface humidities are generally less than 10% RH on such days, humidities of 80% RH are found at 45 cm below the surface, so that animals can maintain water balance by moving downward. Laboratory studies of animals tagged with a gamma-emitting radiolabel and observed and photographed with a gamma scintillation camera indicate that the animals use hygroreceptors located on the antennae to orient in humidity gradients (Edney et al. 1978). Orthokinesis is an important component of the orientation mechanism; the animals are more active in dry than in wet sand, and they aggregate, therefore, at the moist end of a horizontal or vertical humidity gradient.

Observation of mallophagans in subthreshold humidities suggests that they sense humidity as well. The mouthparts are never brought into the absorbing position if the ambient humidity is 2 – 3% below the CEH. Similarly, in the psocid *Psyllopsocus ramburi* desiccated to 80% of the initial weight and subsequently exposed to 76% RH, which is 8% below the CEH, no attempts are made to expose the hypopharyngeal structures to the air. If the RH is increased to 93% RH mouthparts are brought into the absorbing position within a few minutes. However, at 81% RH, 3% below the CEH, the hypopharyngeal structures are brought into the absorbing position, even though their surfaces dry out immediately. When this occurs, the animal retracts the lingual sclerites and closes the salivarium (Rudolph 1982b).

4.3 Physiological Stresses Associated with Localized Sites of Reduced Water Activity

4.3.1 Isolation of Sites of Reduced a_w from Surrounding Tissues

A necessary component of any absorption mechanism is the presence of some sort of barrier to isolate sites of low water activity from the surrounding tissues. For the desert plant *Nolana* (Sect. 3.4) and for the arthropods which utilize external sites of absorption and superficial fluids (Sect. 3.1), this barrier is continuous with the cuticle surrounding the organism. In both plants and terrestrial arthropods the cuticle is a continuous acellular layer which overlies the epidermal cells. Structural integrity of this layer is provided by cross-esterified fatty acids in plants, and by a chitin-protein complex in the arthropods. In both plants and arthropods the waterproof nature of the cuticle is dependent upon superficial deposits of lipids. Maximum waterproofing is provided by long-chain, saturated hydrocarbons with few methyl branches (Hadley 1981).

In *Tenebrio*, the high series resistance provided by the multilayered perinephric membrane presumably isolates the hemolymph from the concentrated solutions of KCl in the lumen of the cryptonephridial tubules. Although it is not known if the tubule cells are in osmotic equilibrium with the fluid in the lumen, the combination of apical microvilli containing mitochondria and long, thin, extracellular spaces between the microvilli might produce sufficiently large unstirred layers to result in a lower permeability than an unfolded membrane (Sect. 3.2.1). If the highly folded apical membrane does not limit water flux from cell to lumen, then the intracellular compartment must also be at a reduced water activity. However, this situation would pose problems for cell volume regulation and the maintenance of an intracellular environment compatible with enzyme function (Sect. 3.2.1).

The nature of the water barrier in the absorption mechanisms of the flea larvae and the thysanurans (Sect. 3.2.3) is even more obscure. Here as well, the luminal plasma membrane is highly folded but nonetheless single-layered, and the lateral and basal plasma membranes are unexceptional. It is unclear, therefore, how the dramatically reduced water activities of the subcuticular space are osmotically isolated from the hemolymph. In other words, why does water not cross the cell membrane down the steep gradient from hemolymph ($a_w = 0.995$) to the subcuticular space ($a_w = 0.40$)? The lowest known permeabilities of cell membranes, found in fresh water protozoans, are in the region of $2-4 \times 10^{-5}$ cm s^{-1}, some four to six orders of magnitude higher than the permeability of insect cuticle. It seems unlikely, therefore, that the cell membranes, apical or basal, can provide the necessary barrier. One possibility, noted above with respect to *Tenebrio*, is that the long, apical microvilli and their enclosed mitochondria greatly increase the resistance to diffusion of water from lumen to cytoplasm. (Sect. 3.2.1).

4.3.2 Implications of High Solute Concentrations for Cell Function

The cellular physiology of the epithelial cells associated with absorption mechanisms has received little attention, and yet some striking adaptations may well be present. Rectal epidermal cells of *Tenebrio,* for example, must adjust to osmotic gradients which change from 0.6 to 6 osmol kg^{-1} as the absorption capability is established after the molt. Similarly, the rectal epithelial cells of the thysanurans and flea larvae may well be exposed to high and changing osmolalities. It seems unlikely that the intracellular osmolality can be adjusted by means of inorganic ions such as K^+ and Na^+, because changes in activities of these ions are known to destabilize intracellular proteins, including enzymes, with detrimental effects on cell function (Yancey et al. 1982). One might anticipate, therefore, the use of nonperturbing osmolytes such as proline for adjustment of intracellular osmolality (Yancey et al. 1982; Somero 1985).

4.4 Hydrophilic Macromolecules – Possible Functions

One common feature of the otherwise quite different absorption mechanisms described above is the presence of a compartment or layer of hydrophilic organic material in close proximity to the site of condensation. Such materials are found in the subcuticular space in the anal or rectal sacs of flea larvae and lepismatids, and in the perirectal space of the tenebrionid cryptonephridial complex. The location and hydrophilic nondiffusible nature of these materials suggests that they function as osmotic couplers between the atmosphere and sites of reduced water activity produced by active transport of underlying cells. In the steady state, water entry into this material as a consequence of WVA is balanced by water removal in response to reduced a_w at the apical surface of the transporting cells beneath. Importantly, this coupling material will buffer adjacent cellular compartments against changes in the rate of water entry in response to fluctuations in ambient humidity.

The protein-containing perirectal fluid in *Tenebrio* is ideally situated to act as an extracellular source of amino acids which could be used for volume regulation by the rectal epidermal cells. During the change from absorbing intermolt animals to nonabsorbing molting animals, osmotic pressures in the rectal complex decline seven fold. The rectal epithelial cells are capable of volume regulation, which presumably prevents them from being damaged by swelling at these times of reduced osmotic pressures in the cryptonephridial complex. In a wide variety of osmotically stressful situations particular organic osmolytes, including some amino acids, are favored for long-term adjustment of intracellular osmotic pressure (Yancey et al. 1982). These adjustments form the basis for the observed cell volume regulation. The solutes employed are termed "compatible osmolytes", reflecting that alterations in their intracellular concentration do not perturb enzyme function (Somero 1985). Alterations in the intracellular concentrations of Na^+ and K^+ do perturb enzyme function, and these ions, in spite of their ready availability, are less useful for long-term adjustments of intracellular osmolyte levels.

A third possible function for the organic macromolecules in rectal uptake sites is suggested by studies of oral uptake in *Arenivaga*. In a broad sense, the bladder cuticle of *Arenivaga* can be thought of as a compartment comprised of hydrophilic macromolecules. In this case, however, the hydrophilic properties of the macromolecules may be cyclically modulated, presumably in response to changes in ionic strength. By extrapolation, it is not inconceivable that the possible secretory and reabsorptive function of the ventral and dorsal cells of the flea larva rectal sac in some way modulates the uptake and release of water by the subcuticular material. We further speculate that changes in ionic strength or external pH mediated by the epithelial cells of the lepismatid anal sac might modulate the hydrophilic nature of the adjacent subcuticular material. For *Tenebrio* such a role would seem superfluous, in view of the elevated osmolalities of the Malpighian tubule cell lumen, which presumably functions as a straightforward osmotic sink. We present these highly speculative suggestions as a reminder of how little is known of the detailed functioning of the cellular mechanisms responsible for WVA. Indeed, the history of this area has shown a succession of controversial ideas which, more often than not, were subsequently refuted. Although tremendous advances in our understanding of the mechanisms of WVA have been made in the last decade, all of the mechanisms described above are tentative.

At the very least, the evidence against solute-dependent WVA in *Arenivaga* provides a caveat for further studies of WVA in arthropods with oral uptake sites. In the absence of direct measurement of elevated solute concentrations in fluids applied to the oral uptake sites, one cannot assume a priori that such fluids are hygroscopic. The studies of WVA in *Arenivaga* support the surprising conclusion that such fluids may play ancillary roles in the absorptive process. Indeed, the wettability of the mouthparts of acarines, psocids, and mallophagans, as indicated by the flow of fluids over them, shows that the cuticle of these regions is modified, as for *Arenivaga* bladder cuticle, from its usual hydrophobic nature. So although the mouthparts of these other species do not show the elaborate arrangements of cuticular hairs found on the hypopharynx of *Arenivaga,* the mouthparts might nonetheless be more than a simple conduit for hygroscopic fluid. The mouthpart cuticle of these species may be as dramatically hydrophilic as that of *Arenivaga*, and absorption and release of water may also be controlled by the cyclical application of nonhygroscopic fluids.

4.5 Do Many Examples of WVA Remain Undiscovered?

The list of species which absorb water vapor has grown rapidly in the last decade, and it seems likely that this trend will continue as new examples are revealed by studies of species which either inhabit appropriate environments or are members of orders or families in which WVA has been demonstrated. For example, the use of a cryptonephridial system for fecal dehydration is widespread among the beetles (Coleoptera), and spezialization of this type of system for WVA is likely to have occurred in genera other than *Tenebrio* and *Onymacris*. It also appears likely that quite different mechanisms will be discovered within the Coleoptera. For example, the threshold humidity for WVA by the cigarette beetle *Lasioderma serricorne* (Anobiidae) is 43% (Knülle and Spadofora 1970), one of the lowest known and well below the value consistent with the concentration of KCl and/or NaCl as occurs in cryptonephridia. Studies of other members of the Anobiidae and Ptinidae (spider beetles) are anticipated (Knülle 1984), because these families are likely to contain other members capable of WVA. Beetles such as the powderpost beetles (Lyctidae) bore into dry and seasoned wood, remaining inside for months, and examination of members of this group for evidence of WVA might also be fruitful.

Dramatically different species and mechanisms may be discovered by further examination of desert biota, particularly burrowing arthropods which can gain access to humid microenvironments beneath the sand surface. It is noteworthy that absorbing species have been found in only two deserts, in the southwest USA (*Arenivaga*) and in Namibia (*Onymacris*). We predict that the analysis of arthropods of the deserts of central Australia, the high Andes plateau, and the Negev and Sinai regions of the Middle East will reveal further examples of WVA. Moreover, because salt glands are characteristic of many halophytic plants, it seems likely that plants other than *Nolana* have developed absorptive abilities.

Acknowledgments. This work was supported by grants from the Natural Sciences and Engineering Research Council (Canada). The authors are grateful for discussions with Prof. G. Needham and M. Sigal concerning absorption by ticks. The technical assistance of G. Lampert and D. Woods is greatly appreciated.

References

Arlian LG, Wharton GW (1974) Kinetics of active and passive components of water exchange between the air and a mite *Dermatophagoides farinae*. J Insect Physiol 20:335–347

Avigad G (1977) DANSYL hydrazine as a fluorometric reagent for thin layer chromatographic analysis of reducing sugars. Chromatography 10:343–347

Baloshov Yu S (1960) Water balance and behaviour of *Hyalomma asiaticum* in desert areas. Medskaya Parasit 29:313–320

Beament JWL (1964) Active transport and passive movement of water in insects. Adv Insect Physiol 2:67–129

Beament JWL (1965) The active transport of water: evidence, models and mechanisms. Symp Soc Exp Biol 19:273–298

Bernotat-Danielowski S, Knülle W (1986) Ultrastructure of the rectal sac, the site of water vapor uptake from the atmosphere in larvae of the oriental rat flea *Xenopsylla cheopis*. Tissue Cell 18:437–455

Brown DB, Machin J (1985) Water-ion coupling in the rectum of an insect: analysis by computer simulation. Physiologist 28:310

Chauvin G, Vannier G (1980) Absorption of water vapor by the larvae of *Tinea pellionella* L. (Lepidoptera: Tineidae). Experientia 36:87–88

Coutchié PA, Machin J (1984) Allometry of water vapor absorption in two species of tenebrionid beetle larvae. Am J Physiol 247:R230–R236

Defay R, Prigogine I (1986) Surface tension and adsorption. Longmans, London

Devine TL (1982a) Neutron activation analysis of saliva from the tick *Amblyomma variegatum*. Experientia 38:676–677

Devine TL (1982b) The dynamics of body water in the booklouse *Liposcelis bostrychophilus* (Badonnell). J Exp Zool 222:335–351

Diamond JM, Bossert WH (1967) Standing gradient osmotic flow. A mechanism for coupling of water and solute transport in epithelia. J Gen Physiol 50:2061–2083

Dunbar BS, Winston PW (1975) The site of active uptake of atmospheric water in larvae of *Tenebrio molitor*. J Insect Physiol 21:495–500

Eckhardt AE, Hayer CE, Goldstein IJ (1976) A sensitive fluorometric method for the detection of glycoproteins in polyacrylamide gels. Anal Biochem 73:192–197

Edney EB (1966) Absorption of water vapor from unsaturated air by *Arenivaga* sp. (Polyphagidae, Dictyoptera). Comp Biochem Physiol 19:387–408

Edney EB (1968) The effect of water loss on the haemolymph of *Arenivaga* sp. and *Periplaneta americana*. Comp Biochem Physiol 25:149–158

Edney EB (1977) Water balance in land arthropods. Springer, Berlin Heidelberg New York

Edney EB, McFarlane J (1974) The effect of temperature on transpiration of the desert cockroach, *Arenivaga investigata* and in *Periplaneta americana*. Physiol Zool 47:1–12

Edney EB, Haynes S, Gibo D (1974) Distribution and activity of the desert cockroach *Arenivaga investigata* (Polyphagidae) in relation to microclimate. Ecology 55:420–427

Edney EB, Franco P, Wood R (1978) The responses of *Arenivaga investigata* (Dictyoptera) to gradients of temperature and humidity in sand studied by tagging with technetium 99 m. Physiol Zool 51:241–255

Epstein HM, Duke KM, Wharton GW (1979) Laser-generated X-ray microradiography applied to entomology. Trans Am Microsc Soc 98:427–436

Evans GD, Sheals JG, MacFarlane D (1961) The terrestrial acari of the British Isles. Alden Press, Oxford

Gerenscer GA (1983) Invertebrate epithelial transport. Am J Physiol 244:R127–R129

Grimstone AV, Mullinger AM, Ramsay JA (1968) Further studies on the rectal complex of the mealworm, *Tenebrio molitor,* L. (Coleoptera, Tenebrionidae). Proc R Soc London Ser B 253:343–382
Gupta BL, Hall TA, Maddrell SHP, Moreton RB (1976) Distribution of ions in a fluid transporting epithelium determined by electron-probe X-ray microanalysis. Nature (London) 264:284–287
Hadley NF (1970) Water relations of the desert scorpion, *Hadrurus arizonensis.* J Exp Biol 53:547–558
Hadley NF (1981) Cuticular lipids of terrestrial plants and arthropods: a comparison of their structure, composition, and waterproofing function. Biol Rev 56:23–47
Hafez M, El-Ziady S, Hefnawy T (1970) Biochemical and physiological studies of certain ticks (Ixoidea). Uptake of water vapor by the different developmental stages of *Hyalomma* (H.) *dromedarii* Koch (Ixodidae) and *Ornithdoros* (O.) *savignyi* (Audouin) (Argasidae). J Parasitol 56:354–361
Hanrahan JW, Phillips JE (1984) KCl transport across an insect epithelium: II. Electrochemical potentials and electrophysiology. J Membr Biol 80:27–47
Harvey WR, Cioffi M, Dow JAT, Wolfersberger MG (1983) Potassium ion transport ATPase in insect epithelia. J Exp Biol 106:91–117
Hill AE, Hill BS (1976) Mineral ions. In: Luttge U, Pitman MG (eds) Transport in plants, vol 2, part B. Tissues and organs. Springer, Berlin Heidelberg New York, pp 225–243
House CR (1974) Water transport in cells and tissues. Arnold, London
Jaworski D, Sauer JR, Williams JP, McNew RW, Hair JA (1984) Age-related effects on water, lipid, hemoglobin, and critical equilibrium humidity in unfed adult lone star ticks (Acari: Ixodidae). J Med Entomol 21:100–104
Katchalsky, A (1954) Polyelectrolyte gels. Prog Biophys Biophys Chem 4:1–59
Kaufman WR, Phillips JE (1973) Ion and water balance in the ixodid tick, *Dermacentor andersoni.* II. Mechanism and control of salivary secretion. J Exp Biol 58:523–536
Knülle W (1965) Die Sorption and Transpiration des Wasserdampfes bei der Mehlmilbe (*Acarus siro* L.). Z Vergl Physiol 49:586–604
Knülle W (1967) Physiological properties and biological implications of the water vapor sorption mechanism in larvae of the oriental rat flea, *Xenopsylla cheopis* (Roths.). J Insect Physiol 13:333–357
Knülle W (1984) Water vapor uptake in mites and insects: an ecophysiological and evolutionary perspective. Acarology 6:71–82
Knülle W, Devine TL (1972) Evidence for active and passive components of sorption of atmospheric water vapor by larvae of the tick *Dermacentor variabilis.* J Insect Physiol 18:1653–1664
Knülle W, Rudolph D (1982) Humidity relationships and water balance of ticks. In: Obenchain FD, Galun RL (eds) Physiology of ticks. Pergamon Press, Oxford New York, pp 43–70
Knülle W, Spadafora RR (1969) Water vapor sorption and humidity relationships in *Liposcelis* (Insecta: Psocoptera). J Stored Prod Res 5:49–55
Knülle W, Spadofora RR (1970) Occurrence of water vapor sorption from the atmosphere in larvae of some stored-produce beetles. J Econ Entomol 4:1069–1070
Knülle W, Wharton GW (1964) Equilibrium humidities in arthropods and their ecological significance. Acarologia 6:299–306
Küppers J, Thurm U (1980) Water transport by electroosmosis. In: Locke M, Smith DS (eds) Insect biology in the future. Academic Press, London New York, pp 125–144
Küppers J, Plagemann A, Thurm U (1986) Uphill transport of water by electroosmosis. J Membr Biol 91:107–119
Lees AD (1946) The water balance in *Ixodes ricinus* L. and certain other species of ticks. Parasitology 37:1–20
Locke M (1974) The structure and formation of the integument in insects. In: Rockstein M (ed) The physiology of insecta, vol 6. Academic Press, London New York, pp 124–213
Machin J (1975) Water balance in *Tenebrio molitor,* L. larvae; the effect of atmospheric water absorption. J Comp Physiol 101:121–132
Machin J (1976) Passive exchange during water vapor absorption in mealworms (*Tenebrio molitor*): a new approach to studying the phenomenon. J Exp Biol 65:603–615
Machin J (1979a) Atmospheric water absorption in arthropods. Adv Insect Physiol 14:1–48
Machin J (1979b) Compartmental osmotic pressures in the rectal complex of *Tenebrio* larvae: evidence for a single tubular pumping site. J Exp Biol 82:123–137

Machin J (1981) Water compartmentalization in insects. J Exp Zool 215:327 – 333
Machin J (1983) Water vapor absorption in insects. Am J Physiol 244:R187 – R192
Machin J (1984) The study of atmospheric water absorption. In: Bradley TJ, Miller TA (eds) Measurements of ion transport and metabolic rate in insects. Springer, Berlin Heidelberg New York, pp 69 – 99
Machin J, O'Donnell MJ, Coutchié PA (1982) Mechanisms of water vapor absorption in insects. J Exp Zool 222:309 – 320
MacLeod GF (1941) X-ray studies of starving mealworm larvae. Ann Entomol Soc Am 34:696 – 701
Maddrell SHP (1971) The mechanisms of insect excretory systems. Adv Insect Physiol 8:199 – 331
Mellanby K (1932) The effect of atmospheric humidity on the metabolism of the fasting mealworm (*Tenebrio molitor* L., Coleoptera). Proc R Soc London Ser B 111:376 – 390
Mooney HA, Gulmon, SL, Ehleringer J, Rundel PW (1980) Atmospheric water uptake by an Atacama desert shrub. Science 209:693 – 694
Mozafar A, Goodin RJ (1970) Vasiculated hairs: a mechanism for salt tolerance in *Atriples holimus*. Plant Physiol 45:62 – 65
Needham GR, Teel PD (1986) Water balance by ticks between bloodmeals. In: Sauer JR, Hair JA (eds) Morphology, physiology, and behavioral biology of ticks. Ellis Horwood, Chichester UK, pp 100 – 151
Noble-Nesbitt J (1970a) Water uptake from subsaturated atmospheres: its site in insects. Nature (London) 225:753 – 754
Noble-Nesbitt J (1970b) Water balance in the firebrat, *Thermobia domestica* (packard). The site of water uptake from the atmosphere. J Exp Biol 52:193 – 200
Noble-Nesbitt J (1977) Active transport of water vapour. In: Gupta BL, Moreton RB, Oschman JL, Wall BJ (eds) Transport of ions and water in animals. Academic Press, London New York, pp 571 – 597
Noirot C, Noirot-Timothee C (1971) Ultrastructure du proctodeum chez le *Lepismodes inquilinus* Newman (= *Thermobia domestica* Packard) II. Le sac anal. J Ultrastruct Res 37:335 – 350
O'Donnell MJ (1977) Site of water vapor absorption in the desert cockroach, *Arenivaga investigata*. Proc Natl Acad Sci USA 74:1757 – 1760
O'Donnell MJ (1978) The site of water vapor absorption in *Arenivaga investigata*. In: Schmidt-Nielsen K, Bolis L, Maddrell SHP (eds) Comparative physiology – Water, ions and fluid mechanics. Cambridge Univ Press, Cambridge, pp 115 – 121
O'Donnell MJ (1980) Water vapor absorption by the desert burrowing cockroach *Arenivaga investigata* (Dictyoptera: Polyphagidae). Ph D Thesis, Univ Toronto
O'Donnell MJ (1981a) Frontal bodies: novel structures involved in water vapor absorption in the desert burrowing cockroach, *Arenivaga investigata*. Tissue Cell 13:541 – 555
O'Donnell MJ (1981b) Fluid movements during water vapor absorption by the desert burrowing cockroach, *Arenivaga investigata*. J Insect Physiol 27:877 – 887
O'Donnell MJ (1982a) Water vapor absorption by the desert burrowing cockroach: evidence against a solute-dependent mechanism. J Exp Biol 96:251 – 262
O'Donnell MJ (1982b) Hydrophilic cuticle – the basis for water vapor absorption by the desert burrowing cockroach, *Arenivaga investigata*. J Exp Biol 99:43 – 60
O'Donnell MJ (1986) Water vapor absorption by arthropods: different sites, different mechanisms. In: Dejours P, Bolis L, Taylor CR, Weibel E (eds) Terrestrial versus aquatic life: contrasts in design and function. Liviana Press (Springer, Berlin Heidelberg New York), pp 500 – 507
Okasha AYK (1972) Water relations in an insect, *Thermobia domestica* II. Relationships between water content, water uptake from subsaturated atmospheres and water loss. J Exp Biol 57:285 – 296
Phillips JE (1964) Rectal absorption in the desert locust, *Schistocerca gregaria* Forskal. I. Water. J Exp Biol 41:15 – 38
Ramsay JA (1964) The rectal complex of the mealworm *Tenebrio molitor*, L. (Coleoptera, Tenebrionidae). Philos Trans R Soc London Ser B 248:279 – 314
Ramsay JA (1976) The rectal complex in the larvae of Lepidoptera. Philos Trans R Soc London Ser B 274:203 – 226
Roth M (1971) Fluorescent reaction for amino acids. Anal Chem 43:880 – 882
Rudolph D (1982a) Occurrence, properties and biological implications of the active uptake of water vapor from the atmosphere in Psocoptera. J Insect Physiol 28:111 – 121

Rudolph D (1982b) Site, process and mechanism of active uptake of water vapor from the atmosphere in the Psocoptera. J Insect Physiol 28:205 – 212
Rudolph D (1983) The water-vapor uptake system of the Phthiraptera. J Insect Physiol 29:15 – 25
Rudolph D, Knülle W (1974) Site and mechanism of water vapor uptake from the atmosphere in ixodid ticks. Nature (London) 249:84 – 85
Rudolph D, Knülle W (1978) Uptake of water vapor from the air: process, site and mechanism in ticks. In: Schmidt-Nielsen K, Bolis L, Maddrell SHP (eds). Cambridge Univ Press, Cambridge, pp 97 – 113
Rudolph D, Knülle W (1979) Mechanisms contributing to water balance in non-feeding ticks and ecological implications. In: Rodriguez JG (ed) Recent advances in acarology, vol 1. Academic Press, London New York, pp 375 – 383
Rudolph D, Knülle W (1982) Novel uptake systems for atmospheric water vapor among insects. J Exp Zool 222:321 – 333
Ryerse JS (1977) Control of mitochondrial movement during development in insect Malpighian tubules. Proc Microsc Soc Can 4:48 – 49
Seely MK, Hamilton WJ (1976) Fog catchment sand trenches constructed by tenebrionid beetles, *Lepidochora*, from the Namib Desert. Science 193:484 – 486
Shih CY, Sauer JR, Eikenbary R, Hair JA, Frick JH (1973) The effects of desiccation and rehydration on the lone star tick. J Insect Physiol 19:505 – 514
Sieber R, Kokwaro D (1982) Water intake by the termite *Macrotermes michaelseni*. Entomol Exp Appl 31:147 – 153
Solomon ME (1966) Moisture gains, losses and equilibria of flour mites, *Acarus siro* L., in comparison with larger arthropods. Entomol Exp Appl 9:25 – 41
Somero GN (1985) Protons, osmolytes, and fitness of internal milieu for protein function. Am J Physiol 251:R197 – R213
Stephens GC, Volk MJ, Wright SH, Backlund PS (1978) Transepidermal accumulation of naturally occurring amino acids in the sand dollar, *Dendraster excentricus*. Biol Bull 154:335 – 347
Tanaka T (1981) Gels. Sci Am 244:124 – 138
Tatchell RJ (1967) Salivary secretion in the cattle tick as a means of water elimination. Nature (London) 213:940 – 941
Tupy J, Machin J (1985) Transport characteristics of the isolated rectal complex of the mealworm *Tenebrio molitor*. Can J Zool 63:1897 – 1903
Wharton GW (1985) Water balance of insects. In: Kerkut GA, Gilbert LI (eds) Comprehensive insect physiology, biochemistry and pharmacology, vol 4. Pergamon Press, Oxford New York, pp 565 – 601
Wharton GW, Devine TL (1968) Exchange of water between a mite, *Laelaps echidnina*, and the surrounding vapor under equilibrium conditions. J Insect Physiol 14:1303 – 1318
Wharton GW, Furumizo RT (1977) Supracoxal gland secretions as a source of fresh water for Acaridei. Acarologia 19:112 – 116
Winston PW, Bates DH (1960) Saturated solutions for the control of humidity in biological research. Ecology 41:232 – 237
Yancey PH, Clark ME, Hand SC, Bowlus RD, Somero GN (1982) Living with water stress: evolution of osomolyte systems. Science 217:1214 – 1222
Zachariassen KE (1979) The mechanism of the cryoprotective effect of glycerols in beetles tolerant to freezing. J Insect Physiol 25:29 – 32

Chapter 3

Nutrient Transport by the Invertebrate Gut

G. A. Ahearn

Contents

1 Introduction

The gastrointestinal tracts of multicellular organisms are often complex and variable in structure and exhibit a variety of physiological roles which aid in the maintenance of stable internal homeostatic conditions as well as participate in the regulation of normal growth and development. Because of medical implications the most detailed studies of gastrointestinal physiology have been conducted on mammals. However, over 95% of living metazoan organisms are not mammals, but are invertebrates which also exhibit highly complex gastrointestinal organs. It is therefore appropriate to examine, in similar detail, the physiological properties of gut function in other animal groups, for only in this way can a complete understanding of the processes involved in gastrointestinal physiology be adequately assessed and their possible adaptive significance ascertained. This review summarizes information concerning organic solute transport by the gastrointestinal tracts of the four most anatomically complex invertebrate phyla: Annelida, Echinodermata, Mollusca, and Arthropoda. Mechanisms of nutrient transport by absorptive organs of each animal group are characterized in general terms, and recent information from the author's laboratory about sugar and amino acid absorption in crustacean gut are discussed in detail.

Department of Zoology, 2538 The Mall, Edmondson 152, University of Hawaii at Manoa, Honolulu, Hawaii 96822, USA

Advances in Comparative and Environmental Physiology. Vol. 2

2 Annelida

The gastrointestinal tract of annelids generally is a simple tube composed of foregut, midgut, and hindgut with regional modifications in structure and function occurring within and between different groups reflecting wide variations in mode of life. Amino acid transport by the esophagus, stomach, and intestine (the first two being foregut, while the latter is largely midgut) has been investigated in the marine lugworm, *Arenicola marina* (Bamford and Stewart 1973a, b), while lipid absorption by the intestine has been examined in the polychaete *Hermodice carunculata* (Marsden 1968).

In *Arenicola* the intestine is the main site for amino acid absorption, exhibiting more than twice the uptake values and concentration ratios than for other gut regions (Bamford and Stewart 1973a). The intestinal epithelium displayed carrier-mediated transport for neutral (alanine) and charged (lysine, aspartate) amino acids. Apparent affinity constants for transapical amino acid transport of these three solutes were 0.63 mM (alanine), 0.43 mM (lysine), and 0.77 mM (aspartate). Both lysine and aspartate were absorbed against a concentration gradient and the metabolic poisons NaCN and 2,4-dinitrophenol significantly inhibited aspartate tissue accumulation (Bamford and Stewart 1973b). On the basis of competition experiments, at least three distinct classes of transport proteins accommodating neutral, basic, and acidic amino acids were suggested for the mucosal border of the annelid intestinal epithelium.

In *Hermodice* products of lipid digestion are absorbed by the anterior intestinal epithelium, where neutral fats are reformed and transported to other parts of the body via the blood or by coelomocytes (Marsden 1968).

3 Echinodermata

The gastrointestinal tracts of echinoderms generally follow two basic designs: (1) a simple tube from mouth to anus which is structurally and functionally differentiated along its length as is found in the echinoids (sea urchins) or holothurians (sea cucumbers); and (2) a gut generally lacking a functional intestine, but instead, exhibiting highly elaborate pyloric and rectal caeca as exemplified by the asteroids (sea stars). Most studies of nutrient absorption in the echinoderms have used forms which exhibit the first gastrointestinal tract design because of its ease of isolation and study in vitro.

3.1 Nutrient Transport by Echinoderms with Tubular Gastrointestinal Tracts

Because intestinal contents of echinoids and holothurians often contain high concentrations of nutrients, while the coelomic fluid surrounding their tubular guts has very low concentrations of the same solutes, early studies with these two groups of echinoderms suggested that the mode of amino acid and sugar movement across this structure was by simple diffusion down concentration gradients

(Farmanfarmaian and Phillips 1962). This suggestion appeared to be confirmed in later studies of sugar transport in isolated intestinal segments of two holothurian species, *Stichopus parvimensis* (Lawrence et al. 1967) and *Thyone briareus* (Farmanfarmaian 1969a, b). In both cases transmural fluxes of sugars were observed, but this transfer was accounted for by simple diffusion. No evidence was presented for carrier transport of sugars in these animals. In contrast, L-alanine and glycine were actively absorbed in *Stichopus,* while AIBA, DL-proline, L-glutamic acid, and DL-lysine were not (Lawrence et al. 1967).

Among the echinoids, segments of intestine from *Paracentrotus lividus* and *Echinus esculentus* displayed active transport of glucose, while segments of stomach wall exhibited both active glucose and galactose transport; fructose was not actively transported by either gut region (Bamford and James 1972; Bamford et al. 1972). Intestinal carrier transport of glucose in *Paracentrotus* displayed an apparent binding constant (K_t) of 0.4 mM, while that of galactose was 0.2 mM. The esophagus, stomach, and intestine of *Echinus* all demonstrated active L-alanine transport with the kinetic constants for transapical transfer being related to expected substrate availability in each gut region (James and Bamford 1974). Furthermore, the stomach preparation of this species displayed active transport of L-alanine, D-alanine, L-leucine, and L-lysine, while L-aspartate was not actively accumulated (Bamford and James 1972).

3.2 Nutrient Transport by Echinoderms with Complex Gastrointestinal Diverticula

The pyloric caeca of asteroids have long been regarded as sites of digestive enzyme production, nutrient absorption, and storage of reserves (Frenzel 1892; Chapeaux 1893; van der Heyde 1923). More recent studies using cytology and histochemistry of these structures confirm their polyfunctional properties (Anderson 1953, 1960, 1962, 1966, 1979; Jangoux 1981, 1982; Jangoux and van Impe 1971, 1977; Jangoux and Perpeet 1972a, b). These modern investigations suggest that food enters the cardiac and pyloric portions of the stomach where preliminary extracellular digestion occurs. Partially digested food is carried to the branches of the pyloric caeca by a flagellated current created by modified stomach cells (Fig. 1). In the lumen of a pyloric caecum, food particles undergo a second extracellular digestion prior to absorption by epithelial cells. Ferguson (1964, 1968) described the transport of a variety of sugars and amino acids by intact pyloric caeca of asteroids which were incubated in a saline medium containing radiolabeled organic solute. Under these conditions uptake of isotope reflected activities of epithelial basolateral membranes and was analogous to movements of substrate from coelomic fluid to caecal tissues in living animals. These studies showed that the basolateral surface of starfish pyloric caeca was capable of accumulating 15 different amino acids from the bathing coelomic fluids by processes which were inhibited by the poison sodium iodoacetate. Inhibitory and stimulatory effects of one amino acid on the transport of another suggested that more than one transport mechanism for this group of compounds existed on the basolateral epithelial pole of this tissue.

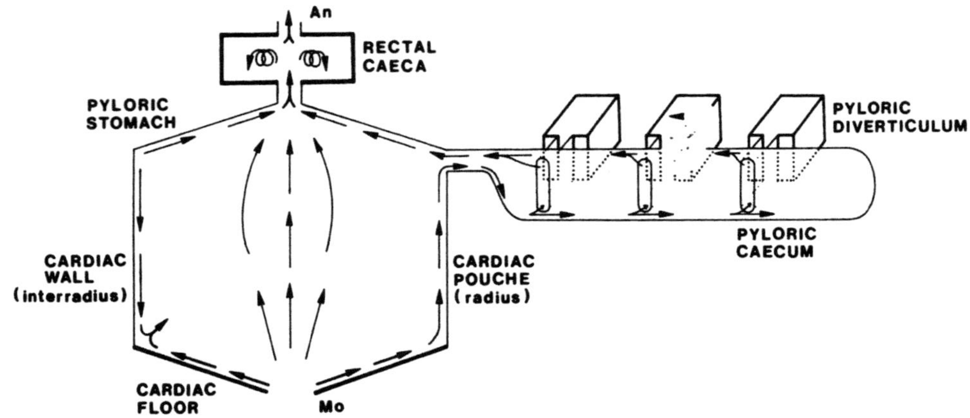

Fig. 1. Direction of water currents inside the gut of *Asterias rubens*. Thin-lined walls indicate transit areas, thick walls indicate areas where extracellular and/or intracellular digestion occurs. Suggested, but unproven, sites of nutrient absorption include the numerous pyloric diverticula and/or the rectal caeca. *An* nus; *mo* mouth. Note lack of functional intestine. (After Jangoux 1981)

At present, no studies have been conducted on apical or transepithelial transport of nutrients by echinoderm pyloric caeca so the physiological characteristics of nutrient absorption by this large organ are still unclear.

4 Mollusca

The gastrointestinal tracts of molluscs are highly diverse in morphology but generally exhibit a tubular conformation with diverticula arising from the wall of the stomach or from the anterior aspect of the intestine. Most mollusks possess an hepatopancreas (i.e., digestive gland, midgut gland, liver), and many often display a blind-ended sac called the caecum. In the cephalopods distinct liver and pancreatic structures have been defined which presumably demonstrate similar functional characteristics as these organs in the vertebrates. Over the last century many of these organs have been ascribed a nutrient absorptive role, although the evidence for this activity has often been indirect.

4.1 Nutrient Transport by the Molluscan Intestine

The intestine appears to be a major site of sugar and amino acid transport among the Polyplacophora (chitons) and the Gastropoda (snails). Using an everted intestinal sac technique, Lawrence and Lawrence (1967) demonstrated active transport of D-glucose and 3-o-methyl-D-glucose in the chiton, *Cryptochiton stelleri,* which was inhibited by anaerobic conditions. D-mannose and D-fructose were not actively absorbed by the intestine in this species. In another study of the same animal, glycine, L-alanine, DL-proline, AIBA, and DL-lysine were actively

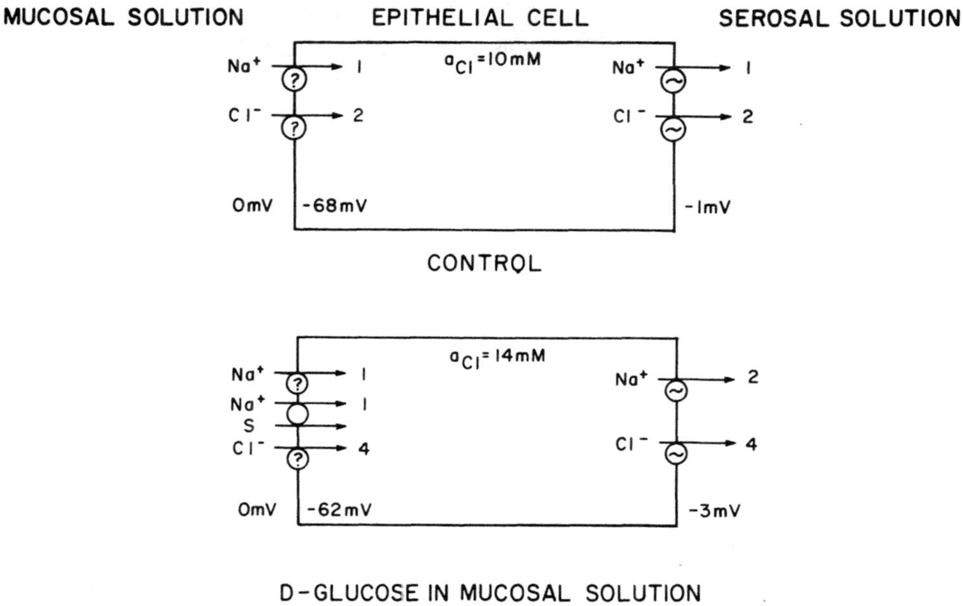

Fig. 2. Models depicting the relationship between the transports of sodium, chloride, and glucose across the brush border of *Aplysia* intestine. *Arrows* and *numbers* at mucosal and serosal membranes designate net movements. The mechanisms at the serosal membrane are active transport processes. In the absence of luminal glucose (or amino acids), Na and Cl enter the epithelium by independent carrier processes at the brush border. Addition of organic solute to the luminal solution, activates a third independent cotransport carrier transferring both luminal Na and glucose to the cytoplasm using the transmembrane Na gradient and transmembrane electrical potential as driving forces for solute accumulation. a_{Cl} intracellular chloride activity; *S* D-glucose. (After Gerencser 1985)

transported by lower intestinal segments, while L-glutamic acid was not (Greer and Lawrence 1967). As with sugar transport, anaerobic incubation significantly depressed the extent of amino acid absorption. Carrier-mediated sugar absorption was characterized for the intestine of the terrestrial snail *Cryptomphalus hortensis*, although in this species active solute transport was sodium-dependent and insensitive to oxygen deprivation (Barber et al. 1975a, b). The intestine of the marine gastropod, *Aplysia californica*, actively transports a variety of sugars and amino acids in conjuction with sodium from luminal solution to blood (Gerencser 1981, 1982, 1985). In this species the net flow of organic solutes and sodium is electrically coupled with that of chloride so that nutrient and salt absorption occur simultaneously (Fig. 2).

4.2 Nutrient Transport by Molluscan Gastrointestinal Diverticula

Many workers regard the hepatopancreas or the caecum as the primary sites of nutrient absorption in bivalves (Potts 1923; Yonge 1926; Morton 1956), chitons (Fretter 1937), gastropods (Millott 1937; Graham 1938; van Weel 1961), and cephalopods (Bidder 1957; Boucaud-Camou et al. 1976). Much of the evidence for this contention is based on cytology and histochemistry, but several workers have used tracer substances to follow the course of dietary components through the gastrointestinal tract. Bidder (1957) used suspensions of carmine dye to demonstrate the absorptive function of the *Octopus* liver (i.e., hepatopancreas). Van Weel (1961) showed that ingested iron saccharate accumulated in the epithelium of the snail (*Achatina fulica*) midgut gland (i.e., hepatopancreas), but not in the intestinal epithelium, suggesting a greater role of the former in absorption. 14C-glycine injected into a living crab before consumption by *Octopus vulgaris* appeared most rapidly and accumulated to the greatest extent in cells of the cephalopod caecum, pancreas, and liver, while virtually none of the isotopic label was recovered from intestinal cells, leading workers to suggest a minimal role of the latter organ in nutrient absorption in this animal (Boucaud-Camou et al. 1976).

Among molluscs, therefore, at least three organ systems may be absorptive and participate in transferring nutrients from lumen to blood: intestine, hepatopancreas, and caecum. The relative contribution of each structure to overall nutrient transfer appears to depend on the particular mollusc and its lifestyle. However, the reasons why similar organs in a variety of molluscs appears to have different absorptive capabilities are presently unclear.

5 Arthropoda

The arthropod digestive system consists of three major divisions: foregut, midgut, and hindgut. The first and last are lined with a chitinous cuticle and arise from the embryonic stomodeum and proctodeum, respectively. The midgut includes the epithelial-lined intestine and an array of anterior and posterior diverticula, all unchitinized and arising from embryonic endoderm. It is largely involved with the secretion of digestive enzymes, the breakdown of polymers under the influence of secreted enzymes, and the absorption of digestive products and water.

5.1 Insecta

Early studies of nutrient transport in insects indicated that sugars and amino acids were absorbed in both the tubular midgut and its anterior diverticula (caeca), with the latter structures transferring the bulk of the solute load to the blood (Treherne 1958, 1959). These studies showed that both types of compounds were passively absorbed from lumen to blood down concentration gradients established by the net flow of ions and water in the same direction. For sugars diffusion was facilitated by the rapid conversion of monosaccharides, such as glucose, into the disaccharide trehalose in the blood, a process which maintained a steep concentration gradient across the gut wall.

More recently, isolated intestinal sheets of lepidopteran larvae (*Hyalophora cecropia*), clamped in modified Ussing chambers, demonstrated active net trans-

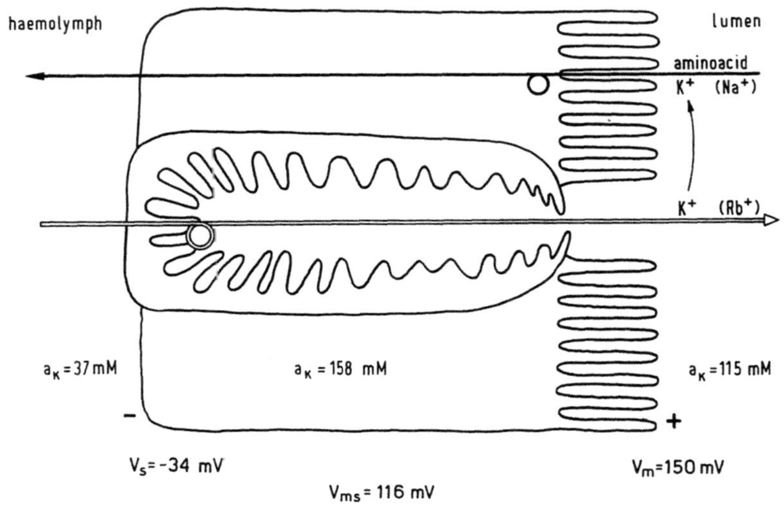

Fig. 3. Model for amino acid absorption across intestinal cells of lepidopteran larvae. A goblet cell between two columnar cells is indicated. The *solid arrow* indicates the amino acid carrier system, while the *open arrow* represents the K-pump. The apical K-pump of the goblet cells maintains the cation at a considerable distance from electrochemical equilibrium across the columnar cell brush border as a result of probable cell-cell coupling, providing a favorable K-diffusion gradient from lumen to cytoplasm and an adequate driving force for cytoplasmic amino acid accumulation by co-transport. V_s transserosal electrical potential difference; V_m transmucosal electrical potential difference; V_{ms} transepithelial electrical potential difference. Since the cell interior is electrically negative with respect to external compartments, polarities are defined as follows: V_s cell to hemolymph; V_m lumen to cell; V_{ms} lumen to hemolymph. Values of electrical potential differences and potassium activities refer to *Bombyx mori* in vivo. (After Giordana et al. 1982)

mural α-aminoisobutyric acid transport from lumen to blood (Nedergaard 1972, 1973). This active amino acid transport was inhibited by lack of oxygen, was independent of the active transport of potassium by the tissue, and was highly sensitive to transmural electrical potential difference.

In the last 6 years a variety of potassium-dependent amino acid transport mechanisms have been disclosed for the brush border membrane of the midgut epithelium in herbivorous lepidopteran larvae (Fig. 3) (Hanozet et al. 1980, 1984; Giordana et al. 1982, 1985). The brush border vesicle preparation from this insect (*Philosamia cynthia*) also demonstrated the capacity for significant D-alanine transport and its inhibition by the L isomer (Hanozet et al. 1984). Furthermore, the D isomer was shown to exhibit active transmural transport across intact intestinal tissue and its net flux was strongly reduced by the presence of L-alanine. Whereas all L-amino acids examined in this species displayed K co-transport, D-alanine transport appeared to illustrate a strong Na dependency. The phytophagous diet of this animal is high in potassium and low in sodium and may explain the occurrence of these unique cotransport systems.

5.2 Crustacea

5.2.1 Anatomical Properties of the Crustacean Intestine

The crustacean intestine runs from the posterior portion of the pyloric foregut to the anterior aspect of the chitinized hindgut. The intestine in crustaceans is a structurally diverse organ ranging from a simple tube in more primative species to a highly complex portion of the gastrointestinal tract having multiple caeca extending from it which often fill the body cavity with their ramified lobes. The intestine is thin-walled in most species and richly supplied with blood vessels from the abdominal artery. The tube is generally lined with a single type of columnar epithelial cell possessing a microvillar apical border and a central or basal nucleus (Yonge 1924; Pillai 1960; Dall 1967; Hootman and Conte 1974; Mykles 1979). Smaller cells in various stages of cell division are often found at the base of the epithelium which are believed to be young cells which give rise to new epithelium (Yonge 1924; Komuro and Yamamoto 1968). An undulating basement membrane and both circular and longitudinal muscle layers occur immediately beneath the epithelium. Food is propelled through the tubular intestine by periodic peristaltic (toward the anus) or antiperistaltic (toward the mouth) waves resulting from circular or longitudinal muscle contractions.

The epithelial cells of crustacean intestine may vary in size along the length of the structure. All cells bear apical microvilli and in some species a fine filamentous coat covering the apical cell border is present that resembles the vertebrate mucopolysaccharide glycocalyx (Talbot et al. 1972). Lateral membranes are generally straight in most species and have a minimum of interdigitation. Zonula adherens and septate desmosomes are common cell-to-cell junction types (Talbot et al. 1972; Cooke and Ahearn 1976; Ahearn and Maginniss 1977). The basal membrane of these cells often exhibits considerable infolding into the cytoplasm and has numerous associated mitochondria, suggesting the occurrence of energy-requiring processes. The basal cytoplasm of decapod intestinal epithelium has an extensive tubular smooth endoplasmic reticulum (Talbot et al. 1972; Mykles 1979) which is apparently continuous with the basal plasma membrane, resembling the basal region of fish chloride cells (Harb and Copeland 1963; Philpott and Copeland 1963).

5.2.2 Nutrient Transport by the Crustacean Intestine

5.2.2.1 Solute Influxes Across the Apical Membrane. During the 1970s a series of studies was performed to determine the extent to which the crustacean intestine could transport sugars and amino acids and whether the transfer mechanisms in this organ were similar to those described for vertebrate gut. Initial experiments were concerned with measuring intestinal amino acid transport in the marine shrimp, *Penaeus marginatus,* and employed longitudinally slit pieces of midgut which were immersed in saline containing an appropriate radiolabeled organic substrate. These early studies indicated that the apical membrane of intestinal epithelial cells exhibited a high affinity, carrier-mediated, sodium-dependent transport process for glycine (Ahearn 1974). Inhibition of mucosal

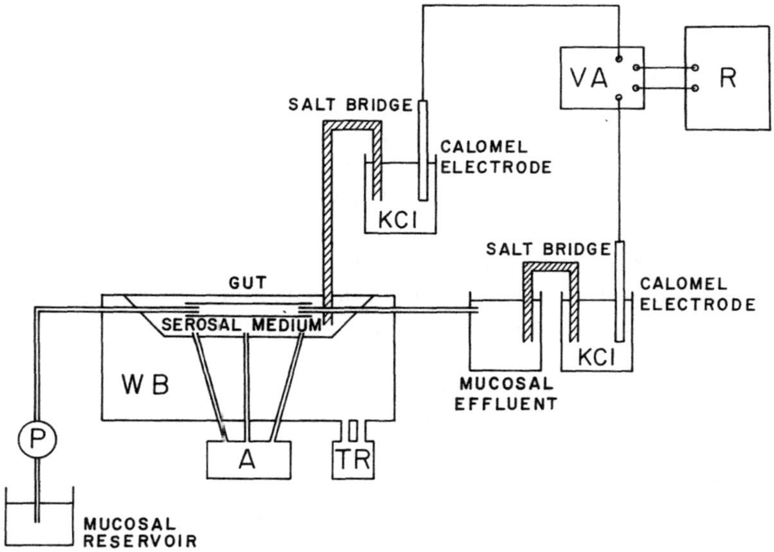

Fig. 4. Experimental apparatus used to measure transmural organic and inorganic solute transport and electrical potential differences across intestine of freshwater prawns. Peristaltic pump (*P*) was used to perfuse saline solution from the mucosal reservoir through the glass capillary tubing to ligated intestine. Following passage through the gut, saline was collected by way of a second glass capillary emptying into a mucosal effluent container. Saturated KCl-agar salt bridges connected both mucosal effluent and serosal medium via 3 M KCl solutions and calomel electrodes to a voltage amplifier (*VA*)-dual pen recorder (*R*) complex for electrical estimations. Serosal medium was kept aerated (*A*) throughout incubation and its temperature was controlled (24 °C) using a surrounding water bath (*WB*) and temperature regulator (*TR*). (After Ahearn 1980)

glycine influx occurred in the presence of both alanine and proline; the former was a noncompetitive inhibitor, whereas the latter was competitive in effect. Later studies on this same animal (Ahearn 1976) examined in detail the role of Na as cosubstrate with glycine in mucosal membrane transport. These investigations showed that both mucosal glycine and Na influxes displayed Michaelis-Menten kinetics and the influx of each solute was markedly dependent upon the presence of the other solute in the bathing medium.

In later studies both amino acid and sugar transport was characterized in the isolated and perfused midgut of the freshwater prawn, *Macrobrachium rosenbergii,* using a lucite chamber that allowed the effective separation of fluids bathing the mucosal and serosal surfaces of the preparation (Fig. 4) (Maginniss 1977; Ahearn and Maginniss 1977; Ahearn 1980, 1982). With this method both transapical influxes and transepithelial fluxes of organic solutes were measured.

Intestinal 3H-L-lysine transport across the mucosal epithelial border in *Macrobrachium rosenbergii* occurred through two parallel carrier-mediated processes, a high affinity system displaying Michaelis-Menten kinetics and a low affinity, nonsaturable mechanism that had a transfer rate which was a linear function of luminal lysine concentration (Brick and Ahearn 1978). The high affinity process was Na-dependent and inhibited by both arginine and iodoacetate. The low affinity mechanism displayed homoexchange diffusion and was Na-independent, inhibited by arginine, and unaffected by iodoacetate. 3H-L-alanine influx across the apical membrane in the perfused freshwater prawn intestine was also Na-dependent (Wyban et al. 1980). 3H-D-Glucose transport across the intestinal apical membrane of the freshwater prawn displayed Michealis-Menten kinetics with no discernible diffusion component (Maginniss 1977; Ahearn and Maginniss 1977). Under Na-free conditions, glucose influx was markedly reduced and was a linear function of luminal sugar concentration, indicative of either slow cellular diffusion or a facilitated process with a low carrier affinity. Phloridzin was a potent competitive inhibitor of glucose influx ($K_i = 3.6 \times 10^{-3}$ M) (Ahearn and Maginniss 1977), galactose was a weak inhibitor (mechanism undetermined), and fructose had no evident effect on glucose uptake.

5.2.2.2 Transepithelial Fluxes of Organic Solutes. Two studies were conducted with perfused freshwater prawn intestine which suggested that minimal luminally-derived sugars or amino acids were transported across this structure to the blood (Ahearn and Maginniss 1977; Wanyban et al. 1980). Transmural transport of both 3H-D-glucose and 3H-L-alanine (1 mM) across the intestine of the freshwater prawn was a linear function of time in both directions after a short 10-min lag period. A significant net flux of labeled sugar toward the blood was observed, but was only 10% or less of that for sugars and amino acids in vertebrate intestine. Unidirectional transmural fluxes of 3H-L-alanine across prawn intestine were not significantly different from one another, suggesting the absence of net amino acid movement across the tissue (Wyban et al. 1980).

Although the transport properties of crustacean intestine for organic solutes have only been described for two species of shrimp, *Penaeus marginatus* and *Macrobrachium rosenbergii*, some tentative conclusions can be made concerning the role of this portion of the gut in total nutrient absorption. Evidence from both transapical solute influxes and transepithelial solute transport suggest a minimal role of the crustacean intestine in total absorption of sugars and amino acids from gut lumen to hemolymph. Table 1 shows a summary of organic solute influx kinetic constants for amino acids, sugars, and vitamins in crustacean and mammalian intestine. It is clear from Table 1 that the crustacean intestine has much higher affinity carrier systems for a particular solute than do those of mammals. These affinity differences may be as great as two orders of magnitude. In addition, maximal transport velocities for each compound are markedly lower in crustaceans. These results suggest that apical carrier processes in crustaceans saturate and deliver considerably less nutrient to the epithelial cytosol per unit of time than do comparable systems of mammals. If reduced quantities of nutrients are transported into cells from the luminal contents, reduced transfer from the cells to the blood can be anticipated. This prediction appears to be verified from the results of transepithelial sugar and amino acid transport experiments (Ahearn and Maginniss 1977; Maginniss 1977; Wyban et al. 1980). A net transmural flux of 1 mM 3H-L-alanine was not observed in prawn intestine owing to extensive metabolism of the amino acid by epithelial cells. Although a significant net flux of 3H-D-glucose was observed across the shrimp intestine, it was less than 10% of that published for mammalian gut (Fisher and Parsons 1953; Fromm and Field 1975; Alvarez et al. 1969). These results confirm that the crustacean

Table 1. Comparison of organic solute influx constants between crustacean and mammalian intestines. (After Ahearn 1982)

Animal	Organic solute	K_t (μM)	J_M (nmol $cm^{-2} h^{-1}$)	Reference
Shrimp	Glycine	360	504	Ahearn (1974, 1976)
Prawn	Alanine	950	–	Wyban et al. (1980)
	Lysine	18	0.30	Brick and Ahearn (1978)
	Glucose	172	76	Ahearn and Maginniss (1977)
	Ascorbic acid	9	0.04	Rogers (1980)
Rabbit	Glycine	22700	4500	Peterson et al. (1970)
	Alanine	14000	12800	Alvarez et al. (1969)
	Lysine	10000	5000	Munck and Schultz (1969)
	Glutamate	7100	1100	Schultz et al. (1970)
	Aspartate	5000	3000	Schultz et al. (1970)
	Methylglucose	18200	4300	Goldner et al. (1969)
Guinea pig	Ascorbic acid	1000	140	Mellors et al. (1977)
Hamster	Folic acid	7	1.2	Rose et al. (1978)

intestine delivers a far lower quantity of amino acid and sugar to the blood per unit surface area than does the mammalian gut.

5.2.3 Anatomical Properties of the Crustacean Hepatopancreas

In crustaceans the largest branching diverticulum of the gastrointestinal tract is the hepatopancreas ("midgut gland"; van Weel 1974). This organ is generally multilobate and possesses secretory ducts leading to the pyloric region of the stomach (proventriculus). The basic unit of the hepatopancreas is a blind tubule containing a single epithelial layer surrounded by a thin muscle network. Many workers consider the primary function of this structure to be secretion, the tubular cells releasing a large complement of digestive enzymes into the gland duct where they pass into the proventriculus and mix with dietary components (Yonge 1924; van Weel 1955, 1960, 1970, 1974; Blandamer and Beechey 1964; Kooiman 1964; DeViliez and Buschien 1967; Sather 1969; Loizzi 1971).

In addition to its well-characterized secretory function, the early literature also suggests that the crustacean hepatopancreas has an absorptive role for nutrients (Yonge 1924; Loizzi 1971; van Weel 1955, 1974; Gibson and Barker 1979; Dall and Moriarty 1983), although the physiological evidence prior to 1985 for this activity is less clear than that relating to secretion.

The histology and ultrastructure of hepatopancreatic tubules from a variety of structurally diverse crustaceans such as shrimp (Pillai 1960; Rigdon and Mensik 1976; Al-Mohanna et al. 1985a, b; Al-Mohanna and Nott 1986), crayfish (Miyawaki and Sasaki 1961; Miyawaki and Tanoue 1961; Miyawaki 1966; Miyawaki et al. 1966; Loizii 1971), crabs (van Weel 1955; Becker et al. 1974; Chen et al. 1974), freshwater isopods (Smith et al. 1975), and terrestrial isopods (Vernon et al. 1974) have been examined and indicate that, although these structures contain only a single layer of epithelium there are several epithelial cell types which may function differentially. Jacobs (1928) reported four distinct cell types in the hepatopancreas of the crayfish, *Astacus leptodactylus*. He classified these cells as E cells (embryonic), which arise by mitotic division at the distal tip of the tubule and differentiate as they migrate proximally; F cells (fibrillar), which have a supranuclear vacuole that grows in size as this cell matures; R cells (resorptive), which contain glycogen, lipid droplets, and may be involved in nutrient absorption; and B cells (blister) which contain a single, immense vacuole bounded by a thin rim of cytoplasm. Since this early study was published, a variety of papers have appeared on several species of crustacean hepatopancreatic tubules which discuss possible interrelationships between the different cell lines (van Weel 1955; Travis 1957; Ogura 1959; Davis and Burnett 1964; Bunt 1968; Stanier et al. 1968; Loizzi 1971; Schultz 1976; Al-Mohanna et al. 1985b; Al-Mohanna and Nott 1986).

Loizzi (1971) presented ultrastructural evidence that, in the crayfish, R and F cells absorb luminal nutrients via molecular and bulk transport, respectively; and B cells, which are derived from F cells, secrete digestive enzymes into the tubule lumen. Figure 5 is a diagrammatic sketch of cells from two distinct regions of the hepatopancreatic tubules of the crayfish, *Orconectes viridis* and *Procambarus clarkii* (Loizzi 1971). Young R cells possess heavy concentrations of glycogen granules and numerous lipid droplets. As the R cells mature, the glycogen granules become more diffuse within the cytoplasm. Smooth endoplasmic reticulum is restricted to the basal region of the cell. Apically, the cell resembles a vertebrate intestinal absorptive epithelium with a dense brush border and an organelle-free region lying below the microvilli. Numerous mitochondria are found below this clear portion of cytoplasm. A surface coat above the microvilli, resembling the vertebrate glycocalyx, is present. Numerous basal invaginations occur in cells of this type.

The F cell possesses extensive rough endoplasmic reticulum with long, narrow, ribosome-studded cisternae (Fig. 5). In contrast to the R cell, the Golgi cisternae in F cells are always dilated and the entire complex is ringed by small, dense vesicles. F cells generally lack glycogen and lipid droplets, have a sparse surface enteric coat, a more uniform distribution of mitochondria, and few basal invaginations. As an F cell matures and migrates proximally, its rough endoplasmic reticulum becomes swollen with amorphous, electron-lucent material. Dilation of the Golgi cisternae increases during this process. The maturation of F cells is manifested also by the growth of the supranuclear vacuole.

Fig. 5. Schematic drawings of transition zone and B-cell zone in the crayfish hepatopancreas tubule showing major cell types occurring in these areas and their differential ultrastructure. Different cell lines are represented by the letters *B, F,* and *R*. *ac,* apical complex; *bi,* basal invaginations; *bl,* basal lamina; *cm,* circular muscle fiber; *cv,* clear vesicle; *dv,* dense-core vesicle; *Fe,* iron granule in supranuclear vacuole; *gly,* glycogen; *gol,* Golgi body; *HEM,* hemolymph surrounding tubule; *ld,* lipid droplet; *lm,* longitudinal muscle fiber; *LU,* lumen of tubule; *myo,* myoepithelial network; *np?,* possible neurosecretory process; *pin,* pinocytic channels and vesicles; *sec,* surface enteric coat; *vac,* vacuole; *z,* circular muscle z-lines. (After Loizzi 1971)

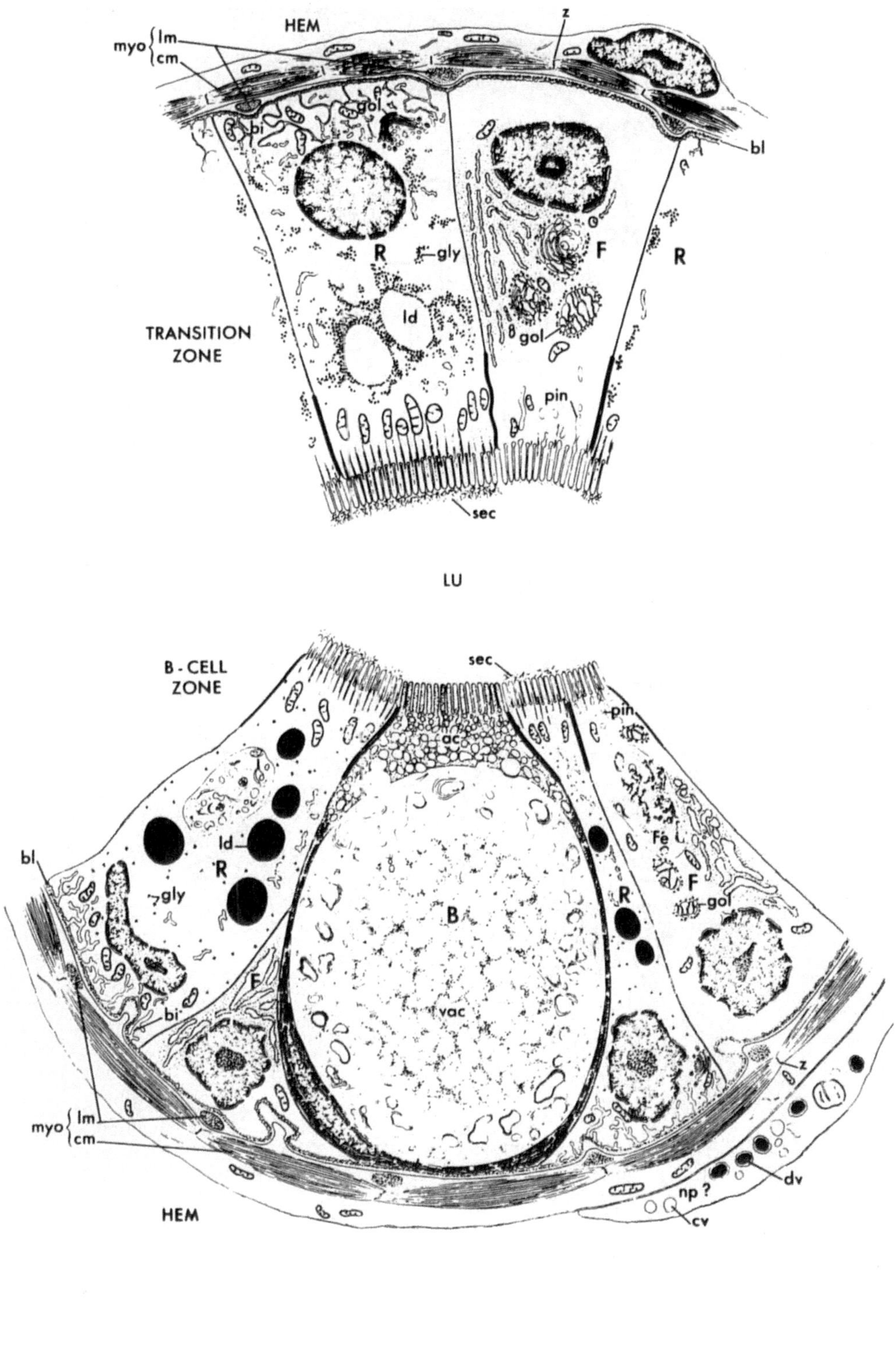
HEM
myo
lm
cm
z
gol
bi
bl
R
gly
F
R
ld
TRANSITION
ZONE
gol
pin
sec
LU
B-CELL
ZONE
sec
pin
ac
ld
R
bl
gly
Fe
F
R
gol
B
F
bi
vac
z
myo
lm
cm
dv
np ?
HEM
cv

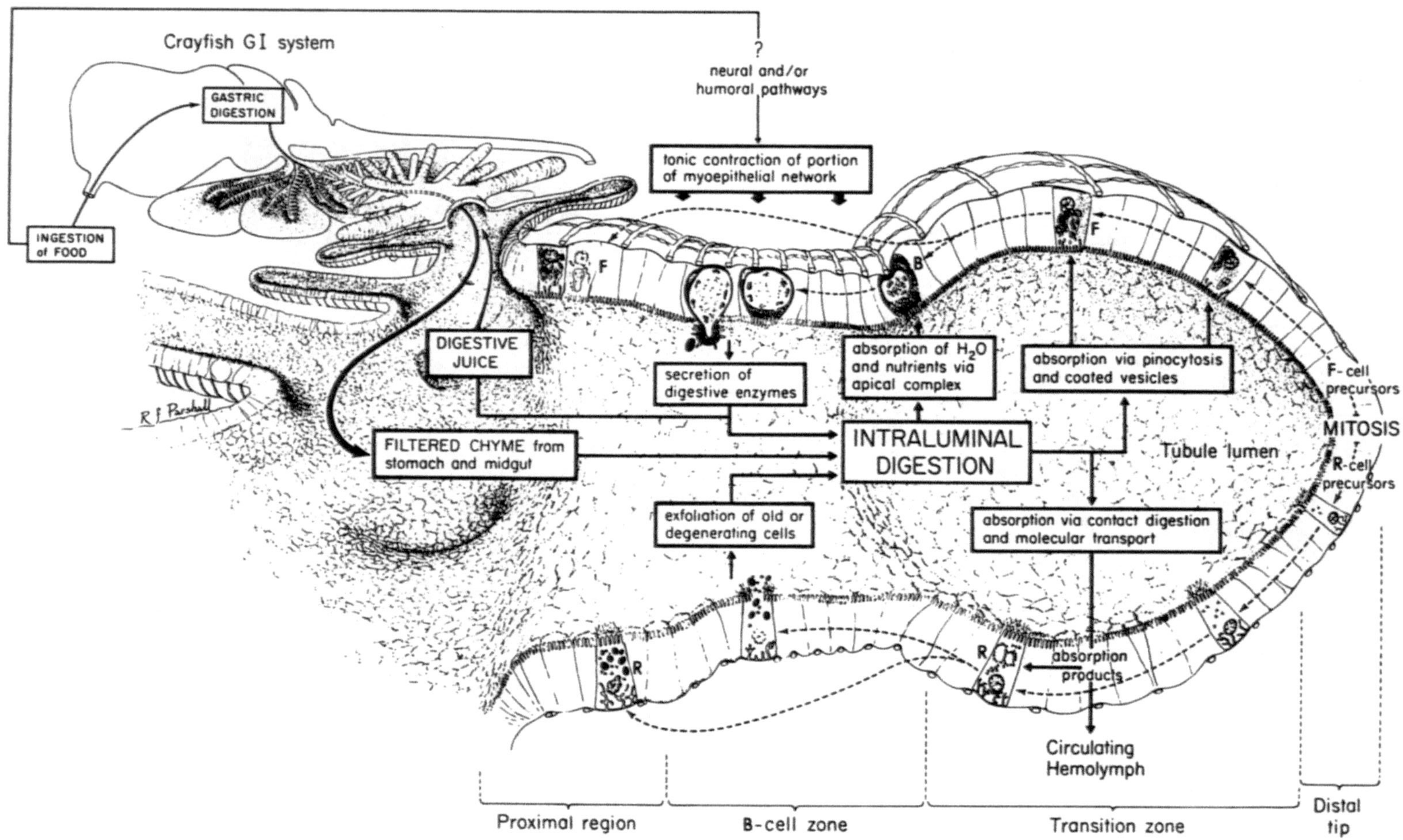
Crayfish GI system
INGESTION of FOOD
GASTRIC DIGESTION
?
neural and/or humoral pathways
tonic contraction of portion of myoepithelial network
DIGESTIVE JUICE
FILTERED CHYME from stomach and midgut
secretion of digestive enzymes
exfoliation of old or degenerating cells
INTRALUMINAL DIGESTION
absorption of H_2O and nutrients via apical complex
absorption via pinocytosis and coated vesicles
absorption via contact digestion and molecular transport
absorption products
Circulating Hemolymph
Tubule lumen
F-cell precursors
MITOSIS
R-cell precursors
F
B
R
R. F. Parshall
Proximal region
B-cell zone
Transition zone
Distal tip

B cells are first found about 0.5 mm from the distal blind end of a tubule and continue for about 60% of the tubule length. These cells contain a single large vacuole enclosed by a thin shell of cytoplasm (Fig. 5). Separating the vacuole from the lumen is an apical complex consisting of a microvillar border, dense cytoplasm, small mitochondria, and a meager surface enteric coat. No glycogen granules or lipid droplets are present. On the basis of general similarities between mature F and B cells, Loizzi (1971) believes that the former is transformed into the latter via enlargement of the supranuclear vacuole. B cell secretion is believed to occur when all or part of the apical complex pinches off in an apocrinelike fashion, allowing spillage of the vacuolar contents into the lumen.

Loizzi (1971) provided a scheme of digestive and absorptive physiology of crayfish hepatopancreas based on his ultrastructural observations (Fig. 6). This scheme proposes the distoproximal differentiation of two cell lines from precursor cells in the tubule apex. Differentiating cells pass through a transition zone in the tubule where they become either F or R cells. An F cell found in the transition zone may then change into a secretory cell laden with digestive enzymes in the B cell zone. Finally, differentiation is complete in the proximal region which is composed mainly of mature B and R cell types. According to this author, nutrients are absorbed in two ways: R cells apparently use contact digestion at the surface enteric coat, followed by molecular transport through the microvillar membrane; F cells appear to absorb nutrients and water by pinocytosis, a process that contributes to growth of the supranuclear vacuole. Bulk transport is continued by the apical complex after transformation of F cells to B cells and could be related to secretion of digestive enzymes by this cell line.

5.2.4 Nutrient Transport by the Crustacean Hepatopancreas

As suggested by Loizzi (1971) and several other investigators for much of this century, as a result of histological and ultrastructural studies, the hepatopancreas appears to be a major site of nutrient absorption in crustaceans, with the tubular intestine playing a minor role in many species (Yonge 1924; van Weel 1974; Gibson and Barker 1979; Dall and Moriarty 1983). Until recently, there have been no direct measurements of solute transfer across either the hepatopancreatic brush border membrane or the epithelial cell layer to clarify organ function. In order to directly define apical and basolateral nutrient transfer properties of crustacean hepatopancreatic epithelial cells, membrane vesiculation techniques developed by mammalian gastrointestinal and renal physiologists were applied to this invertebrate organ.

Fig. 6. Diagrammatic scheme of digestive physiology in the crayfish hepatopancreas tubule. Precursor cells in the distal tip of each tubule for F cells and R cells undergo extensive mitotic division and migrate proximally. Differentiation into these two cell lines occurs in the transition zone. In the B-cell zone, F cells transform into B cells by the enlargement of the supranuclear vacuole. Mature R and B cells are presumed functionally differentiated into absorptive and secretory roles, respectively. Differentiations of R and F cell types (*dashed lines*) interact with movements of nutrients distally and digestive juices proximally (*solid lines*). Relative zone lengths are not drawn to scale. (After Loizzi 1971)

5.2.4.1 Methods of Producing Purified Hepatopancreatic Epithelial Brush Border Membrane Vesicles (BBMV). Isolated and purified brush border membrane vesciles (BBMV) of lobster (*Homarus americanus*) hepatopancreatic epithelial cells were prepared using a magnesium precipitation technique originally developed for mammalian epithelia by Kessler et al. (1978) and Biber et al. (1981). These centrifugation techniques produced a final purified membrane pellet of hepatopancreatic brush border vesicles that was composed of large numbers of sealed, osmotically active transporting units that ranged in diameter from 0.1 to 0.8 μm (Fig. 7; Ahearn 1987). This size range was approximately the same as that of purified vesicles from vertebrate gut or renal epithelia (Kinne et al. 1975; Kinne and Murer 1976). Minimal visual contamination by organelles or membranes from other cellular locations appears present in electron micrographs taken of final purified BBMV preparations.

The biochemical purity of hepatopancreatic BBMV was assessed by comparing the enrichments of membrane enzyme activity in final vesicle samples with enzyme activity in the original tissue homogenate prior to differential centrifugation. Results indicate enrichments of alkaline phosphatase, Na/K-ATPase, and cytochrome c oxidase of 15.0-, 1.0-, and 0.2-fold, respectively, in the purified membrane fraction, suggesting a high enrichment of brush border membranes and minimal contamination from basolateral or organelle sources (Table 2).

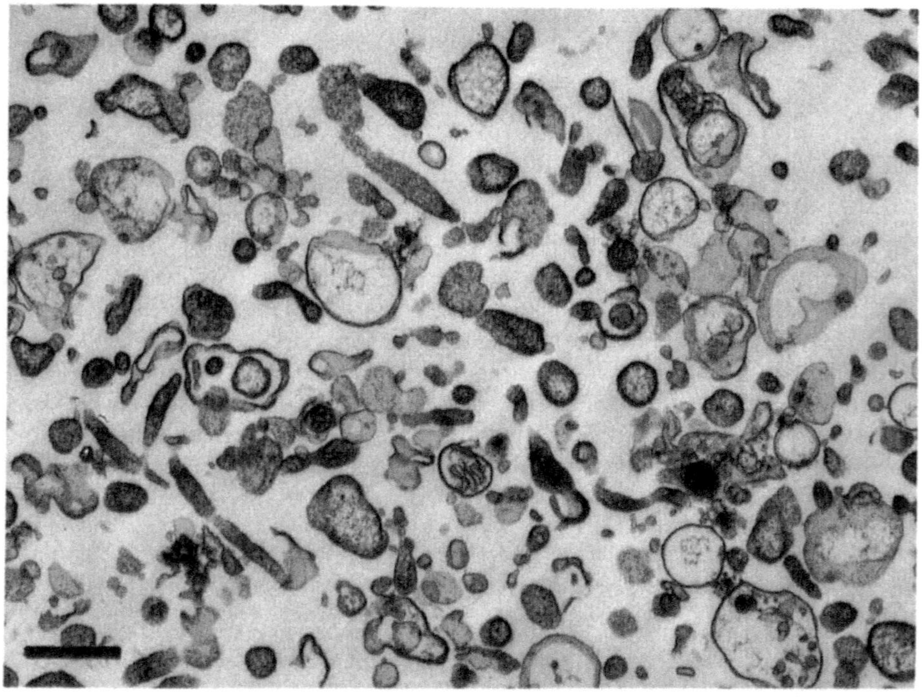

Fig. 7. Electron micrograph of lobster (*Homarus americanus*) hepatonpancreatic brush border membrane vesicles produced using the preparative scheme described in Ahearn et al. (1985) and discussed in the text. Length of bar corresponds to 0.5 μm (×25000)

Table 2. Enzyme characterization of lobster hepatopancreatic brush border membrane vesicles.[a] (After Ahearn et al. 1985)

Enzyme	Pellet 2	Pellet 7	Purification factor	Enzyme recovery (%)	Protein recovery (%)	Animal number
	activity (μmol mg^{-1} h^{-1})					
Alkaline phosphatase	1.28 ± 0.37	10.32 ± 2.26	15.26 ± 4.65	30.47 ± 7.85	3.37 ± 0.76	8
Na/K-ATPase	0.41 ± 0.10	0.29 ± 0.07	1.04 ± 0.42	5.00 ± 2.52	4.99 ± 0.79	7
Cytochrome c oxidase	42.26 ± 4.36	7.80 ± 1.51	0.19 ± 0.04	0.47 ± 0.15	2.60 ± 0.69	5

[a] Values are means ± SE. Enzyme activities are in μmol inorganic phosphate released per mg protein per hour (alkaline phosphatase and Na/K-ATPase) and in μmol reduced cytochrome c oxidized per mg protein per hour (cytochrome c oxidase). Purification factors are means of individual pellet 7 activities divided by individual pellet 2 activities. Enzyme and protein recoveries represent comparisons between contents of pellets 2 and 7.

5.2.4.2 Sugar Transport by Hepatopancreatic BBMV. Transport studies using lobster hepatopancreatic BBMV were conducted using the Millipore filtration technique of Hopfer et al. (1973). The first series of studies using purified brush border vesicles involved the characterization of D-glucose transport. Vesicles were loaded with 300 mM mannitol and incubated in a series of external media of varying composition for time intervals up to 60 min. Incubation of BBMV in a medium containing 150 mM NaCl and 0.1 mM 3H-D-glucose resulted in rapid uptake of the sugar into the vesicles with a maximal accumulation occurring by 1 min, followed by a slow reduction in sugar content to equilibrium at 60 min (Fig. 8). Maximal transient glucose uptake (overshoot) by these vesicles was generally three for four times the equilibrium values. The overshoot phenomenon of glucose accumulation was not observed when incubation medium either contained 150 mM KCl, 300 mM mannitol, or 150 mM NaCl + 0.5 mM phloridzin. Apparent glucose influx rates, estimated at 15-s incubations were:

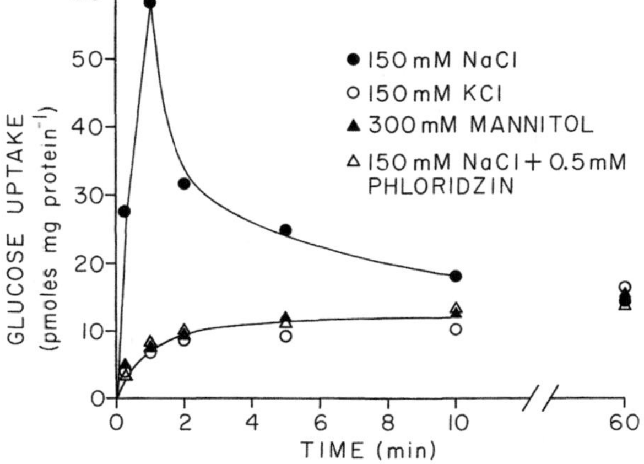

Fig. 8. Time course of 0.1 mM 3H-D-glucose uptake by lobster hepatopancreatic BBMV loaded with 300 mM mannitol at pH 7.4 and incubated in media of various compositions at the same external pH. (After Ahearn et al. 1985)

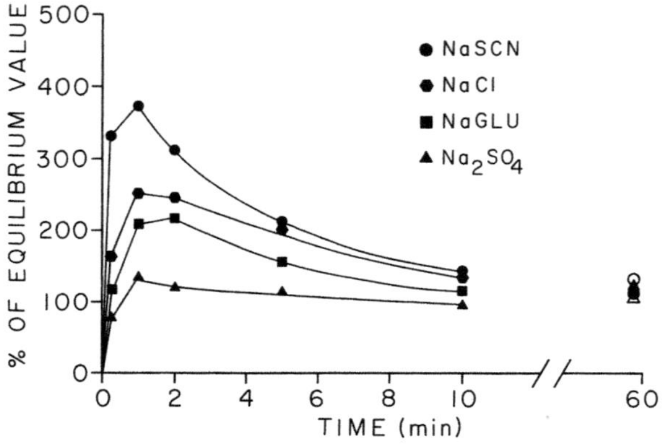

Fig. 9. Effect of external anions on the time course of Na-dependent 0.1 mM 3H-D-glucose uptake by lobster BBMV loaded with 300 mM mannitol at pH 7.4 and incubated in media of the same pH containing the respective sodium salts at 150 mM. (After Ahearn et al. 1985)

27.6 (NaCl) and 3.9 (KCl) pmol mg^{-1} protein 15 s^{-1}, suggesting a sevenfold greater entry rate in sodium medium. These results indicate a clear Na dependency of glucose uptake by BBMV and its significant inhibition by the drug, phloridzin.

Additional experiments characterizing Na-dependent glucose transport by hepatopancreatic BBMV suggested that the transfer mechanism was sensitive to transmembrane electrical potential. 3H-D-glucose uptake was measured in mannitol-loaded vesicles incubated in media containing a variety of sodium salts (NaCl, NaSCN, $NaSO_4$, and Na-gluconate) composed of anions of known different rates of diffusional transfer across other membranes. Greatest apparent glucose influxes and highest overshoot values were obtained when sodium salts with highly permeant anions were used in the external media (e.g., NaSCN, NaCl) (Fig. 9). This experiment suggested that Na-dependent glucose uptake by lobster BBMV was electrogenic and responded to variations in transmembrane potential induced by differential anion fluxes.

Gastric contents of crustaceans are acidic (Gibson and Barker 1979) and are periodically flushed through hepatopancreatic tubules, where they come into contact with apical surfaces of absorptive cells. A series of experiments was conducted to determine the influence of luminal pH on glucose transport by hepatopancreatic BBMV. Vesicles were loaded with mannitol adjusted to pH 6.0, 7.4, and 8.0 and were incubated in media of the same pH containing NaCl and 0.1 mM 3H-D-glucose. At each pH an Na-dependent glucose uptake overshoot at 1-min incubation occurred, but the initial rate of uptake at 15 s and the extent of overshoot were markedly greater at pH 6.0 than at either of the other two pH conditions (Ahearn et al. 1985).

To ascertain how pH exerted its effect on glucose transport, 15-s sugar influx as a function of external sugar concentration was assessed at two pH extremes, pH 6.0 and 8.0. Vesicles were loaded with mannitol and were incubated in a series of media containing NaCl and glucose concentrations ranging from 0.1 to 20 mM. As shown in Fig. 10, glucose influx was a curvilinear function of glucose concentration at each pH and was greater at every external sugar concentration at pH 6.0 than at pH 8.0. Glucose influx for both conditions was described as the sum of at least two independent processes operating simultaneously: (1) a

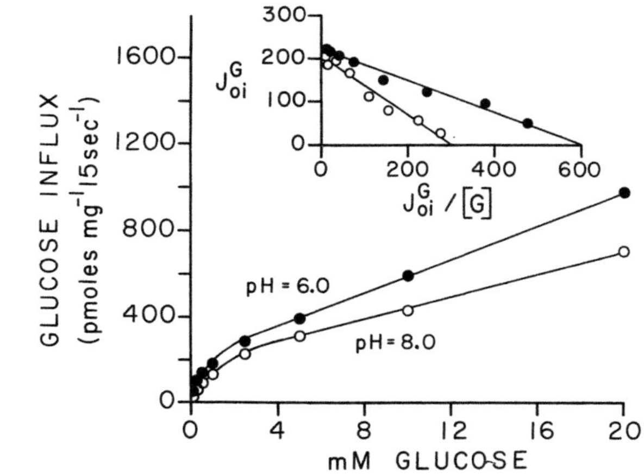

Fig. 10. Influence of bilateral pH conditions on concentration dependence of glucose influx (J_{oi}) (15-s exposures) in lobster BBMV. Vesicles were loaded with 300 mM mannitol at each pH and were exposed to identical external pH conditions in a 150 mM NaCl solution containing respective 3H-D-glucose concentrations. *Inset* is an Eadie-Hofstee plot of calculated carrier influxes after subtracting nonsaturable entry. Half-saturation constants and maximal transfer velocities were obtained from the slopes and vertical intercepts, respectively. (After Ahearn et al. 1985)

Michaelis-Menten carrier mechanism illustrating saturation kinetics, and (2) a linear entry system having a rate that was proportional to the external glucose concentration. These two processes acting together at each pH follow the equation:

$$J_{oi} = ((J_M [Glu])/(K_t + [Glu])) + P [Glu] \quad , \tag{1}$$

where J_{oi} is total glucose influx in pmol mg^{-1} protein 15 s^{-1}; J_M is the apparent maximal carrier-mediated influx; K_t is the apparent glucose concentration resulting in 1/2 J_M; [Glu] is the external glucose concentration, and P is the rate constant of the linear entry process, which can operationally be defined as an apparent diffusional permeability coefficient.

Apparent diffusional sugar influx was subtracted from total glucose entry at each external concentration providing an estimate of the carrier transport component at each pH. These calculated carrier-mediated glucose influx values were graphed in Eadie-Hofstee plots in the inset of Fig. 10 to yield estimates of the apparent transport constants K_t and J_M. Results indicated that an alteration in pH from 8.0 to 6.0 stimulated glucose influx because it significantly enhanced the apparent sugar-binding affinity (reduced K_t) and increased the apparent membrane diffusional permeability (P). Maximal glucose influx was virtually unaffected by pH (Ahearn et al. 1985).

5.2.4.3 Na-Independent Amino Acid Transport by Hepatopancreatic BBMV. Two recent studies of alanine (Ahearn et al. 1986) and lysine (Ahearn and Clay 1987a) transport by lobster hepatopancreatic BBMV indicate that in crustaceans these luminal amino acids are transported largely, if not exclusively, by Na-in-

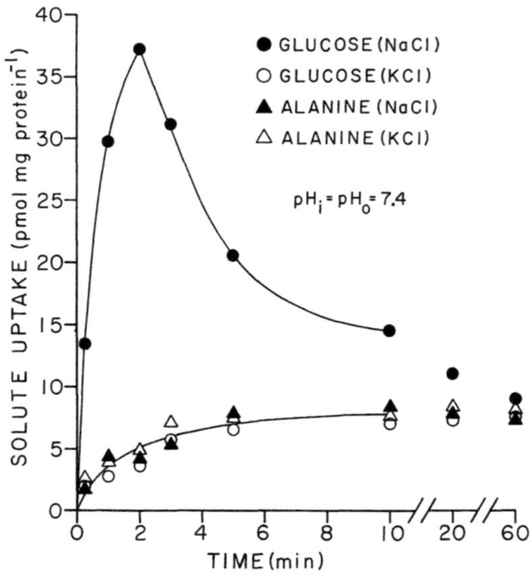

Fig. 11. Effects of external ion composition on the time course of 0.05 mM 3H-D-glucose or 0.05 mM 3H-L-alanine uptake by lobster BBMV. Vesicles were loaded with 200 mM mannitol at pH 7.4 and incubated in either 100 mM NaCl or 100 mM KCl at pH 7.4. (After Ahearn et al. 1986)

dependent carrier mechanisms. In Fig. 8 it was previously shown that glucose transport across lobster hepatopancreatic BBMV was Na-gradient dependent and that K did not substitute for Na in stimulating sugar transfer (Ahearn et al. 1985). Figure 11 shows that 3H-L-alanine uptake by these vesicles did not illustrate the same Na dependence as was exhibited by sugar transport.

In order to determine whether L-alanine uptake by hepatopancreatic BBMV was influenced by any exogenous ions, a series of experiments was conducted that was designed to test the effect of a single transmembrane ion gradient on 0.05 mM 3H-L-alanine accumulation by BBMV. The results of this test indicated that individual cation gradients (H, Na, or K) had no influence on apparent vesicular alanine influxes (15-s uptake) or the extent of amino acid accumulation against a concentration gradient (Fig. 12). In contrast, a transmembrane Cl gradient, directed inward, in the presence of an acidic external environment led to a fourfold increase in initial alanine uptake rate and a transient amino acid uptake overshoot were vesicles contained approximately 2.5 times the equilibrium alanine content. Similar results were also obtained with lysine transport across lobster BBMV (Ahearn and Cay 1987a).

Both the transports of L-alanine and L-lysine were electrogenic in the absence of cation gradients at acidic external pH and were stimulated by a transmembrane electrical potential established either by varying external anions as discussed for sugar transport, or in response to a valinomycin-induced K diffusion potential directed outwardly (electrically negative vesicular interior) (Ahearn et al. 1986; Ahearn and Clay 1987a). In order to establish whether anions in the external medium could influence lysine uptake rates, as previously shown for the transport of alanine (Fig. 12), BBMV were loaded with 200 mM mannitol at pH 5.5 and were incubated in media at pH 5.5 containing 100 mM K-gluconate, KCl, or KSCN, and the radiolabeled amino acid (Ahearn and Clay 1987a). Results indicated that the species of external anion had a significant effect on the time

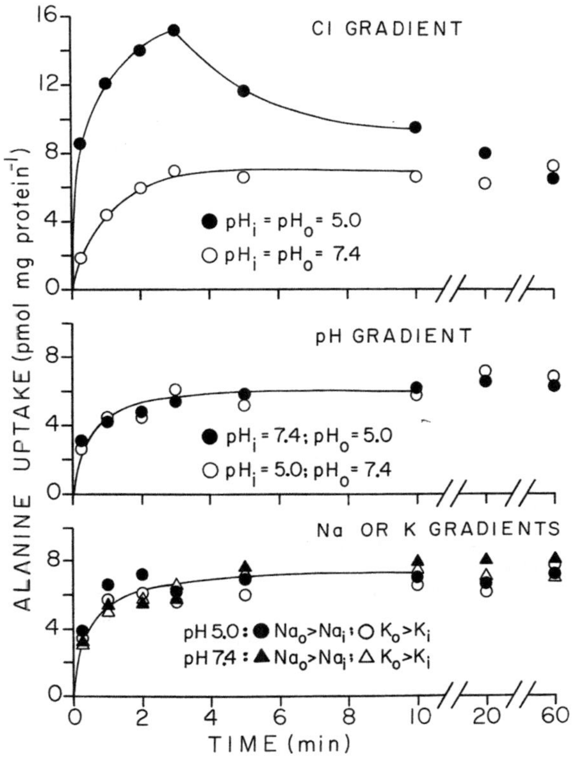

Fig. 12. Effects of a single transmembrane ion gradient on the time course of 0.05 mM 3H-L-alanine uptake by lobster BBMV. In the *upper panel* vesicles were loaded with 100 mM Na-gluconate at either pH 5.0 or 7.4 and were incubated in 100 mM NaCl at identical external pH. In the *central panel* vesicles were loaded with, and incubated in, 100 mM NaCl, and had proton gradients directed inwardly or outwardly. In the *lower panel* vesicles were loaded with 100 mM choline chloride at either pH 5.0 or 7.4 and were incubated in 100 mM NaCl or KCl at identical external pH. (After Ahearn et al. 1986)

course of Na-independent amino acid uptake by these membranes at acidic pH in much the same way as they influenced Na-dependent glucose transport at neutral pH (Fig. 9; Ahearn et al. 1985). Apparent lysine influxes displayed the following responses to differential anion gradients: SCN > Cl > gluconate (15.3 ± 0.3; 10.6 ± 0.3; 6.3 ± 0.5 pmol mg^{-1} protein 15 s^{-1}, respectively).

Figure 13 illustrates the results of experiments conducted to confirm the electrogenic nature of lysine transport in the absence of sodium, but at an acidic pH (Ahearn and Clay 1987a). Similar results were obtained for alanine transport (Ahearn et al. 1986). Figure 13 (left) shows that in the presence of valinomycin, an outwardly directed, potassium diffusion potential (inside negative) accelerated uptake of lysine at this pH compared to that illustrated by vesicles lacking the ionophore. Figure 13 (right) indicates that a transmembrane SCN gradient, in the absence of valinomycin, stimulated lysine uptake compared to that displayed by vesicles short-circuited by the ionophore.

pH had a strong effect on the BBMV transports of alanine and lysine (Ahearn et al. 1986; Ahearn and Clay 1987a). In both cases a reduction in pH from 7.4 to 4.0 significantly increased both initial rates of transport (influx) as well as the extent of amino acid accumulation within vesicles against a concentration gradient. These results suggest that a drop in pH protonated amino acid carboxyl groups largely converting alanine and lysine into Na-insensitive, cationic species responsive to transmembrane electrical potential.

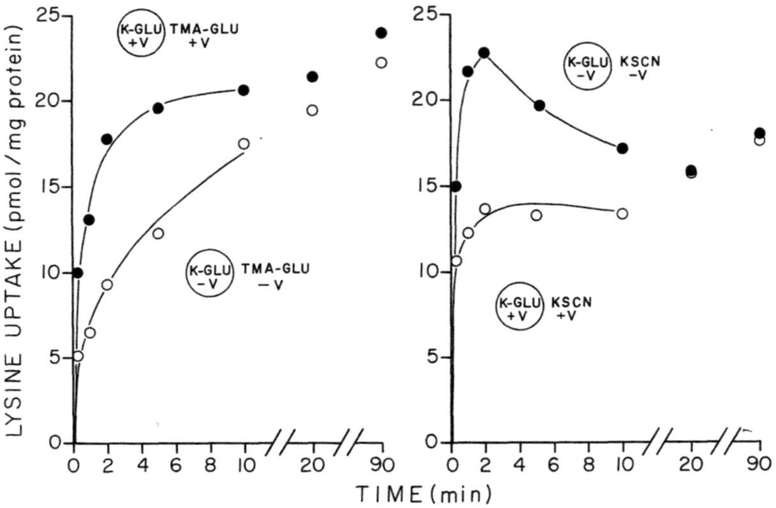

Fig. 13. Effects of transmembrane electrical potential on the time course of 5.5 μM 3H-L-lysine uptake by lobster BBMV. *Left:* Vesicles were loaded with 100 mM K-gluconate at pH 5.5 and were incubated in a medium containing labeled lysine and 100 mM TMA-gluconate at pH 5.5. One group of vesicles (*solid circles*) had 50 μM valinomycin added to internal and external media, while the other group (*open circles*) lacked the ionophore. *Right*: Vesicles were loaded with 100 mM K-gluconate at pH 5.3 and were incubated in a medium containing labeled lysine at 100 mM KSCN at pH 5.3. One group of vesicles (*open circles*) was incubated with 50 μM valinomycin (both sides), while the other group (*solid circles*) lacked the ionophore. (After Ahearn and Clay 1987a)

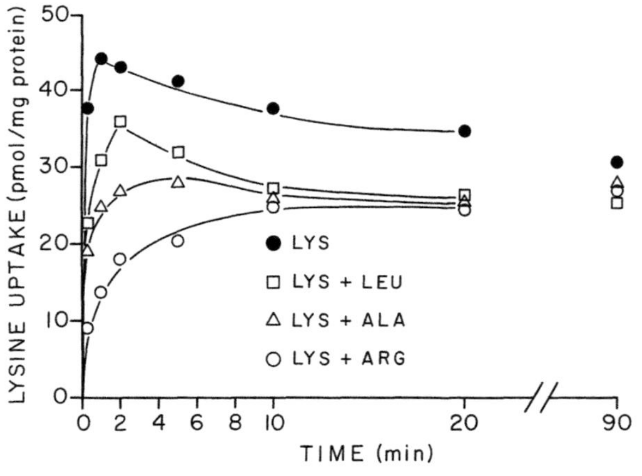

Fig. 14. Inhibitory effects of external amino acids on time course of 5.5 μM 3H-L-lysine uptake by lobster BBMV. Vesicles were loaded with 100 mM K-gluconate at pH 5.0 and were incubated in media containing the labeled amino acid, 100 mM KSCN at pH 5.0 and 5 mM L-leucine (*squares*), L-alanine (*triangles*), or L-arginine (*open circles*). Under control conditions (*solid circles*) no inhibitor was added to the incubation medium. (After Ahearn and Clay 1987a)

Influxes of 3H-L-alanine and 3H-L-lysine into hepatopancreatic BBMV were both curvilinear functions of external amino acid concentration and yielded composite curves similar in appearance to that shown in Fig. 10 for 3H-D-glucose entry into the same preparation (Ahearn et al. 1986; Ahearn and Clay 1987a). Both amino acids displayed at least one carrier-mediated influx component that exhibited saturation kinetics and a linear process that could not be distinguished from simple diffusion. At pH 5.0 the apparent binding affinity constant (K_t) for alanine was 1.74 mM (Ahearn et al. 1986), while that disclosed for lysine was 0.06 mM (Ahearn and Clay 1987a). While a decrease in external pH enhanced apparent binding affinity of 3H-D-glucose transport, elevated proton concentration strongly influenced both apparent alanine binding affinity as well as maximal transfer rate with a predominant effect on the latter transport property.

Amino acids known to be transported by several distinct mechanisms in other tissues were examined for their potential inhibitory effects on lysine transport in lobster BBMV. The time course of 5.5 μM 3H-L-lysine uptake into vesicles loaded with 100 mM K-gluconate at pH 5.0 and incubated in media containing the labeled amino acid, 100 mM KSCN at pH 5.0, and 5 mM L-leucine, L-alanine, or L-arginine was followed. Figure 14 indicated that the control condition, that is, 5.5 μM 3H-L-lysine uptake in the absence of other amino acids, produced the most rapid influx rate (37.6 ± 2.6 pmol mg^{-1} protein 15 s^{-1}) and greatest accumulation of label against a concentration gradient (overshoot). Addition of 5 mM L-arginine strongly reduced the apparent rate of lysine influx (9.0 ± 0.9 pmol mg^{-1} protein 15 s^{-1}) and abolished the overshoot properties of lysine accumulation. Adding 5 mM L-leucine or L-alanine to the external media led to apparent lysine influxes that were intermediate between the two extremes. These results suggest that all three amino acids may have inhibitory effects on lysine uptake by BBMV.

In order to determine the mechanism of inhibition by external amino acids on lysine influx, 8-s 3H-L-lysine uptakes, measured at 5.5, 10, 20, 40, 100, 150, and 250 μM lysine and corrected for lysine binding to the external vesicle surface at each concentration, were measured in the presence of 500 μM, L-alanine, L-arginine, or L-leucine. Carrier-mediated lysine influx in the presence and absence of each inhibitory amino acid is displayed in Fig. 15 as Eadie-Hofstee plots. Both L-alanine and L-arginine significantly ($P < 0.01$) decreased lysine influx K_t (reduced apparent binding affinity), without having significant ($P > 0.05$) effects on maximal lysine influx (J_M). In contrast, L-leucine did not have a significant ($P > 0.05$) influence on lysine influx K_t, but did significantly ($P < 0.05$) lower maximal lysine entry rate. These results suggest that alanine and arginine were competitive inhibitors of lysine influx, arginine being the most potent, while leucine was a noncompetitive inhibitor of lysine entry.

The interactions between alanine and lysine discussed above suggest that they appear to share the same Na-independent, membrane potential-sensitive carrier mechanism in lobster BBMV. This carrier may be analogous to the classical y+ system described for mammalian cells (Christensen 1964, 1975). The nature of the noncompetitive effect of L-leucine on L-lysine influx (Fig. 15) is presently unclear, but may either be the result of amino acid interactions at a single membrane transport protein such as the y+ system, or may be due to competition for the membrane potential between protonated amino acids using two distinct Na-independent, membrane potential-sensitive carrier proteins (e.g., y+ and L systems).

Recent unpublished observations on 3H-L-proline and 3H-L-leucine transport by hepatopancreatic BBMV suggest that these two amino acids are also transferred across the luminal membrane of hepatopancreatic epithelial cells by

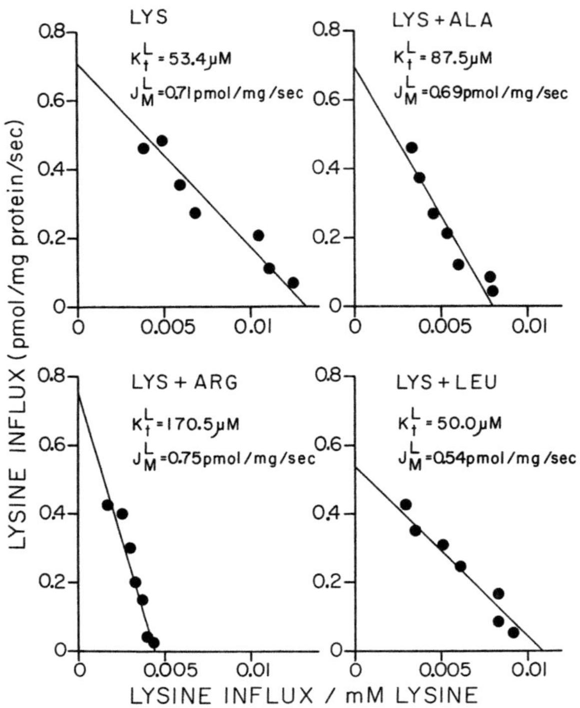

Fig. 15A–D. Mechanisms of amino acid inhibition of 3H-L-lysine influx into lobster BBMV. Vesicles were loaded with 100 mM K-gluconate at pH 5.0 and were incubated for 8 s in media containing 100 mM KCl at pH 5.0, the labeled amino acid at concentrations of 5.5, 10, 20, 40, 60, 150, and 250 μM, and 500 μM L-alanine (*B*), L-arginine (*C*), or L-leucine (*D*). The control condition (*A*) lacked external inhibitory amino acids. Lysine influx was corrected for binding and nonsaturable entry. Each *point* is the mean of triplicate samples. (After Ahearn and Clay 1987a)

Na-independent carrier processes which appear to be different from the process described above for alanine and lysine transport. Figure 16 shows the results of a preliminary experiment illustrating the Na independence of 3H-L-proline uptake by these membrane preparations. Vesicles were loaded with 200 mM mannitol at pH 4.0 and were incubated in media at the same pH containing 0.05 mM 3H-L-proline, 100 mM KCl, KSCN, KNO_3, or 200 mM mannitol. All three anions produced proline uptake overshoots in the absence of external sodium, where the intravesicular amino acid concentration transiently exceeded that at equilibrium. The greatest overshoot was observed in vesicles incubated in NO_3 medium. These data suggest that at acidic pH proline enters vesicles largely in the cationic form and is electrically coupled to the diffusional entry of permeable anions.

An additional experiment with L-proline uptake by BBMV in the absence of external sodium tested the effects of various preloaded amino acids on trans-stimulation of 10-s proline influx. Vesicles were loaded with 200 mM mannitol at pH 4.0 and contained, in addition, a series of potentially stimulatory amino acids at 0.5 mM. The internal organic solutes were L-proline, hydroxy-L-proline, glycine, L-lysine, and mannitol as the control. The preloaded vesicles were incubated in a medium of the same pH containing 0.05 mM 3H-L-proline and 100 mM KNO_3. Vesicles containing L-proline, hydroxy-L-proline, and glycine significantly trans-stimulated the 10-s uptake of 3H-L-proline, while those loaded with L-lysine did not exert a significant effect on 3H-L-proline transport. These results suggest that an Na-independent imino acid transport protein shared

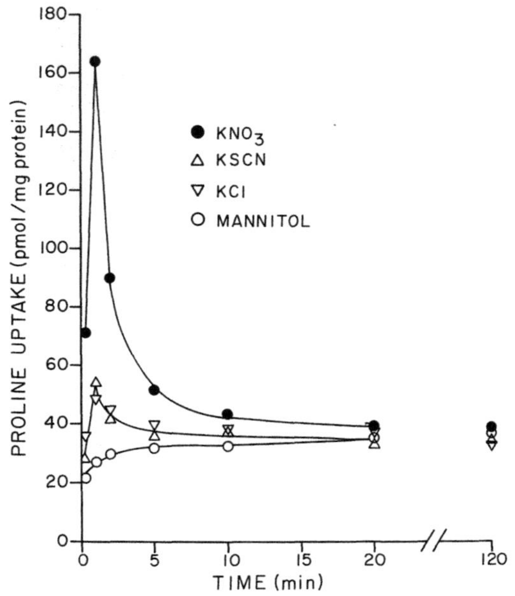

Fig. 16. Effects of inwardly directed anion gradients on the time course of Na-independent proline uptake by lobster BBMV. Vesicles were loaded with 200 mM mannitol at pH 4.0 and were incubated in media of the same pH containing 100 mM KNO_3, KSCN, KCl, or 200 mM mannitol, and 0.05 mM 3H-L-proline

by the above stimulatory amino acids may occur in these membranes which reresents a separate system than the Na-independent process used by L-lysine and L-alanine.

The noncompetitive inhibitory effects of L-leucine on 3H-L-lysine influx in the absence of external sodium discussed above (Fig. 15) suggested that L-leucine may employ a separate Na-independent transport system than that used by L-lysine and L-alanine. A preliminary trans-stimulation experiment was conducted to establish the substrate specificity of this putative L-leucine carrier process. Vesicles were loaded with a variety of amino acids and mannitol (control) at 0.5 mM and 200 mM mannitol at pH 4.0 and were incubated for 10 s in a medium of the same pH containing 0.05 mM 3H-L-leucine and 100 mM KNO_3. Significant trans-stimulation of 3H-L-leucine influx occurred when L-leucine, L-methionine, and L-alanine were added to the internal medium. Intravesicular L-lysine and L-proline did not significantly stimulate Na-independent 3H-L-leucine uptake by these membrane vesicles. These results suggest the possible occurrence of a third Na-independent transport process for L-leucine and other amino acids with nonpolar side chains, which appears to be a separate protein class from the two previously discussed Na-independent transfer systems used by L-lysine/L-alanine or by L-proline/glycine. Additional experiments must be performed to confirm the occurrence and characteristics of both the L-proline and L-leucine transport mechanisms in hepatopancreatic BBMV.

5.2.4.4 Na-Cl-Glutamate Cotransport by Hepatopancreatic BBMV. A recent study of L-glutamate transport by lobster hepatopancreatic BBMV suggested that, in contrast to the Na independence of L-alanine and L-lysine transport, and the Na dependence of D-glucose transport, simultaneous cotransport of this

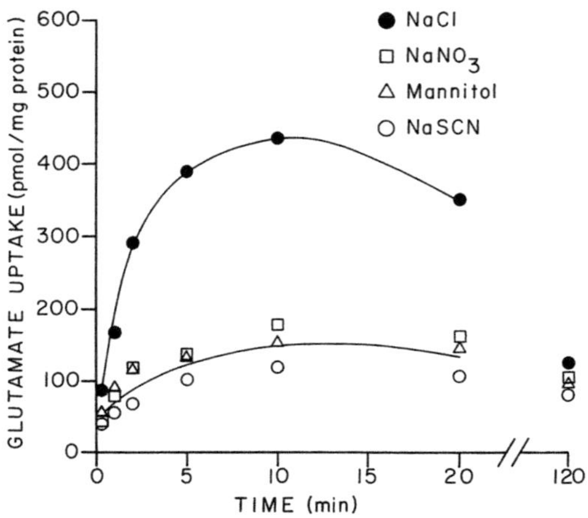

Fig. 17. Effects of inwardly directed anion gradients on the time course of glutamate uptake by lobster BBMV. Vesicles were loaded with 200 mM mannitol at pH 4.0 and were incubated in media of the same pH containing 100 mM NaCl, NaSCN, $NaNO_3$, or 200 mannitol, and 0.05 mM 3H-L-glutamate. (After Ahearn and Clay 1987b)

amino acid with both sodium and chloride ions occurred (Ahearn and Clay 1987b). Sodium specificity of the cotransport system was shown by the results of an experiment where vesicles were loaded with 200 mM mannitol at pH 5.0 and were incubated in media at the same pH containing either 100 mM NaCl, KCl, or TMA-Cl and 0.05 mM 3H-L-glutamate. Vesicles incubated in NaCl exhibited significantly ($P < 0.05$) greater apparent amino acid influxes (15-s uptake) than those exposed to the other two media, and the former displayed a marked uptake overshoot where the intravesicular amino acid concentration transiently exceeded that at equilibrium by a factor of three.

The requirement of the glutamate transport system on external Cl was demonstrated by the results in Fig. 17. In these experiments vesicles were loaded with 200 mM mannitol at pH 4.0 and were incubated in media of the same pH, but which contained 100 mM NaCl, NaSCN, $NaNO_3$, or 200 mM mannitol and 0.05 mM 3H-L-glutamate. Transient 3H-L-glutamate accumulation against a concentration gradient only occurred in the presence of an inwardly directed NaCl gradient, all other conditions led to slow hyperbolic amino acid uptake. These results and those mentioned above, suggest that concentrative 3H-L-glutamate transport by hepatopancreatic BBMV at acidic pH required inwardly directed gradients of both Na and Cl, and that in the absence of either ion, transient accumulation against a concentration gradient was abolished and apparent unidirectional transport to the vesicular interior was markedly reduced.

In order to determine whether an outwardly directed K gradient enhanced glutamate uptake in the presence of inwardly directed gradients of Na and Cl, vesicles were preloaded with either 200 mM mannitol, 100 mM KNO_3, or 100 mM KCl at pH 4.0 and were incubated in media containing 0.05 mM 3H-L-glutamate and either 100 mM NaCl or 100 mM $NaNO_3$ at the same pH. Under these conditions 3H-L-glutamate uptake was strongly stimulated in the presence of an inwardly directed gradient of NaCl as indicated earlier. When an outwardly directed gradient of KNO_3 or KCl was added to inwardly directed NaCl or $NaNO_3$ gradients significantly reduced uptake of the amino acid occurred. These results provide support for the dependence of 3H-L-glutamate uptake on both

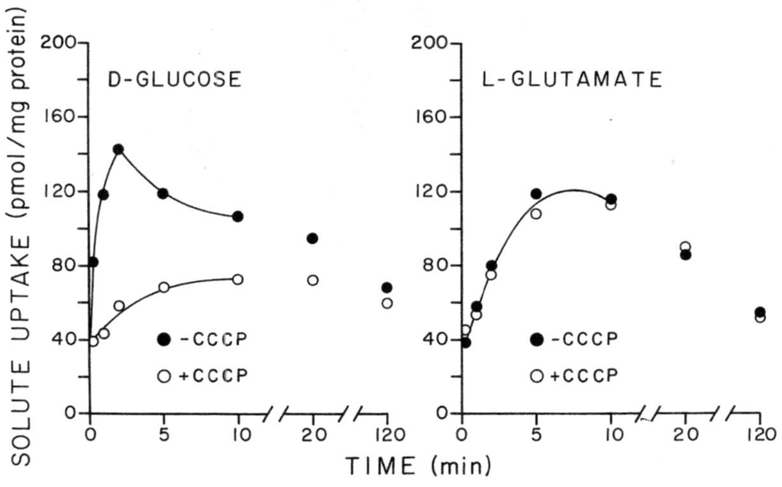

Fig. 18. Effects of protonophore, CCCP, on the time course of 0.05 mM 3H-D-glucose or 3H-L-glutamate uptake by lobster BBMV. Vesicles were loaded with 200 mM mannitol at pH 5.0 and were incubated in media of the same pH containing 100 mM NaCl and either radiolabeled solute. One group of vesicles for each organic solute was preincubated in 50 µM CCCP for 30 min prior to experimentation. (After Ahearn and Clay 1987b)

external Na and Cl, and clearly show that intravesicular K or KCl do not have a stimulating effect on the entry of the amino acid.

As with the hepatopancreatic transport of all organic solutes studied to date, transmembrane transfer of 3H-L-glutamate was strongly stimulated by a decrease in pH. Vesicles were loaded with 200 mM mannitol at pH 4.0, 5.0, 7.0, and 8.0 and were incubated in media of the same pH containing 100 mM NaCl and the labeled amino acid. Only at pH 4.0 and 5.0 was there concentrative 3H-glutamate accumulation which exceeded that at the respective 90-min equilibrium uptake interval (Ahearn and Clay 1987b). In addition, apparent 15-s amino acid influxes significantly increased as pH was lowered from 8.0 to 4.0. These data suggest that NaCl-dependent 3H-L-glutamate transport was stimulated by a drop in pH and that transient vesicular glutamate accumulation only occurred when pH was in the acidic range.

In order to define the electrogenic nature of NaCl-dependent 3H-L-glutamate transport, the effects of the protonophore, CCCP, on the uptake of glutamate in the presence of an inwardly directed NaCl gradient at pH 5.0, were compared to the effects of the drug on Na-dependent glucose transport in the same preparation. Vesicles were loaded with 200 mM mannitol at pH 5.0 and were incubated in media of the same pH containing 100 mM NaCl and either 0.05 mM 3H-D-glucose or 0.05 mM 3H-L-glutamate. One group of vesicles for each organic solute was preincubated in 50 µm CCCP for 30 min prior to experimentation. Figure 18 shows that in the absence of the ionophore 3H-D-glucose uptake exhibited a strong NaCl dependency, as reported earlier (Fig. 8), which was abolished when the vesicles were short-circuited. In contrast, CCCP had no significant effect ($P > 0.05$) on 3H-L-glutamate uptake. These results suggest that

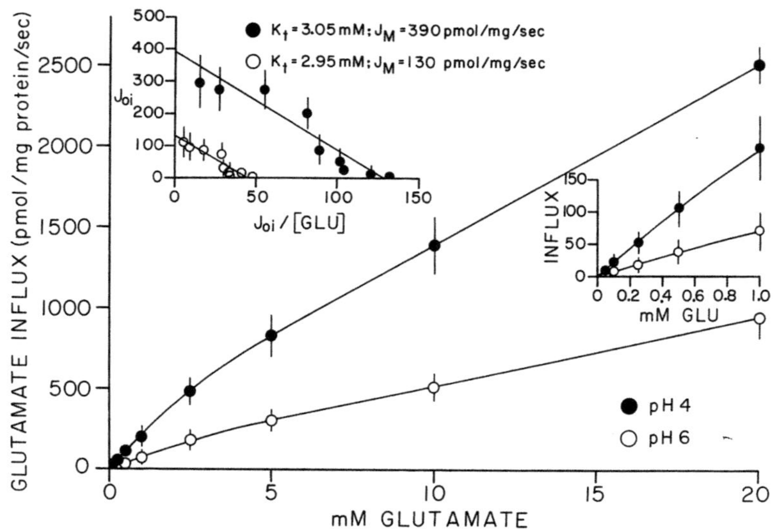

Fig. 19. Effects of external glutamate concentration of 3H-L-glutamate influx (10-s uptake) into lobster BBMV under two pH conditions. Vesicles were loaded with 200 mM mannitol at either pH 4.0 or 6.0 and were incubated in media of the same pH containing 100 mM NaCl and a selected amino acid concentration ranging from 0.05 to 20 mM. *Inset to the right* is an expansion of influx data from 0.05 to 1.0 mM glutamate concentrations. *Inset to the left* is an Eadie-Hofstee plot of calculated carrier-mediated 3H-L-glutamate influx after subtraction of the apparent diffusion component of amino acid influx from total transfer in the main body of the figure. Displayed kinetic constants were derived from slope (K_t) and vertical intercept (J_M) of Eadie-Hofstee plot using linear regression analysis. (After Ahearn and Clay 1987b)

NaCl-dependent L-glutamate transport was electroneutral, whereas that of Na-dependent glucose uptake was strongly electrogenic.

Figure 19 shows the effects of pH on 10-s 3H-L-glutamate influx from incubation medium to the vesicular interior for external glutamate concentrations ranging from 0.05 to 20 mM. In all cases vesicles were loaded with 200 mM mannitol at the respective pH and were incubated in media containing 100 mM NaCl and the selected amino acid concentration. Over this concentration range, influx was a curvilinear function of external amino acid concentration at both pH values and could be described as the sum of at least one Michaelis-Menten carrier mechanism illustrating saturation kinetics and a linear system having a rate that was proportional to the external amino acid concentration and probably represented simple diffusion. At either pH these two processes operating together could be described by Eq. (1) presented earlier for glucose entry into the same vesicle preparation. Nonsaturable 3H-L-glutamate influx was subtracted from total influx at each amino acid concentration yielding an estimate of carrier transport. Calculated carrier-mediated influxes were graphed in the Eadie-Hofstee inset to provide estimates of the transport constants K_t and J_M.

The data displayed in Fig. 19 show that the pH had a significant effect on the influx of 3H-L-glutamate across hepatopancreatic BBMV. The major impact of a drop in pH on amino acid entry was to triple both the maximal carrier-

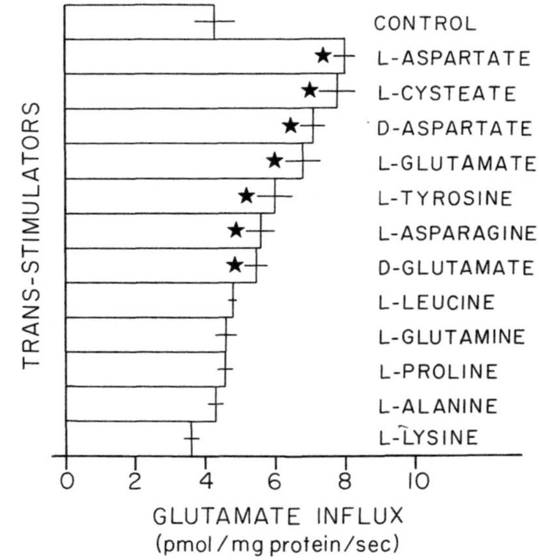

Fig. 20. Trans-stimulation of 0.05 mM 3H-L-glutamate influx by preloaded amino acids in lobster BBMV. Vesicles were preincubated for 90 min on ice prior to experimentation with 200 mM mannitol at pH 4.0 and one of 12 different substrates at 0.5 mM. Following preincubation, 10-s uptake of 0.05 mM 3H-L-glutamate from an external medium containing 100 mM NaCl at pH 4.0 was measured. Extent of *bar* represents the mean of five replicates per treatment, the *horizontal lines* signify the standard error values, and the *stars* represent those results that were significantly different from the control which was loaded with 0.5 mM mannitol. (After Ahearn and Clay 1987b)

mediated transport component (J_M at pH $4.0 = 390 \pm 72$; J_M at pH $6.0 = 130 \pm 47$ pmol mg^{-1} s^{-1}) as well as the apparent diffusional permeability of the membrane to this compound (P at pH $4.0 = 111 \pm 23$; P at pH $5.0 = 42 \pm 17$ pmol mg^{-1} s^{-1} mM^{-1}). The apparent half-saturation constant for amino acid association with transport proteins in the membrane was not significantly ($P > 0.05$) influenced by a reduction in pH (K_t at pH $4.0 = 3.05 \pm 0.41$; K_t at pH $6.0 = 2.95 \pm 0.25$ mM).

A series of trans-stimulation experiments was performed to ascertain the relative substrate specificity of the glutamate carrier process. Vesicles were preincubated for 90 min on ice prior to influx experiments with 200 mM mannitol at pH 4.0 and one of 12 organic solutes at 0.5 mM. Following the loading process, 10-s influxes of 0.05 mM 3H-L-glutamate from an external medium containing 100 mM NaCl at pH 4.0 was measured. Figure 20 indicates that vesicles preloaded with L-aspartate, L-cysteate, D-aspartate, L-glutamate, L-tyrosine, L-asparagine, and D-glutamate significantly ($P < 0.05$) stimulated 3H-L-glutamate influx compared to vesicles containing mannitol (control). In contrast, 3H-L-glutamate influx into vesicles preloaded with 0.5 mM L-leucine, L-glutamate, L-proline, L-alanine, and L-lysine was not significantly ($P > 0.05$) different than the rate of entry under control conditions. These results generally suggest that organic molecules that structurally resemble 3H-L-glutamate at acidic pH tend to trans-stimulate the uptake of the labeled substrate most likely as a result of common usage of a membrane carrier mechanism. Many of these trans-stimulating compounds are the same as those cis-inhibiting L-glutamate influx in mammalian renal brush border membrane vesicles (Schneider et al. 1980). The lack of significant trans-stimulation by the second group of substrates may reflect the independence of these compounds from the L-glutamate transport process.

It was previously mentioned that 3H-L-glutamate uptake shows a strong dependency upon both external Na and Cl which was not satisfied by other

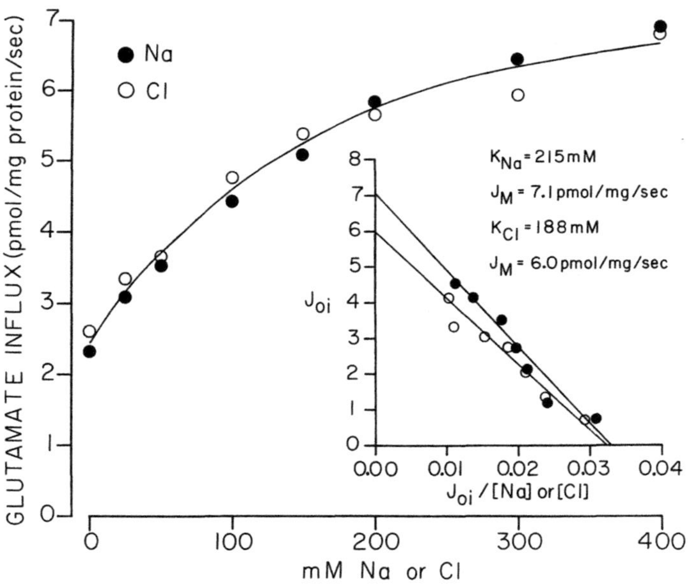

Fig. 21. Effects of variable external Na or Cl concentrations on carrier-mediated 3H-L-glutamate influx in lobster BBMV. In the Na experiments, vesicles were loaded with 800 mM mannitol at pH 4.0 and were incubated for 10 s in media of the same pH containing a fixed Cl concentration (400 mM), 0.05 mM 3H-L-glutamate, and variable concentrations of Na (TMA replacing Na) from 0 to 400 mM. In the Cl experiments, conditions were the same except that a fixed Na concentration (400 mM) and variable concentrations of Cl (NO_3 replacing Cl) from 0 to 400 mM were used. *Inset* shows Eadie-Hofstee plots of data obtained from the main body of the figure after subtraction of the respective influxes occurring at 0 mM concentrations of the respective ions. Kinetic constants were obtained from the slopes (K_t) and vertical intercepts (J_M) of the Eadie-Hofstee plots. (After Ahearn and Clay 1987b)

cations or anions. Additional experiments were conducted to evaluate the nature of these specific ion dependencies. In one experiment 3H-L-glutamate influx at fixed amino acid and Cl concentrations was measured as a function of variable external Na concentration. In the second experiment, 3H-L-glutamate influx was determined at fixed amino acid and Na concentrations using variable concentrations of external Cl. Figure 21 indicates that amino acid influx was a hyperbolic function of both external Na or Cl concentrations at fixed concentrations of the respective counterion. The Eadie-Hofstee inset shows the half-saturation constants for apparent Na and Cl binding to the amino acid carrier, the maximal Na-stimulated 3H-L-glutamate influx, and the maximal Cl-stimulated 3H-L-glutamate influx. Neither apparent half-saturation constants nor maximal transport velocities were significantly different from one another ($P > 0.05$). These results suggest that the Na and Cl dependencies of glutamate uptake are likely due to cotransport of the cation, anion, and amino acid on a shared carrier process. Furthermore, because of the hyperbolic nature of the relationships between variable external ion concentrations and glutamate influx, the stoichiometry of coupled entry is likely 1 Na : 1 Cl : 1 glutamate (Turner 1983).

5.2.4.5 Summary of Nutrient Transport Mechanisms in Lobster Hepatopancreatic Brush Border Membrane. The sugar and amino acid transfer mechanisms summarized in this review clearly indicate that the crustacean hepatopancreatic epithelium possesses a complement of brush border membrane transport proteins that facilitate the movement of nutrients from tubular lumen to epithelial cytosol in vivo. Some of the transport characteristics demonstrated for each compound resemble those reported for the same substance in vertebrate intestinal or renal epithelia, but because of the acidic nature of the crustacean hepatopancreatic absorption site other transfer properties in the crustacean depart markedly from the vertebrate paradigm.

A current model of lobster hepatopancreatic brush border membrane showing the sugar and amino acid transport systems described in this review, the in vivo net electrical charges expected on the transported substrates at acidic pH, and the respective driving forces for each transfer mechanism are shown in Fig. 22. Five distinct nutrient transport carriers are proposed for this membrane. These are: (1) an Na-dependent sugar transporter which can accumulate glucose against a concentration gradient and which uses the combined driving forces of a

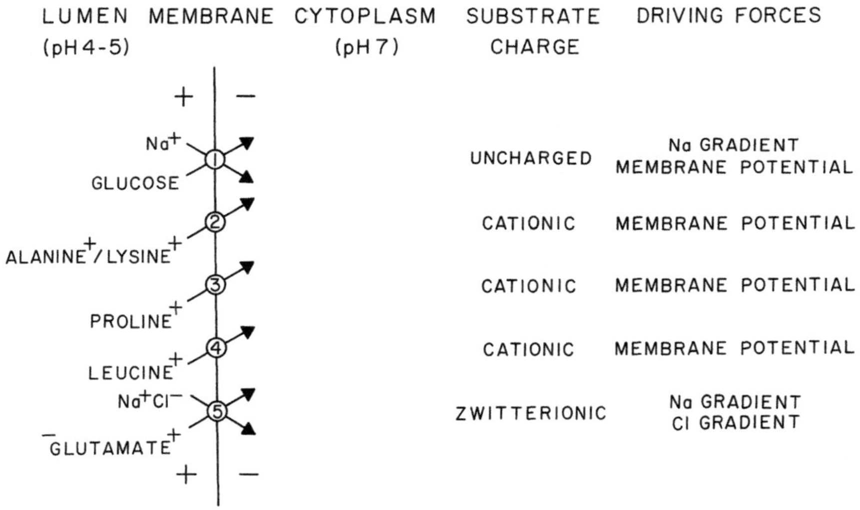

Fig. 22. Nutrient transport mechanisms of the crustacean hepatopancreatic epithelial brush border membrane. At least three distinct carrier proteins for sugars and amino acids reside on this membrane. Two additional Na-independent transport systems for specific groups of amino acids may also occur in this location. Predicted substrate charge at acidic pH and the respective driving forces for organic solute accumulation are shown on the figure for each transfer process. Direction of *arrows* indicates proposed mode of transfer for each solute, that is, movement against or down a concentration gradient. Electroneutral sugars employ Na-coupled transport resulting in an electrically positive transported complex which responds to both the transmembrane Na gradient as well as the membrane potential. Amino acids transported as cationic species at physiologically acidic pH appear to be Na-independent in their transfer across vesicle membranes and respond solely to the transmembrane electrical potential. Zwitterionic amino acids at physiologically acidic pH move across the vesicular wall in conjunction with both Na and Cl as electroneutral complexes. These complexes use both the inwardly directed transmembrane Na and Cl gradients to bring about nutrient accumulation within the vesicle interior

transmembrane sodium concentration gradient as well as a transmembrane electrical potential difference to bring about net hexose transfer; (2) an Na-independent amino acid transport system which uses the transmembrane potential alone to accumulate positively charged amino acids such as alanine and lysine at physiologically acidic pH; (3) a putative second Na-independent carrier process which also uses the transmembrane potential to accumulate imino acids such as L-proline; (4) a putative third Na-independent mechanism employing the membrane potential to accommodate lipophilic amino acids such as L-leucine; and (5) an NaCl-coupled zwitterionic transport process that transfers electrically silent amino acids at physiologically acidic pH using the combined transmembrane ion gradients of both Na and Cl for nutrient accumulation.

Glucose uptake by the proposed Na-dependent mechanism is quite similar to that described in mammalian intestine or kidney epithelial brush border membranes except that protons stimulate the transfer process in crustaceans and inhibit it in the vertebrate system (Toggenburger et al. 1978). The competitive interactions between the transports of L-alanine and L-lysine suggest that they largely employ the same Na-independent membrane protein for transfer from tubular lumen to cytoplasm. The properties of this system are similar to those of the y+ dibasic transport protein characterized for ascites tumor cells (Christensen 1964, 1975) and found to occur widely in mammalian intestinal and renal epithelia (Busse 1978; Hilden and Sacktor 1981; Munck 1981; Stevens et al. 1982; Hammerman 1982; Stieger et al. 1983). The NaCl-dependent, zwitterionic transporter of L-glutamic acid exhibited a substrate specificity similar to that shown in mammals for the X_{AG}, "acidic" amino acid, transfer system (Schneider et al. 1980; Christensen 1984). However, because the crustacean absorption site is acidic, L-glutamate is protonated in these animals and is transported as an electrically neutral substrate in combination with Na and Cl ions. Furthermore, in contrast to the mammalian situation (Burckhardt et al. 1980; Schneider and Sacktor 1980; Sacktor et al. 1981; Corcelli and Storelli 1983), cytoplasmic K does not seem to enhance NaCl stimulation of L-glutamate uptake.

At least two additional Na-independent amino acid transporters may occur on lobster hepatopancreatic brush border as suggested by preliminary observations with this system, serving the imino acids such as L-proline and the hydrophobic amino acids that resemble L-leucine, respectively. At physiologically acidic pH both of these Na-independent systems would likely transfer their respective substrates as cations in the same manner as described for the L-alanine/L-lysine-shared y+ system. To date in lobster hepatopancreatic BBMV at acidic pH, no amino acids have illustrated Na-dependent/Cl-independent transport with properties resembling the classical A or ACS paradigms (Stevens et al. 1982).

6 Overview of Invertebrate Nutrient Absorption

This review has focused on the mechanisms of nutrient transport which occur in the gastrointestinal tracts of the four most anatomically advanced invertebrate phyla: Annelida, Echinodermata, Mollusca, and Arthropoda. Carrier-mediated, membrane-bound transport mechanisms for sugars and amino acids were disclosed for gastrointestinal organs from each group which displayed many properties in common to one another and to vertebrate absorptive epithelia as well. All groups appear to have distinct transport proteins accommodating sugars as a class and amino acids as a class with little overlap between the substrate classes. However, several different transfer proteins appear to be responsible for transmembrane flow of amino acids, specificity largely appearing to depend on charge and side group configuration. A common feature among certain molluscan and crustacean intestinal preparations was sodium dependency of sugar and amino acid transport (Gerencser 1985; Ahearn 1982). In these cases cellular nutrient accumulation appeared to follow the classic Na-gradient model for organic solute uptake first advanced for vertebrate epithelia (Crane 1962, 1977; Schultz and Curran 1970).

Several instances of departure from the vertebrate nutrient transport paradigms appear to have occurred in invertebrate gut. Perhaps the most striking dissimilarity between vertebrate and invertebrate gut absorption of organic solutes is that among the vertebrates the small intestine is the center for nutrient transfer from gut lumen to blood, while a variety of organs appear to perform this function among the invertebrates. Evidence has accumulated over the last century that invertebrate stomach (Bamford and James 1972), intestine (Hanozet et al. 1980; Ahearn 1982; Gerencser 1985), hepatopancreas (van Weel 1974; Ahearn 1987), and caecum (Boucaud-Camou et al. 1976) may all perform certain nutrient absorptive processes.

Where examined in detail the molecular characteristics of nutrient absorption in some invertebrate gastrointestinal organs also strongly depart from processes described for the same substances in vertebrate gut. The paradigm for transapical transport of many sugars and amino acids in mammalian intestine is that the uphill transfer of certain organic solutes is coupled to the downhill movement of the sodium ion on a shared carrier process using the energy inherent in the transmembrane ion gradient to accumulate the nutrient intracellularly to considerably higher concentrations than in the gut lumen. The transapical sodium gradient is maintained through energy expenditure on the basolateral Na/K-ATPase. In certain lepidopteran larvae which consume large concentrations of K in their diet, many amino acids appear to be cotransported into the gut epithelial cells in conjunction with K instead of Na (Fig. 3) (Hanozet et al. 1980; Giordana et al. 1982, 1985). Here, a favorable transapical electrochemical gradient for K entry is maintained by an electrogenic apical K pump which actively transfers intracellular K to the lumen as it enters simultaneously with organic solutes.

Another paradigm which has been established for vertebrate intestinal transport is that nutrient transfer occurs into epithelial cells from a luminal solution with a near-neutral pH. At this pH luminal amino acids occur as positively charged, negatively charged, or zwitterionic species and appear to be transported across the luminal membrane largely according to their charge properties. In crustacean hepatopancreas, where much of the bulk of nutrient absorption appears to occur, the luminal pH in vivo is between pH 4.0 – 5.0. At this pH many of the zwitterionic amino acids in mammals are transported as cations, and the anionic amino acids in mammals appear as zwitterions. As a result the number of independent gastrointestinal amino acid transport systems and the substrate specificity of each system may be markedly different between mammals and crustaceans.

As more invertebrates are examined for gut epithelial transport properties, particularly those species inhabiting extreme environments or those with unusual dietary preferences, the more we may come to realize that the mammalian paradigms for nutrient absorption may not necessarily be relevant to all other types of animals. Characterization of novel transport processes for nutrients and other important substances among the vast array of invertebrate organisms will be one of the more exciting challenges of the next generation of comparative physiologists.

Acknowledgments. This work was funded by NSF grant No. DCB85-11272 which is gratefully acknowledged.

References

Ahearn GA (1974) Kinetic characteristics of glycine transport by the isolated midgut of the marine shrimp, *Penaeus marginatus.* J Exp Biol 61:677 – 696

Ahearn GA (1976) Co-transport of glycine and sodium across the mucosal border of the midgut epithelium in the marine shrimp, *Penaeus marginatus.* J Physiol (London) 258:499 – 520

Ahearn GA (1980) Intestinal electrophysiology and transmural ion transport in freshwater prawns. Am J Physiol 239:C1 – C10

Ahearn GA (1982) Water and solute transport in crustacean gastrointestinal tract. In: Podesta RB (ed) Membrane physiology of invertebrates, chap 8. Dekker, New York Basel, pp 261 – 339

Ahearn GA (1987) Nutrient transport by the crustacean gastrointestinal tract: recent advances with vesicle techniques. Biol Rev 62:45 – 63

Ahearn GA, Clay LP (1987a) Membrane-potential-sensitive, Na-independent lysine transport by lobster hepatopancreatic brush border membrane vesicles. J Exp Biol 127:373 – 387

Ahearn GA, Clay LP (1987b) Na-Cl-glutamate cotransport by lobster hepatopancreatic brush border membrane vesicles. J Exp Biol 130:175 – 191

Ahearn GA, Maginniss LA (1977) Kinetics of glucose transport by the perfused mid-gut of the freshwater prawn, *Macrobrachium rosenbergii.* J Physiol (London) 271:319 – 336

Ahearn GA, Grover ML, Dunn RE (1985) Glucose transport by lobster hepatopancreatic brush border membrane vesicles. Am J Physiol 248:R133 – R141

Ahearn GA, Grover ML, Dunn RE (1986) Effects of Na+, H+, and Cl− on alanine transport by lobster hepatopancreatic brush border membrane vesicles. J Comp Physiol B 156:537 – 548

Al-Mohanna SY, Nott JA (1986) B-cells and digestion in the hepatopancreas of *Penaeus semisulcatus* (Crustacea: Decapoda). J Mar Biol Assoc UK 66:403 – 414

Al-Mohanna SY, Nott JA, Lane DJW (1985a) M-"midgut" cells in the hepatopancreas of the shrimp, *Penaeus semisulcatus* De Haan, 1844 (Decapoda, Natantia). Crustaceana 48:260 – 268

Al-Mohanna SY, Nott JA, Lane DJW (1985b) Mitotic E- and secretory F-cells in the hepatopancreas of the shrimp *Penaeus semisulcatus* (Crustacea: Decapoda). J Mar Biol Assoc U K 65:901 – 910

Alvarez O, Goldner AM, Curran PF (1969) Alanine transport in rabbit jejunum. Am J Physiol 217:946 – 950

Anderson JM (1953) Structure and function of the pyloric caeca of *Asterias forbesi.* Biol Bull 105:47 – 61

Anderson JM (1960) Histological studies on the digestive system of a starfish, *Henricia,* with notes on Tiedemann's pouches in starfishes. Biol Bull 119:371 – 398

Anderson JM (1962) Studies on visceral regeneration in seastars. I. Regeneration of pyloric caeca in *Henricia leviuscula.* Biol Bull 122:321 – 342

Anderson JM (1966) Aspects of nutritional physiology. In: Boolootian RA (ed) Physiology of Echinodermata. Wiley-Interscience, New York, pp 329–357

Anderson JM (1979) Histological studies on the pyloric stomach and its appendages in *Oreaster reticulatus* (L.) (Asteroidea). Biol Bull 156:1–18

Bamford DR, James D (1972) An in vitro study of amino acid and sugar absorption in the gut of *Echinus esculentus*. Comp Biochem Physiol 42A:579–590

Bamford DR, Stewart MG (1973a) Absorption of neutral amino acids by the gut of *Arenicola marina*. J Comp Physiol 82:291–304

Bamford DR, Stewart MG (1973b) Absorption of charged amino acids by the intestine of *Arenicola marina*. Comp Biochem Physiol 46A:537–547

Bamford DR, West B, Jeal F (1972) An in vitro study of monosaccharide absorption in echinoid gut. Comp Biochem Physiol 42A:591–600

Barber A, Jordana R, Ponz F (1975a) Effect of anaerobiosis, dinitrophenol and fluoride on the active intestinal transport of galactose in snail. Rev Esp Fisiol 31:125–130

Barber A, Jordana R, Ponz F (1975b) Sodium dependence of intestinal active transport of sugars in snail (*Cryptomphalus hortensis* Muller). Rev Esp Fisiol 31:271–276

Becker GL, Chen C, Greenawalt JW, Lehninger AL (1974) Calcium phosphate granules in the hepatopancreas of the blue crab *Callinectes sapidus*. J Cell Biol 61:316–326

Biber J, Stieger B, Haase W, Murer H (1981) A high yield preparation for rat kidney brush border membranes. Different behaviour of lysosomal markers. Biochim Biophys Acta 647:169–176

Bidder AM (1957) Evidence for an absorptive function of the "liver" of *Octopus vulgaris* Lam. Pubbl Stn Zool Napoli 29:139–150

Blandamer A, Beechey RB (1964) The identification of an α-amylase in aqueous extracts of the hepatopancreas of *Carcinus meanus*, the common shore crab. Comp Biochem Physiol 13:97–105

Boucaud-Camou E, Boucher-Rodoni R, Mangold K (1976) Digestive absorption in *Octopus vulgaris* (Cephalopoda: Octopoda). J Zool 179:261–271

Brick RW, Ahearn GA (1978) Lysine transport across the mucosal border of the perfused midgut in the freshwater prawn, *Macrobrachium rosenbergii*. J Comp Physiol 124:169–179

Bunt AH (1968) An ultrastructural study of the hepatopancreas of *Procambarus clarkii* (Girard) (Decapoda, Astacidea). Crustaceana 15:282–288

Burckhardt G, Kinne R, Stange G, Murer H (1980) The effects of potassium and membrane potential on sodium-dependent glutamic acid uptake. Biochim Biophys Acta 599:191–201

Busse D (1978) Transport of L-arginine in brush border vesicles derived from rabbit kidney cortex. Arch Biochem Biophys 191:551–560

Chapeaux M (1893) Sur la nutrition des Echinodermes. Bull Acad R Belg Ser 3 26:227–232

Chen C, Greenawalt JW, Lehninger AL (1974) Biochemical and ultrastructural aspects of Ca^{2+} transport by mitochondria of the hepatopancreas of the blue crab *Callinectes sapidus*. J Cell Biol 61:301–315

Christensen HN (1964) A transport system serving for mono- and diamino acids. Proc Natl Acad Sci USA 51:337–344

Christensen HN (1975) Biological transport, 2nd edn. Benjamin, Reading, Mass

Christensen HN (1984) Organic ion transport during seven decades. The amino acids. Biochim Biophys Acta 779:255–269

Cooke WJ, Ahearn GA (1976) Comparative cell morphology of freshwater and marine shrimp midgut epithelia. Am Zool 16:225

Corcelli A, Storelli C (1983) The role of potassium and chloride ions on the Na^+ acidic amino acid cotransport system in rat intestinal brush-border membrane vesicles. Biochim Biophys Acta 732:24–31

Crane RK (1962) Hypothesis of mechanisms of intestinal active transport of sugars. Fed Proc 21:891–895

Crane RK (1977) The gradient hypothesis and other models of carrier-mediated active transport. Rev Physiol Pharmacol 78:99–159

Dall W (1967) The functional anatomy of the digestive tract of a shrimp *Metapenaeus bennettae* Racek and Dall (Crustacea: Decapoda: Penaeidae). Aust J Zool 15:699–714

Dall W, Moriarty DJW (1983) Functional aspects of nutrition and digestion. In: Mantel LH (ed) The biology of Crustacea. Internal anatomy and physiological regulation, vol 5. Academic Press, London New York, pp 215–261

Davis LE, Burnett AL (1964) A study of the growth and cell differentiation in the hepatopancreas of the crayfish. Dev Biol 10:122–153
DeViliez E, Buschien K (1967) Survey of tryptic digestive enzyme in various species of Crustacea. Comp Biochem Physiol 21:541–546
Farmanfarmaian A (1969a) Intestinal absorption and transport in *Thyone*. I. Biological aspects. Biol Bull 137:118–131
Farmanfarmaian A (1969b) Intestinal absorption and transport in *Thyone*. II. Observations on sugar transport. Biol Bull 137:132–145
Farmanfarmaian A, Phillips JH (1962) Digestion, storage, and translocation of nutrients in the purple sea urchin (*Strongylocentrotus purpuratus*). Biol Bull 123:105–120
Ferguson JC (1964) Nutrient transport in starfish. II. Uptake of nutrients by isolated organs. Biol Bull 126:391–406
Ferguson JC (1968) Transport of amino acids by starfish digestive glands. Comp Biochem Physiol 24:921–931
Fisher RB, Parsons DS (1953) Glucose movements across the wall of the rat small intestine. J Physiol (London) 119:210–223
Frenzel J (1892) Beiträge zur vergleichenden Physiologie und Histologie der Verdauung. I. Mittheilung. Der Darmkanal der Echinodermen. Mullers Arch Anat Physiol 1892:81–114
Fretter V (1937) The structure and function of the alimentary canal of some species of polyplacophora (Mollusca). Trans R Soc Edinburgh 59:119–163
Fromm D, Field M (1975) Effects of theophylline on Na and alanine transport across isolated rabbit ileum. Am J Physiol 229:683–688
Gerencser GA (1981) Effects of amino acids on chloride transport in *Aplysia* intestine. Am J Physiol 240:R61–R69
Gerencser GA (1982) Membrane physiology of molluscs. In: Podesta RB (ed) Membrane physiology of invertebrates. Dekker, New York, pp 409–442
Gerencser GA (1985) Transport across the invertebrate intestine. In: Gilles R, Gilles-Baillien M (eds) Transport processes, iono- and osmoregulation. Springer, Berlin Heidelberg New York, pp 251–264
Gibson R, Barker PL (1979) The decapod hepatopancreas. Oceanogr Mar Biol 17:285–346
Giordana B, Sacchi FV, Hanozet GM (1982) Intestinal amino acid absorption in lepidopteran larvae. Biochim Biophys Acta 692:81–88
Giordana B, Parenti P, Hanozet GM, Sacchi VF (1985) Electrogenic K^+-basic amino-acid cotransport in the midgut of lepidopteran larvae. J Membr Biol 88:45–53
Goldner AM, Schultz SG, Curran PF (1969) Sodium and sugar fluxes across the mucosal border of rabbit ileum. J Gen Physiol 53:362–383
Graham A (1938) The structure and function of the alimentary canal of aeolid molluscs, with a discussion on their nematocysts. Trans R Soc Edinburgh 59:267–307
Greer ML, Lawrence AL (1967) The active transport of selected amino acids across the gut of the chiton (*Cryptochiton stelleri*). 1. Mapping determinations and effects of anaerobic conditions. Comp Biochem Physiol 22:665–674
Hammerman MR (1982) Na^+-independent L-arginine transport in rabbit renal brush border membrane vesicles. Biochim Biophys Acta 685:71–77
Hanozet GM, Giordana B, Sacchi VF (1980) K-dependent phenylalanine uptake in membrane vesicles isolated from the midgut of *Philosamia cynthia* larvae. Biochim Biophys Acta 596:481–486
Hanozet GM, Giordana B, Parenti P, Guerritore A (1984) L- and D- alanine transport in brush border membrane vesicles from lepidopteran midgut: evidence for two transport systems. J Membr Biol 81:223–240
Harb JM, Copeland DE (1963) Fine structure of the pseudobranch of the founder *Paralichthys lethostigma*. Z Zellforsch Mikrosk Anat 101:167–174
Heyde HC van der (1923) Petites contributions à la physiologie comparée. II. La résorption chez les Echinodermes. Arch Neerl Physiol 8:118–147
Hilden SA, Sacktor B (1981) L-arginine uptake into renal brush border membrane vesicles. Arch Biochem Biophys 210:289–297
Hootman SR, Conte FP (1974) Fine structure and function of the alimentary epithelium in *Artemia salina* nauplii. Cell Tissue Res 155:423–436
Hopfer U, Nelson K, Perrotto J, Isselbacher KJ (1973) Glucose transport in isolated brush border membrane from rat small intestine. J Biol Chem 248:25–32

Jacobs W (1928) Untersuchungen über die Cytologie der Sekretbildung in der Mitteldarmdrüse von *Astacus leptodactylus.* Z Zellforsch 8:1 – 62
James DW, Bamford DR (1974) Regional variation in alanine absorption in the gut of *Echinus esculentus.* Comp Biochem Physiol 49A:101 – 113
Jangoux M (1981) Etude structuale et fonctionnelle du tube digestif d'*Asterias rubens* L. Thalassia Jugosl 12:181 – 186
Jangoux M (1982) Digestive systems: Asteroidea. In: Jangoux M, Lawrence JM (eds) Echinoderm nutrition, chap 11. Balkema, Rotterdam, pp 235 – 272
Jangoux M, Impe E van (1971) Etude comparative de activités phosphomonoesterasiques alcalines du tube digestif de plusieurs espéces d'astérodes (echinodermes), précédeeé d'une note anatomique. Cah Biol Mar 12:405 – 418
Jangoux M, Impe E van (1977) The annual pyloric cycle of *Asterias rubens* L. (Echinodermata: Asteroidea). J Exp Mar Biol Ecol 30:165 – 184
Jangoux M, Perpeet C (1972a) Etude comparative de la structure fine des caecums pyloriques de trois espèces d'Asteriidae (Echinodermata, Asteroidea). Cah Biol Mar 13:401 – 420
Janoux M, Perpeet C (1972b) Etude histochimique de l'effet du jeune sur les organes digestifs de l'ètoile de mer *Asterias rubens* L. C R Acad Sci 274:2587 – 2589
Kessler M, Acuto O, Storelli C, Murer H, Muller H, Semenza G (1978) A modified procedure for the rapid preparation of efficiently transporting vesicles from small intestinal brush border membranes. Their use in investigating some properties of D-glucose and choline transport systems. Biochim Biophys Acta 506:136 – 154
Kinne R, Murer H (1976) Polarity of epithelial cells in relation to transepithelial transport in kidney and intestine. In: Robinson JWL (ed) Intestinal ion transport. Univ Park Press, Baltimore, Maryland, pp 79 – 95
Kinne R, Murer H, Kinne-Saffran E, Thees M, Sachs G (1975) Sugar transport by renal plasma membrane vesicles. Characterization of the systems in the brush-border microvilli and basal-lateral plasma membranes. J Membr Biol 21:375 – 395
Komuro T, Yamamoto T (1968) Fine structure of the epithelium of the gut in the crayfish (*Procambarus clarkii*) with special reference to the cytoplasmic microtubules. Arch Histol Jpn 30:17 – 32
Kooiman P (1964) The occurrence of carbohydrases in digestive juice and in hepatopancreas of *Astacus fluviatillis* (Fabr.) and of *Homarus vulgaris* (M-E.). J Cell Comp Physiol 63:197 – 201
Lawrence AL, Lawrence DC (1967) Sugar absorption in the intestine of the chiton, *Cryptochiton stelleri.* Comp Biochem Physiol 22:341 – 357
Lawrence DC, Lawrence AL, Greer ML, Mailman D (1967) Intestinal absorption in the sea cucumber, *Stichopus parvimensis.* Comp Biochem Physiol 20:619 – 627
Loizzi RF (1971) Interpretation of crayfish hepatopancreas function based on fine structural analysis of epithelial cell lines and muscle network. Z Zellforsch 113:420 – 440
Maginniss LA (1977) Glucose transport by the perfused midgut of the freshwater prawn, *Macrobrachium rosenbergii.* Ph D Diss, Univ Hawaii, Honolulu
Marsden JR (1968) The digestion and storage of fat by the amphinomid polychaete, *Hermodice carunculata.* Can J Zool 46:615 – 618
Mellors AJ, Nahrwold DL, Rose RC (1977) Ascorbic acid flux across mucosal border of guinea pig and human ileum. Am J Physiol 233:E374 – E379
Millot N (1937) On the morphology of the alimentary canal, process of feeding, and physiology of digestion of the nudibranch mollusc, *Forunna tomentosa* (Cuvier). Philos Trans R Soc London Ser B 228:173 – 217
Miyawaki M (1966) Eyestalk hormones and P^{32} incorporation of the hepatopancreas cells in the crayfish, *Procambarus clarkii.* Ann Zool Jpn 39:137 – 141
Miyawaki M, Sasaki N (1961) A preliminary report on uptake of Ca^{45} by the hepatopancreas of crayfish, *Procambarus clarkii.* Kumamoto J Sci 5:170 – 172
Miyawaki M, Tanoue S (1961) Electron microscopy of the hepatopancreas in the crayfish, *Procambarus clarkii.* Kumamoto J Sci 6:1 – 4
Miyawaki M, Arimura H, Ukeshima A (1966) The relationship between eyestalk hormones and the ultrastructure of the hepatopancreas cells in the crayfish, *Procambarus clarkii.* Kumamoto J Sci 8:1 – 11

Morton JE (1956) The tidal rhythm and action of the digestive system of the lamellibranch *Lasaea rubra*. J Mar Biol Assoc U K 35:563–586
Munck BG (1981) Intestinal absorption of amino acids. In: Johnson LR (ed) Physiology of the gastrointestinal tract, vol. 2. Raven Press, New York, pp 1097–1122
Munck BG, Schultz SG (1969) Lysine transport across isolated rabbit ileum. J Gen Physiol 53:157–182
Mykles DL (1979) Ultrastructure of alimentary epithelia of lobsters, *Homarus americanus* and *H. gammarus*, and crab, *Cancer magister*. Zoomorphologie 92:201–215
Nedergaard S (1972) Active transport of α-aminoisobutyric acid by the isolated midgut of *Hyalophora cecropia*. J Exp Biol 56:167–172
Nedergaard S (1973) Transport of amino acids in cecropia midgut. In: Ussing HH, Thorn NA (eds) Transport mechanisms in epithelia. Academic Press, London New York, pp 372–381
Ogura K (1959) Midgut gland cells accumulating iron or copper in the crayfish *Procambarus clarkii*. Ann Zool Jpn 32:133–142
Peterson SC, Goldner AM, Curran PF (1970) Glycine transport in rabbit ileum. Am J Physiol 219:1027–1032
Philpott CW, Copeland DE (1963) Fine structure of chloride cells from three species of *Fundulus*. J Cell Biol 18:389–404
Pillai RS (1960) Studies on the shrimp *Caridina laevis* (Heller). 1. The digestive system. J Mar Biol Assoc India 2:57–74
Potts FA (1923) The structure and function of the liver of *Teredo*, the shipworm. Proc Cambridge Philos Soc 1:1–17
Rose RC, Koch MJ, Nahrwold DL (1978) Folic acid transport by mammalian small intestine. Am J Physiol 235:E678–E685
Rigdon RH, Mensik DJ (1976) Gastrointestinal tract of *Penaeus aztecus* Ives, 1891 (Decapoda, Natantia), a histologic study. Crustaceana 30:164–168
Sacktor B, Rosenbloom IL, Liang CT, Cheng L (1981) Sodium gradient- and sodium plus potassium gradient-dependent L-glutamate uptake in renal basolateral membrane vesicles. J Membr Biol 60:63–71
Sather BT (1969) A comparative study of amylases and proteinases in some decapod Crustacea. Comp Biochem Physiol 28:371–379
Schneider EG, Sacktor B (1980) Sodium gradient-dependent L-glutamate transport in renal brush border membrane vesicles. Effect of an intravesicular > extravesicular potassium gradient. J Biol Chem 255:7645–7649
Schneider EG, Hammerman MR, Sacktor B (1980) Sodium gradient-dependent L-glutamate transport in renal brush border membrane vesicles. Evidence for an electroneutral mechanism. J Biol Chem 255:7650–7656
Schultz SG, Curran PF (1970) Coupled transport of sodium and organic solutes. Physiol Rev 50:637–718
Schultz SG, Yu-Tu L, Alvarez OO, Curran PF (1970) Dicarboxylic amino acid influx across brush border of rabbit ileum. Effects of amino acid charge on the sodium-amino acid interaction. J Gen Physiol 56:621–639
Schultz TW (1976) The ultrastructure of the hepatopancreatic caeca of *Gammarus minus* (Crustacea, Amphipoda). J Morphol 149:383–400
Smith JM, Nadakavukaren MJ, Hetzel HR (1975) Light and electron microscopy of the hepatopancreas of the isopod *Acellus intermedius*. Cell Tissue Res 163:403–410
Stanier JE, Woodhouse MA, Griffin RL (1968) The fine structure of the hepatopancreas of *Carcinus meanas* (L.) (Decapoda, Brachyura). Crustaceana 14:56–66
Stevens BR, Ross HJ, Wright EM (1982) Multiple transport pathways for neutral amino acids in rabbit jejunal brush border vesicles. J Membr Biol 66:213–225
Stieger B, Stange G, Biber J, Murer H (1983) Transport of L-lysine by rat renal brush border membrane vesicles. Pfluegers Arch gesamte Physiol Menschen Tiere 397:106–113
Talbot P, Clark WH, Lawrence Al (1972) Fine structure of the midgut epithelium in the developing brown shrimp, *Penaeus aztecus*. J. Morphol 138:467–486

Toggenburger G, Kessler M, Rothstein A, Semenza G, Tannenbaum C (1978) Similarity in effects of Na^+ gradients and membrane potentials on D-glucose transport by, and phlorizin binding to, vesicles derived from brush borders of rabbit intestinal mucosal cells. J Membr Biol 40:269–290

Travis F (1957) The molting cycle of the spiny lobster, *Panulirus argus* Latreille. IV. Postecdysial histochemical changes in the hepatopancreas and integumental tissues. Biol Bull 113:451–479

Treherne JE (1958) The absorption of glucose from the alimentary canal of the locust *Schistocerca gregaria* (Forsk.). J Exp Biol 35:297–306

Treherne JE (1959) Amino acid absorption in the locust (*Schistocerca gregaria* Frosk.). J Exp Biol 36:533–545

Turner RJ (1983) Quantitative studies of cotransport systems: models and vesicles. J Membr Biol 76:1–15

Vernon GM, Herold L. Witkus ER (1974) Fine structure of the digestive tract epithelium in the terrestrial isopod, *Armadillidium vulgare*. J Morphol 144:337–360

Weel PB van (1955) Processes of secretion, restitution, and resorption in gland of midgut (glandula media intestini) of *Atya spinipes* Newp. Physiol Zool 28:40–54

Weel PB van (1960) On the secretion of digestive enzymes by the marine crab *Thalamita crenata*. Z Vergl Physiol 43:567–577

Weel PB van (1961) The comparative physiology of digestion in molluscs. Am Zool 1:245–252

Weel PB van (1970) Digestion in Crustacea. In: Florkin M, Scheer BT (eds) Chemical Zoology, vol 5. Academic Press, London New York, pp 97–115

Weel PB van (1974) Hepatopancreas? Comp Biochem Physiol 47A:1–9

Wyban JA, Ahearn GA, Maginniss LA (1980) Effects of organic solutes on transmural PD and Na transport in the intestine of freshwater prawns. Am J Physiol 239:C11–C17

Yonge CM (1924) Studies on the comparative physiology of digestion. II. The mechanism of feeding, digestion, and assimilation in *Nephrops norvegicus*. J Exp Biol 1:343–390

Yonge CM (1926) The digestive diverticula in the lamellibranchs. Trans R Soc Edinburgh 56:703–718

Chapter 4

Nutrient Transport Across Vertebrate Intestine

W. H. Karasov

Contents

1 Introduction

1.1 Chapter Overview

Modern vertebrates are diverse in their trophic characteristics. All higher classes of vertebrates, for example, include species which feed at some point in their life on diets with relatively high proportions of carbohydrate (e.g., herbivores and omnivores) and species which feed on nearly carbohydrate-free diets (e.g., carnivores). Vertebrates also differ tremendously in the amounts of foods they eat daily. For example, daily metabolic rates and hence food requirements of endothermic mammals and birds are generally more than an order of magnitude greater than those of ectothermic fish, amphibians, and reptiles.

This diversity in the types and amounts of foods eaten among vertebrates represents a major challenge to the digestive system. The gross differences in digestive system morphology among mammals with different diets (e.g., Chivers

Department of Wildlife Ecology, University of Wisconsin, Madison, WI 53706, USA

Advances in Comparative and Environmental Physiology. Vol. 2

and Hladik 1980) are just one testament to the fact that a single "standard model" digestive tract cannot effectively process the myriad types of foods eaten by representatives of that class. But morphology is just one of several features upon which natural selection has operated in the evolution of different digestive systems. Other features might include the particular enzymes involved in the breakdown and absorption of nutrients.

In this chapter I review adaptation of intestinal transport of nutrients in vertebrates, principally sugars and amino acids. Research on intestinal transport has traditionally focused on the mechanisms of nutrient absorption in vertebrates, especially laboratory mammals, and many excellent reviews are available (for a list see Hopfer 1987). This review differs from most of those in its focus on two questions important in comparative and environmental physiology: (1) Do vertebrates share the same pathways for transporting nutrients from intestinal lumen into blood? and (2) Are transport rates themselves modulated to fit the particular requirements set by a vertebrate's feeding or environmental circumstances? These questions are inherently interesting but answers to them are also important for our effective use of animal models in the study of nutrient absorption and its regulation as it relates to human health and disease.

The chapter is organized in three parts. Part one (Sect. 2) briefly surveys digestive tract morphology in vertebrates as it relates to sugar and amino acid transport. Part two (Sect. 3) is devoted to a discussion of the transport pathways and their occurrence among vertebrates. It reviews the mechanisms and signals by which nutrient transport rates can be varied within and between species. Part three (Sect. 4) is devoted to the patterns of adaptation of nutrient transport rates in response to variation in diet composition, nutrient intake rate, and environmental conditions. Where possible distinctions are made between phenotypic (within species) and genotypic (between species) adaptations.

1.2 Terms

Certain terms are used throughout and are best defined here. Carrier-mediated transport refers to processes which display saturation kinetics. Mediated transport can occur against a substrate's electrochemical potential gradient (usually called active transport), in which case it is directly or indirectly dependent on cellular metabolic energy. When carrier-mediated transport occurs down an electrochemical gradient it is considered to be facilitated diffusion. Simple diffusion occurs down an electrochemical gradient and involves neither a carrier nor cellular metabolic energy. Intestinal absorption involves movement of solutes by these processes in at least two steps: uptake from the gut lumen into the enterocyte across its apical membrane, and exit from enterocyte to blood across the basolateral membrane.

1.3 Technical Issues

Analyses of adaptation of intestinal nutrient transport are frequently dependent upon comparisons of nutrient transport rates. However, a simple comparison of rates measured in different laboratories is rarely enlightening and often not even possible. Reasons for this include (1) that a variety of preparations have been used in different laboratories and the type of preparation affects measured values of kinetic constants (Thomson and Dietschy 1980a), and (2) transport rates are normalized to many different measures of amount of intestine.

Type of Preparation. There are at least two kinds of differences between preparations which confound comparisons. The most basic difference is in the process actually measured, e.g., brush border uptake, basolateral efflux, or some composite of these. Some methods for measuring transport (e.g., everted sacs, intestinal rings, in vivo absorption) do not allow extraction of information concerning the function of either the brush border or basolateral membranes, though they may be suited for the demonstration of active transport (e.g., analysis of tissue/medium concentration ratios in sacs and rings) (Munck 1981). For most analyses of transport *rates* in this chapter I have tended to rely, whenever possible, on studies which applied methodology which does specify a particular cell membrane and process. The largest body of such data is for uptake in intact tissue in vitro in preparations which isolate the mucosal and serosal surfaces.

Even when the same membrane and transport step are studied, there may still be significant differences in rates derived with different methods. An example of this is given in Section 3.1.2. Thus, I have concluded that values such as kinetic constants for transport or transport rates can only be directly compared if the same methodology was employed. This requirement, and the previous consideration, guided my selection of data for the analyses which follow.

Expression of Nutrient Flux Rates. These rates can be normalized to intestinal length, mass, several measures of surface area, protein content, DNA content, and so on. The decision of what to normalize transport rates to may depend upon the particular focus of the study but even then can affect one's conclusions about the nature of adaptation. Illustration of these points can be found in several sections of the chapter (Sects. 4.2.1 and 4.3.1). I find normalization to intestinal length or surface area most useful for comparisons of species. It is meaningful physiologically, especially when the length or area of the intestine is known and one can then calculate an integrated or summed uptake rate over the entire intestine. Comparisons of transport rates normalized to intestinal mass are complicated by the fact that the proportions of the mass which are mucosa and underlying muscle layers are not known for most species. Normalization to intestinal mass, tissue protein, or DNA content is most useful when studying treatment effects on transport rates using adjacent tissues from within the same animal.

2 Where Sugar and Amino Acid Absorption Occurs

2.1 Location in the Gastrointestinal Tract

Monosaccharides, amino acids, and di- and tripeptides are the breakdown products of dietary carbohydrate and protein which are absorbed in the gut. Vertebrates differ in their gut morphology and in the regions where most nutrient absorption occurs.

Intestines of some vertebrates are a single, unsectioned tube and have transport activity all along the length. Examples of this pattern occur in fish (Ferraris and Ahearn 1983), amphibians (e.g., frogs; Karasov unpublished data), reptiles (e.g., snakes; Karasov unpublished data), birds (Karasov et al. 1986a), and mammals (Karasov et al. 1985a). In these species, and in species with distinct small intestine, the sugar and amino acid transport rate often varies with position. Usually, uptake rates are greatest in the proximal or midintestine, possibly due to induction by unabsorbed solutes (Karasov and Diamond 1983b). Regardless of the mechanistic basis for this gradient in transport activity, the importance of this for the comparative physiologist is that in order to make interspecies comparisons at the tissue level, one must therefore specify where along the intestine uptake measurements were made.

In many vertebrates the intestine is separated from the caecum and colon by a valve. In these cases the proximal region (small intestine) is probably the major site of absorption. The existence of carrier-mediated uptake of sugars and amino acids by the apical membrane of colon epithelium in mammals is disputed (Henriques de Jesus et al. 1979; Ilundain and Naftalin 1981; Robinson et al. 1963), but when it appears to occur the uptake rates are usually less than 10% of those in small intestine (Ilundain and Naftalin 1981; Karasov et al. 1985b). In the mammals and reptiles which have been studied, the total uptake capacity of the colon is only 1 – 8% of that of the small intestine (Karasov et al. 1985b). Colon transport in birds has been little studied. In chicken colon it appears that carrier-mediated uptake of sugars and amino acids occurs, but only in early stages of development (Holdsworth and Wilson 1967; Lerner et al. 1975).

In some fish species the most important region of nutrient absorption may not be the intestine, but a series of blind appendages emptying into the proximal intestine called pyloric caeca (Garg 1978; Buddington and Diamond 1987). In cod and rainbow trout, for example, the caeca contribute about 70% of the total postgastric area of the gut and the same percentage of the total glucose and proline uptake capacity of the combined caeca-intestine. The function of the proximally located caeca in fish appears to be enzymatic digestion and absorption in contrast to the distally located caeca of mammals, reptiles, and some birds in which microbial fermentation occurs.

Whether or not significant nutrient absorption occurs in the caeca of mammals, reptiles, and birds is of considerable importance in understanding the role of this organ in the nutrition of nonruminant herbivores. Bacterial protein synthesis from urea occurs in the caeca, for example, but the utility of this for the animal is limited in the absence of an absorptive mechanism or some form of coprophagy (Hornicke and Bjornhaag 1980). Amino acid absorption appears to be negligible in the caecum of adult rabbits (Hoover and Heitmann 1975), in chickens past the first few days of life (Lerner et al. 1975), and in captive willow ptarmigan (*Lagopus lagopus*; Mortensen and Tindall 1981). This issue needs to be further studied, especially in light of a recent report of carrier-mediated absorption of sugar in proximal portions of caecum from 8-week-old chickens (Vinardell et al. 1986).

2.2 Solute Concentrations at the Site of Absorption

Solute concentrations in the intestine presumably reflect the relative rates of hydrolysis and absorption in vivo, and possibly provide insight as to the significance of passive vs active absorption. There are probably few generalizations for all vertebrates with respect to these issues.

Judging by the relatively low free amino acid concentrations in the gut of man after a milk meal, for example, one concludes that amino acid absorption was quite rapid compared with protein hydrolysis, whereas in the fish absorption may be more rate-limiting since free amino acids appear to accumulate in the lumen (Table 1). A comparison of humans digesting milk or bovine protein for 3 h suggests that the particular protein or carbohydrate being digested is a factor. In humans absorption of glucose from sucrose or maltose occurs at a slower rate than hydrolysis and glucose molecules derived from maltose actually accumulate intraluminally, whereas hydrolysis is rate-limiting in the digestion of lactose (Gray and Santiago 1966).

In the digestion of a discrete meal one might expect that at any one position in the gut concentrations might rise and then fall with increasing time since inges-

Table 1. Concentration (mM) of free amino acids in the digestive tract of fish, reptiles, birds, and mammals

Amino acid	Common carp[a]	Rainbow trout[b]	Cayman[c]	Duck[d]	Man[e]	Man[f]	Man[g]	Rat[h]
Phenylalanine	4.67	6.59	1.20	2.3	5.34	1.76	0.19	2.83
Tyrosine	3.68	5.57	1.28	–	5.56	1.69	0.23	2.37
Leucine	8.95	11.89	1.80	4.4	14.20	3.02	0.19	4.98
Isoleucine	5.98	7.12	0.36	–	2.00	0.89	0.08	1.68
Methionine	4.89	5.33	0.48	1.0	0.80	0.60	0.06	1.04
Valine	6.29	7.30	0.60	2.5	6.62	1.69	0.15	2.18
Cystine	1.67	0.53	–	–	–	–	0.11	0.16
Alanine	6.31	10.69	1.78	5.2	5.58	1.88	0.16	3.89
Glycine	4.50	6.60	1.68	–	0.38	0.74	0.20	2.14
Glutamic acid	5.58	11.75	1.40	4.9	4.80	2.62	0.58	–
Proline	4.11	13.68	–	–	0	0.85	0.13	3.45
Serine	5.71	9.61	0.15	3.0	4.16	1.05	0.16	1.99
Threonine	6.45	8.07	1.20	–	3.78	1.14	0.13	1.87
Aspartic acid	3.56	4.77	0.12	–	2.04	1.37	0.14	1.88
Arginine	6.76	9.18	0.40	–	1.97	2.06	0.35	4.46
Histidine	2.71	2.16	0.29	–	2.12	1.90	0.18	0.60
Lysine	11.18	13.36	0.84	3.3	8.91	5.63	0.70	4.50
Total	93.00	134.50	13.58	–	57.26	29.29	3.72	40.02

[a] Dabrowski (1983); average for entire gut 3 h after feeding fish-meal based ration.
[b] Dabrowski and Dabrowska (1981); midgut 14 h after feeding fish-meal based ration.
[c] Coulson and Hernandez (1970); average concentration in four caymens 18 h after feeding fish.
[d] Crompton and Nasheim (1969); midgut of ducks fed soybean ration ad libitum; assumes 80% of digesta wet mass was water.
[e] Chung et al. (1979); jejunum 2 h after feeding bovine serum albumin.
[f] Adibi and Mercer (1973); jejunum 3 h after feeding bovine serum albumin.
[g] Nixon and Mawer (1970); jejunum 3 h after feeding milk meal.
[h] Steiner and Gray (1969); jejunum 45 min after 5 g rat chow.

tion, and this appears to be the case for man eating a test meal containing bovine serum albumin (Table 1). And of course nutrients decrease in concentration distally as they are absorbed along the intestine; luminal glucose concentration in fed rats was 104.4 mM in the jejunum and 1.59 mM in ileum (Debnam 1982a). Thus, nutrient transporters may or may not be saturated by their substrates in vivo depending upon time since the last meal, position in the gut, and the carbohydrate or protein source of monosaccharides and amino acids. Based on the available measurements of intraluminal nutrient concentration both passive and carrier-mediated nutrient transport are important pathways for absorption in vivo.

Amino acid absorption can occur via carrier-mediated transport of amino acids and di- and tripeptides. In jejunum of humans 2 – 3 h after a meal 70 – 80% of the total alpha-amino nitrogen in the lumen was associated with the peptide fraction (Adibi and Mercer 1973; Chung et al. 1979). The fraction in pigs was smaller ($\leq 37\%$) and depended on diet (Low 1979). In carp the free amino acid fraction made up 4 – 94% of the total nitrogen in the gut depending upon the particular amino acid considered, the part of the digestive tract, and the time since feeding (Dabrowski 1983). Thus, while high peptide fractions have suggested to some that peptides are quantitatively a more important substrate for absorption sites in the intestine of man than are amino acids (Adibi and Kim 1981), one cannot presently extend this conclusion to vertebrates in general.

3 Pathways for Sugar and Amino Acid Absorption and Mechanisms of Regulation

This section is devoted to a short summary of the pathways in intestine for sugar and amino acid absorption, mechanistic explanations for differences in their rates of absorption, and a discussion of some of the signals involved in the regulation of their absorption rates.

3.1 Pathways for Uptake Across the Brush Border

3.1.1 Description and Occurrence of the Pathways

Sugars. Sugars cross the luminal surface passively and also through carrier-mechanisms-facilitated diffusion or active transport. The latter Na^+-dependent uptake process has been identified in the intestine of vertebrates from every class (cf. Ferraris and Ahearn 1984, for fish; Gunter-Smith et al. 1982, for amphibians; Karasov et al. 1985b, for reptiles and mammals; Kimmich and Randles 1984, Lerner 1984, for birds). There is evidence for multiple glucose carriers in hamster (Honegger and Semenza 1973), rats (Debam and Levin 1976), rabbits (Thomson et al. 1982; Kaunitz et al. 1982), cows (Kaunitz and Wright 1984), and mice (Ferraris and Diamond 1986) though extensive studies in chick intestinal cells (Kimmich 1981) have not provided evidence for more than one type of transporter. Hopfer (1987) provides a review and analysis of the evidence for and against multiple carriers for aldohexoses.

A distinct Na^+-independent facilitative uptake pathway for fructose has been demonstrated in many vertebrates (Albus et al. 1983; Schultz and Strecker 1970; Honegger and Semenza 1973; Bode et al. 1981; Csaky and Fischer 1984; Solberg and Diamond 1987).

Amino Acids. Analysis of amino acid uptake is complicated by the existence of numerous transporters with sometimes overlapping specificities. The transporters are operationally defined in terms of their Na^+ dependence and the kinetic criteria of saturation and competition among different amino acids.

Several reviews are available that summarize information in this complicated field (e.g., Munck 1981, 1983; Stevens et al. 1984). In both mammals and birds (Lerner 1984) there appear to be at least four distinct Na^+-dependent mechanism for the four classes of amino acids: alpha-aminomonocarboxylic acids (neutral amino acids), cationic (basic) amino acids, anionic (acidic) amino acids, and imino acids. These pathways are not necessarily specific for their major substrates. The Na^+-dependent neutral pathway, for example, appears to transport the imino acid proline in several species. Beyond these four mechanisms there appear to be additional mechanisms present in some species but not in others [e.g., two carriers for lysine in rabbit but only one in rats (Munck 1983); a Na^+-dependent pathway for phenylalanine and methionine in rabbits but not in mice (Karasov et al. 1986c)]. Carrier-mediated, Na^+-independent uptake of amino acids can also be demonstrated in the brush border membrane.

What can be said about the existence of multiple amino acid transport pathways in ectothermic vertebrates? There would seem to be little doubt that amino acids from the four groups are transported in a Na^+-dependent fashion, based on either direct demonstrations or observations of their transport against concentration gradients in both reptiles [Moin-Un-Nisa et al. (1970) for all four classes of amino acids] and fish [e.g., Mepham and Smith (1966) for neutral amino acids; Hokazono et al. (1979) for basic amino acid; T. H. Wilson (1957) and Collie (1985) for imino acids]. Also, amino acid competition studies in fish have confirmed that multiple carriers exist (reviewed in Ferraris and Ahearn 1984; Storelli et al. 1986). However, the paucity of studies of mutual inhibition of uptake by amino acids precludes one from evaluating whether the same distinct carrier systems exist in ectothermic vertebrates as in mammals and birds.

3.1.2 Characterization of Passive Uptake

The carrier-mediated transport processes cannot be properly studied until one has subtracted the passive component from the total flux. Several procedures are generally used to measure passive uptake of sugar across the apical membranes of enterocytes. Thomson et al. (1982) compared the following methods for determining passive D-glucose uptake in four mammal species and found them all to yield similar estimates: (1) the slope of the concentration curve over a range of concentrations which saturate glucose carrier(s), (2) equating total L-glucose up-

Table 2. Kinetic constants for D-glucose uptake and Na^+ dependence of L-proline uptake across the brush border membrane of vertebrate intestine

Species	D-glucose uptake				L-proline uptake	Method[a]	Reference[b]
	V_{max} (nmol min^{-1} cm^{-2})	K_m^* (mM)	K_d^* (µl min^{-1} cm^{-2})	Fraction of total uptake at 50 mM which is passive	Fraction of total uptake at 25–50 mM which is Na^+-dependent		
Mammals							
Lab mouse fed high-carbohydrate ration	654	6.2	0.73	0.06	0.24	a	1
Lab mouse fed carbohydrate-free ration	337	3.2	0.57	0.08		a	1
Desert woodrat	190	1.3	0.14	0.04	0.23	a	2
Hamster	418	1.5	4.3	0.35	0.18	a	2, 4
Belding's ground squirrel	121	0.4	0.14	0.06	–	a	6
Kangaroo rat	244	1.6	0.85	0.15	0.26	a	4
Lab rat	264	4.0	0.77	0.14	0.79	a	5
Prairie vole	–	–	1.1	0.12	0.67	a	3
Green monkey	–	–	0.8	0.19	–	a	3
Birds							
Chicken	306	4.9	–	–	0.45	a	4
Hummingbird	411	4.3	~0	~0		a[c]	6
American robin	–	–	1.11	0.52	0.45	a	8
Starling	–	–	1.21	0.87	0.17	a	8
Cedar waxwing	–	–	1.42	0.44	0.33'	a	8
Reptiles and Amphibians							
Desert iguana	40	0.7	0.19	0.19	0.35	a	2
Chuckwalla	130	0.8	0.45	0.15	0.24	a	2
Box turtle					0.44	a	2
Bullfrog	49	0.6	–	–		a	4

Fish							
Coho salmon, fresh water					0.35	a	9
carp					0.55	a	4
Mammals							
Hamster	62	2.0	2.58	0.68		b	7
Guinea pig	42	5.0	0.66	0.46		b	7
Lab rat	11	2.8	1.41	0.87		b	7
Rabbit	21	0.8	0.64	0.61		b	7

[a] All results are for intact midgut in vitro and uptake rates are normalized to nominal surface areas. The in vitro methods measured unidirectional uptake across the brush border and utilized stirring to minimize and control the unstirred layers. *Methods*: (a) everted sleeve (Karasov and Diamond 1983a); (b) intestinal disks (Lukie et al. 1974). Kinetic constants derived using the same methods can be compared.

[b] References: (1) Diamond and Karasov (1984); (2) Karasov et al. (1985b); (3) Karasov (unpubl. observations); (4) Karasov et al. (1985a); (5) Karasov and Debnam (1987); (6) Karasov et al. (1986a); (7) Thomson et al. (1982); (8) Levey, Martinez del Rio, and Karasov (unpubl. observations); (9) Collie (1985).

[c] In this instance the serosal surface was not isolated and solute entry was not necessarily solely across the apical membrane.

take with D-glucose passive uptake; and (3) the slope of the concentration curve of D-glucose uptake in Na^+-free buffer. To this list one can add (4) D-glucose uptake in the presence of phloridzin. Methods 1 and 3 could also be used with amino acids. Munck (1983) considered that estimates of passive permeation of sugars or amino acids with mannitol were too high, though in brush border membrane vesicle preparations the passive permeability coefficient of mannitol is intermediate among a selection of amino acids and D-glucose (Stevens et al. 1984).

What proportion of sugar and amino acid uptake occurs by passive as opposed to carrier-mediated mechanisms, and are there any important differences between the classes of vertebrates? A priori one might wonder whether ectotherms might take advantage of their much lower nutrient requirements to absorb most of their nutrients by slower but energetically free passive uptake. A comparison of mammals and reptiles suggests that this is not the case for glucose or proline. In three reptile and three mammal species eating the same ration passive uptake accounted for less than 10% of the total D-glucose uptake even at a high concentration (50 mM), with the exception of the hamster for which the value was 35% (Karasov et al. 1985 b). The proportion would be smaller at lower concentrations. Na^+-independent proline uptake, which includes the passive component and possibly also a saturable component, accounted for 6 – 53% of total L-proline uptake at 50 mM, the exact proportion varying with intestinal position and with species but only slightly higher in reptiles. Scantier data for other classes also suggest no gross differences between endotherms and ectotherms (Table 2).

Care must be exercised when making this calculation and comparison with measurements made using different intestinal preparations. The proportion of total glucose or amino acid uptake by the passive route will appear higher in those preparations where carrier-mediated uptake is low, perhaps due to decreased functional integrity of the tissue. An example of this can be found in the comparison (Table 2) of unidirectional glucose uptake kinetics in rat and hamster determined by two different techniques: the everted sleeve method (Karasov and Diamond 1983 a) and the intestinal disk method (Lukie et al. 1974). The latter method tends to yield lower uptake rates for active transport and so even though apparent passive permeability coefficients (K_d^*, apparent because uncorrected for unstirred layer effects) were similar by the two methods, glucose uptake by the passive route was proportionally greater when measured by the latter method.

Interestingly, both methods showed hamster intestine to have a high K_d^*, compared with other mammals. Thus, differences between species in passive permeation may exist and might have physiological significance. Hummingbird intestine, for example, had an immeasurably low glucose K_d^*, possibly as an adaptation to prevent solute loss from the blood in the face of high fluid transit rates (< 1 h) through the intestine (Karasov et al. 1986 a). Fruit-eating birds (e.g., robin and waxwing) which also have rapid digesta transit times (< 1 h), however, had K_d^* values more similar to those of other vertebrates (Table 2).

3.1.3 Characterization of Carrier-Mediated Uptake

The carrier-mediated transport processes can be described kinetically by their maximal transport rate V_{max}, which reflects the number of transport sites and the translocation rate of the nutrient-carrier complex, and their apparent Michaelis constant K_m^*, (apparent because uncorrected for unstirred layer effects) which varies inversely with the nutrient-carrier affinity. Anyone who reviews the literature on kinetics of glucose or amino acid uptake in different species will be impressed by the range of values reported for the parameters. This is even apparent when focusing solely on in vitro measurements, sometimes made by the same technique (cf. Table 2). In many (perhaps most) cases, it is probably incorrect to conclude that differences in Michaelis constants represent real differences in the solute-binding characteristics of the sugar or amino acid carriers. At least two other possibilities discussed below, both related to the effects of unstirred layers (USL), must be considered as sources of variation before invoking phenotypic or genotypic differences in carrier-solute "affinity" as an explanation for observed differences in K_m^*.

Effects of Unstirred Layers. Unstirred layers (USL) have a dramatic effect on determinations of K_m^* but do not necessarily introduce error into V_{max} determinations (Winne 1973; P.H. Barry and Diamond 1984; Thomson and Dietschy 1984). According to Winne (1977) though, even V_{max} can be underestimated if uptake is measured in the presence of an USL and the carrier-mediated portion is calculated by subtraction of passive uptake from total uptake. The magnitude of the error in K_m^* increases with USL resistance (R) and with V_{max} according to the following relationship:

$$K_m^* = K_m + 0.5\, V_{max}\, R = K_m + 0.5\, (\delta/DA)\, V_{max} \,, \tag{1}$$

where K_m is the true K_m for the solute, A is the effective surface area of the USL for the particular solute used, δ its thickness, and D the solute's diffusion coefficient. In practice, then, variation in both USL resistance and V_{max} can be sources of variation in measures of K_m^*, as illustrated by the following examples.

The effects of USL thickness on K_m^* determinations can be illustrated experimentally by varying shaking or stirring rates in tissue uptake studies (Thomson and Dietschy 1980b; Lherminier and Alvarado 1981) and also by a comparison of K_m^* values for glucose absorption in different preparations which vary in their USL thicknesses. Thus, K_m^* in rat jejunum determined in vivo (high USL thickness; Winne 1976) ranges between 20 – 30 mM (Debnam and Levin 1975, 1976; Debnam 1982b), whereas in intact tissues in vitro (reduced USL thickness; Thomson and Dietschy 1980b; Karasov and Diamond 1983a) it ranges between 0.4 – 6 mM (Table 2), and in vesicle preparations (presumably negligible USL) values fall in the micromolal range (Hopfer 1987). In uptake studies in vitro with intact tissue USL resistance can be minimized and controlled (e.g., Thomson and Dietschy 1980b; Karasov and Diamond 1983a) but the USL cannot be entirely eliminated. Unfortunately, there is no straightforward way to measure intestinal USL resistance for a nutrient because different solutes may be transported over different fractions of the villus area; one cannot measure δ and A for one solute and use them to correct measured kinetic constants for another solute (P.H. Barry and Diamond 1984).

The effects of variaton in V_{max} on K_m^* determined in the presence of an USL is illustrated by the correlation between the two when many determinations using the same methodology are compared. Equation (1) implies that if one measured transport in several animals differing in V_{max} but with identical unstirred layers and K_m's, the measured values of K_m^* would vary in parallel with V_{max} because of the USL artifact. Karasov et al. (1985a) tested this possibility in mice by measuring glucose K_m^* and V_{max} under several experimental conditions which altered V_{max}. For all mouse data combined, V_{max} and K_m^* were positively correlated, as predicted by Eq. (1). Furthermore, for all determinations made in 12 species of vertebrates by the everted sleeve method (which minimizes and controls the USL) V_{max} and K_m^* were positively correlated. For the everted sleeve data reported in Table 2 V_{max} and K_m^* are also positively correlated ($P < 0.01$). Additionally, an earlier review (Karasov and Diamond 1983b) found that numerous studies of phenotypic variation in D-glucose kinetics reported that V_{max} and K_m^* change in the same direction.

Differences between species in K_m^* for glucose are probably not any greater in magnitude than could be explained by USL effects. Differences between species in K_m^* for particular amino acids (when determined by the same methods) are also not great and can probably be explained similarly. For example, K_m^* for proline uptake at the brush border of intestinal sleeves was as follows for the following five species: mouse, 3.9 mM; woodrat, 4.2 mM; chicken, 5.3 mM; desert iguana, 0.6 mM; salmon in fresh water, 4–9 mM (Karasov et al. 1985a, b, 1986c; Collie 1985).

Thus, perhaps phenotypic and genotypic changes in the binding constants for glucose or amino acids do not occur. On the other hand, a significant difference between species in K_m^* for monosaccharide uptake (e.g., Wolffram et al. 1986) was observed using brush border membrane vesicles, and one might think that this preparation would be insensitive to USL artifacts. Interestingly, in this example V_{max} and K_m^* changed in the same direction.

The glucose transporter can also be characterized by its specificity for aldohexoses. Early studies with mammalian intestine established that the structure of the D-pyranose ring, C_2-OH in the equatorial position and C_6 are minimally required for sugar-carrier interaction (T.H. Wilson and Landau 1960; Crane 1960). Subsequent studies have shown that these structural requirements can roughly be applied to vertebrates from every vertebrate class (Shimada and Hoshi 1986). Carrier specificity is not absolute but the affinity scale of sugars for the Na^+-dependent D-glucose carrier system is similar or identical when comparing species from all five higher vertebrate classes (cf. Brot-Laroche and Alvarado 1983; Shimada and Hoshi 1986; Storelli et al. 1986). Thus, the proposal that a phylogenetically identical carrier mechanism for sugar transport is present in the intestine of most vertebrates (Alvarado and Monreal 1967) is entirely reasonable.

3.2 Pathways for Transport Across the Basolateral Membrane

Transport across the basolateral membrane has been effectively studied using isolated enterocytes and membrane vesicles. In both intact cells (Kimmich 1981) and membrane vesicles (Stevens et al. 1984) glucose basolateral transport occurs by simple and facilitated diffusion. The specificity of the basolateral mediated pathway differs from that for Na^+ cotransport in brush borders as exemplified by fructose's interaction with the former but not the latter (Kimmich 1981). In basolateral membrane vesicles amino acid transport occurs by passive diffusion and the Na^+-independent (*L*) and Na^+-dependent (*A, ASC*) carriers found almost universally in nonepithelial membranes (Stevens et al. 1984; Hopfer 1987). Inasmuch as there have been relatively few studies of basolateral transport, an interspecies survey or comparison is not yet possible.

3.3 Mechanisms of Adaptation of Nutrient Absorption Rates

Rates of nutrient absorption by the small intestine have been found to vary between species, and within a species in response to many factors. Here, I introduce five possible mechanistic bases for these variations in transport rates, and in Section 4 specific examples of adaptation and their particular mechanistic bases are discussed. The possible mechanisms can sometimes be partially distinguished experimentally by how specific they are in their effect. Some instances of intestinal adaptation, however, may represent hybrid responses involving more than one of these mechanisms of regulation.

3.3.1 Anatomic Adaptations

The actual mucosal area of the gut of vertebrates is several orders of magnitude greater than that of the corresponding smooth bore tube (whose area I call the nominal area of the gut). Pfeiffer (1968) described four general types of folds and four general forms of villi in vertebrates. Almost all measurements of the contributions of these folds and villi, and of the microvilli, to total surface area have been made in mammals (Table 3).

Anatomic adaptation occurs at the gross level, villous level, and microvillous level. Examples and time courses of such adaptations within a species to factors such as lactation, starvation, diet, resection, and diabetes are summarized in Karasov and Diamond (1983b). Changes in surface area often cause parallel changes in the transport of all nutrients which are studied. In pregnancy and lactation, for example, transport rates of sugars, amino acids, copper, zinc, and iron increase in rats, hamsters, and mice (Penzes and Simon 1968; Musacchia and Hartner 1970; Linder and Munro 1973; Cripps and Williams 1975; Davies and Williams 1976, 1977; Karasov 1988).

To what extent can differences in surface area between species account for differences in transport rates? One simple relation one might expect to see is a correspondence between surface area per unit nominal area and apparent passive

Table 3. Contribution of villi and microvilli to intestinal surface area in vertebrate species

Species	Mass (kg)	Intestine nominal area (cm^2)	Multiplication due to villi	Multiplication due to microvilli	References[a]
Mammals					
Big brown bat	0.018	16	7.3		1
Mole-shrew	0.018	35	4.6		1
Long-tailed deermouse	0.025	39	3.9		1
Woodland deermouse	0.020	24	4.2		1
Northern jumping mouse	0.020	23	4.04		1
Prairie vole	0.030	23	4.1		1
Eastern vole	0.04	36	3.1		1
Laboratory mouse	0.035	42	5.1	48	2, 3
Desert woodrat	0.121	44	11.5	80	4, 3
Egyptian fruit bat	0.125			57	14
Laboratory rat	0.295	112 – 142	4.8 – 6.3	18 – 27.2	5, 6, 12, 14
		100	5.8	20	7
Flying-fox (bat)	0.63	215	16.3	77	8
Rabbit	2.6	958	21.1	28.0	9, 10, 11
Cat	1.95	330	12.9		5
Dog	12.00	1910	8.5	26	12
Human	70.0	3300	6.6	30	13, 15
Birds					
Pigeon	0.300	85	8.1		9
Reptiles					
Desert iguana	0.073	18	3.4	82	4, 3

[a] *References:* (1) Barry (1976); (2) Diamond and Karasov (1984); (3) Ferraris, Lee and Diamond (pers. comm.); (4) Karasov et al. (1985b); (5) Wood (1944); (6) Palay and Karlin (1959); (7) Mayhew and Middleton (1985); (8) Tedman and Hall (1985); (9) Verzar and McDougall (1936); (10) Westergaard and Dietschy (1974); (11) Chivers and Hladik (1980); (12) Anderson and Taylor (1973); (13) Marsh and Swift (1969); (14) Keegan and Modinger (1979); (15) Wilson (1962).

permeability per unit nominal area. This simple relation, however, does not hold when one compares surface areas and glucose K_d^*'s among lab mice, lab rats, desert woodrats, and desert iguanas (Tables 2 and 3). Woodrats have by far the greatest microvillous surface area ($11.5 \times 80 = 1280$ cm^2 per cm^2 nominal surface area) but only an intermediate K_d^*, whereas lab rats have the highest K_d^* but the lowest surface area. Possible explanations include that passive uptake of glucose does not occur primarily over the entire microvillous surface area of the mucosa (perhaps instead at cell junctions), that there are significant differences between these species in the permeability of a unit area of membrane, and that true K_d's are similar among the species but that USL resistances (which are additive to membrane resistance) differ even though stirring rates were similarly high for the measurements on all four species.

There are several reasons to expect poor correspondence between total surface area per unit nominal area and carrier-mediated uptake rate per unit nominal area. Some of them are discussed below: densities of transporters per unit membrane area can vary, and carrier-mediated transport may not occur at all along the lower two-thirds of villi.

In summary, there is little doubt that a change in surface area is the mechanistic basis for adaptative change in transport rate in certain examples of phenotypic adaptation. But, when comparing species the relationship at the tissue level between surface area and transport rate does not appear either simple or predictable.

3.3.2 Changes in Specific Transport Systems

In some instances carrier-mediated transport changes in one direction for one solute with no change or a change in the opposite direction for other solutes. Evidence for specific regulation of nutrient transport exists for the sugars glucose and fructose, the dipeptide carnosine, the amino acids proline, aspartate, histidine, lysine, leucine, alanine, and methionine, two vitamins (thiamine and ascorbic acid) and three minerals (calcium, iron, phosphate) (Diamond and Karasov 1987).

Only for glucose (Ferraris and Diamond 1986) and iron (Linder and Munro 1977) has regulation of transporter site densities been assessed directly by binding studies. In both cases up-regulation of activity was associated with increased site density, and in the former case the factorial increases in V_{max} of transport (Diamond and Karasov 1984) and in phloridzin binding site density (Ferraris and Diamond 1986) were identical. This indicates that the increase in transport activity is due entirely to increased number of functional copies of the transporter. Comparisons of V_{max} and binding site density in mice, woodrats, and desert iguanas indicated that there might be differences between the species in turnover number ($= V_{max}$/binding site density) of the glucose carrier (Ferraris, Diamond, and Lee, unpubl. observations).

Another example of a specific adaptation would be the expression of a new transport mechanism. A Na^+-dependent transport mechanism for lysine, for example, was expressed when rats were fed a high protein ration (Wolffram et al. 1984).

3.3.3 Changes in the Electrochemical Gradient for Na^+

The magnitude of the electrochemical gradient for Na^+ is an important component of the total driving force for active solute movement across the brush border membrane (Carter-Su and Kimmich 1980). Changes in this gradient might have partially specific effects, affecting any Na^+-coupled nutrient transport such as for sugars and amino acids. The mutual dependence of sugar and amino acid transport on the Na^+ gradient is evidenced by the cross-inhibition of uptake between sugars and amino acids which has been observed in intestinal tissue of mammals (Munck 1981), chickens (Tucker and Kimmich 1973), and fish (Albus et al. 1983).

One example of adaptation of intestinal transport possibly accounted for by changes in the electrochemical gradient for Na^+ is the increased brush border transport of both sugars and amino acids which occurs in diabetes (reviewed in

Karasov and Diamond 1983b). Debnam et al. (1988) found that in jejunum of rats 7 days diabetic, the brush border membrane was hyperpolarized. Carrier-mediated (i.e., Na^+-dependent) glucose uptake in vitro was enhanced as was total L-proline uptake at a low concentration where uptake was largely (79%) Na^+-dependent. But diabetes was without significant effect on total proline uptake at a high concentration where uptake was predominantly (79%) Na^+-independent. These observations are in accord with the suggestion by Hopfer (1975) that a decrease in the Na^+ conductance of the brush border membrane in diabetic rats may result in an increase in the electrochemical gradient of Na^+ across that membrane.

As yet, there are no examples of differences between species in Na^+-dependent uptake which are accounted for by this mechanism. Mammals may have slightly higher sugar and amino acid uptake rates at the tissue level than fish (Sect. 4.2.2). Baseline potential differences across the mucosal membrane (V_m), however, appear to be more hyperpolarized in goldfish (−53.9 mV; Albus et al. 1983) than rats (−45.5; Debnam and Thompson 1984; Debnam et al. 1988).

3.3.4 Changes in Membrane Lipids

A change in membrane lipids affecting membrane fluidity might affect passive permeability or carrier-mediated transport (by affecting translocation rates of nutrient-carrier complexes). There are no well-studied examples of adaptation of intestinal nutrient transport by this mechanism, though it may modulate sugar transport in plasma membranes (Pilch et al. 1980). In fish decreases in acclimation temperature (Smith and Kemp 1971) and increases in salinity (Leray et al. 1984) have been found to cause an increased degree of unsaturation and (for the salinity case) increased membrane fluidity in intestinal membrane. Increased unsaturation was correlated with increased apical uptake of Na^+ and Cl^- (DiCostanzo et al. 1983). Smith and Kemp (1971) suggested that in temperature-acclimated goldfish basolateral efflux of amino acids was affected by changes in fatty acid composition of the membrane.

3.3.5 Changes in Enterocyte Differentiation

As enterocytes migrate up the villi from the crypts they mature and express progressively more transport function. Thus, sugar (Kinter and Wilson 1965), amino acid (King et al. 1981), and dipeptide (Cheeseman 1986) uptake is achieved mostly by epithelial cells that are closer to the villus tip. Details of enterocyte maturation have been studied in a number of mammalian species (Smith et al. 1984), as well as the role of changes in maturation in adaptation of intestinal nutrient transport to change in diet (Smith 1985; Cheeseman 1986), temperature (Smith 1985), and resection (Menge et al. 1983).

When rats were fed low protein rations, cell migration rates up the villus were slower but enterocyte life span and the time needed for an enterocyte to first express the capacity to absorb amino acids and dipeptides (i.e., latency) was relatively constant (Smith 1985; Cheeseman 1986). Villus height was shorter in low protein rats but a greater proportion of the enterocytes along the villus transported amino acids and dipeptides. The underlying mechanisms causing changes in transport rates with enterocyte age may have to do with age-related changes in V_m, microvillous length, and/or rates of synthesis or insertion of new carriers (Cremaschi et al. 1982; Cheeseman 1986).

3.3.6 Changes at the Basolateral Membrane

Regulation of glucose transport at the basolateral membrane has been demonstrated directly and indirectly. Rats made hyperglycemic for at least 5 h by jugular infusions of glucose (Csaky and Fischer 1981), or which were perfused at least 4 h in the distal ileum with glucose solution (Debnam 1985), have enhanced carrier-mediated glucose absorption in jejunum. When brush border carrier-mediated glucose uptake was measured in vitro in intact jejunum (Karasov and Debnam 1987) or brush border membrane vesicles (Maenz and Cheeseman 1986) there was no significant change in uptake, suggesting that the rapid change in glucose absorption in vivo was mediated by changes at the level of the basolateral membrane. Indeed, in basolateral membrane vesicles of 5-h hyperglycemic rats glucose uptake was significantly enhanced. Further evidence for enhanced transport by basolateral transporters includes observations by Csaky and Fischer (1981) that the enhanced transepithelial transport of sugar in hyperglycemic rats was Na^+-independent and inhibited by phloretin but not phloridzin. Considering the many examples of regulation of transport by plasma membranes (e.g., Shotwell et al. 1983; Kilberg et al. 1985), one might expect to see more examples in the future of adaptation of intestinal nutrient transport mediated by changes at the basolateral membrane.

3.4 Proximate Signals for Adaptation

In the discussion so far I have mentioned that intestinal transport changes within a species with numerous conditions such as pregnancy and lactation, diet, temperature, diabetes and hyperglycemia, and resection. What are the steps involved in connecting changes in these factors with changes in nutrient transport? There are no examples of intestinal adaptation for which we know the answer to this question. The two most likely proximate signals for adaptation appear to be the transported substrates themselves or intermediate signals such as hormones (Karasov and Diamond 1983b).

3.4.1 The Solutes Themselves

Some of the best evidence that a transported solute can act as a signal for its own regulation is provided by studies on the regulation of glucose transport in cultured kidney cells (LLC-PK_1). These cells transport glucose in a Na^+-dependent fashion, as do intestinal cells. The concentration of glucose in the growth medium regulates the Na^+-coupled hexose transport (Moran et al. 1983, 1984). Metabolizable but nontransported sugars, such as fructose or mannose, mimic the regulatory effect of glucose, though not to the same extent as glucose, whereas the nonmetabolized sugar alpha-methyl glucose does not. Thus, the regulatory signal is not a direct effect of the hexose on its carrier, but involves hexose metabolism.

This conclusion is somewhat similar to the situation in rat small intestine where luminal or dietary glucose stimulates hexose transport more effectively than does fructose, but nonmetabolizable analogs apparently do not stimulate hexose transport in a specific manner (Bode et al. 1981; Debnam and Levin 1976; Roy and Dubois 1972; Richter et al. 1983; Kotler et al. 1981; Debnam 1985). In mice, replacement of dietary glucose with nonmetabolizable 3-O-methylglucose caused hyperphagia and mucosal hypertrophy thereby nonspecifically increasing uptake of sugars and amino acids, but only metabolizable sugars induced the brush border glucose carrier(s) specifically and fructose was the best inducer of brush border fructose transport (Solberg and Diamond 1987).

While only metabolizable sugars appear to induce the brush border glucose carrier, the same may not be true for the basolateral carrier. Nonmetabolizable 3-O-methylfructose apparently induced basolateral fructose transport in rats when it was infused into the jugular vein for 5 h (Casky and Fischer 1984).

A stimulation of mucosal growth occurs following intestinal infusion of certain amino acids and amino acid mixtures (Spector et al. 1977, 1981) and induction of amino acid transport by amino acids has been studied recently. Studies have used parenterally nourished rats whose intestines were perfused with amino acids (Levine 1986) and mice fed rations differing in amino acid composition (Stein et al. 1987). Because there are several amino acid carriers, one question has been whether amino acids induce their own transport. In mice, but not in rats, the acidic amino acid aspartate induced its own transport and in rats valine induced its own transport. Dietary proline did not induce proline transport in mice. In these studies there were many examples of amino acids of one class inducing the transport of amino acids of another class; e.g., lysine was a good inducer of the transport of acidic, basic, and neutral amino acids in the rat, and aspartate was a good inducer of lysine transport in the mouse.

For most of these examples of induction of intestinal nutrient transport by a luminal or dietary substrate one cannot rule out hormones or nerves as mediating signals. A possible exception is the induction of fructose transport by dietary fructose. The body rapidly converts fructose into glucose yet dietary fructose stimulated fructose transport more so than did dietary glucose in both mice (Solberg and Diamond 1987) and rats (Mavrias and Mayer 1973; Bode et al. 1981).

3.4.2 Hormonal Signals

Evidence that nervous or hormonal signals are involved in regulation of absorption rates comes from studies in which changes in mucosa structure or function are observed in the absence of direct contact with luminal nutrients. Examples of this include: (1) intestinal mucosal growth was stimulated in one member of a parabiotic pair of rats when the other member underwent small bowel resection (Williamson et al. 1978; Williamson and Malt 1981); (2) infusion of nutrients into the ileum of rats led to changes in proximal sections of the intestine such as hyperplasia when amino acids were infused (Spector et al. 1977) and increased carrier-mediated glucose absorption when sugars were infused (Debnam 1985); (3) hyperplasia and/or enhanced glucose transport have been observed in Thiry-Vella fistulas of lactating rats (Elias and Dowling 1973) and of rats with intestinal resections (Hanson et al. 1977).

Experimental administration of some hormones has been found to alter intestinal structure and/or function with times scales ranging from fractions of an hour to several weeks. Intravenous administration of somatostatin reduced glucose and amino acid absorption in vivo in humans within an hour, and adsorption rates returned to pretreatment levels within an hour after administration ceased (Krejs et al. 1980). The underlying mechanism was not determined. Studies with dogs have also suggested that absorption rates are inhibited following somatostatin administration (Ross and Shaffer 1983), but other in vivo and in vitro studies in rats showed no effect (Daumerie and Henquin 1982; F. A. Wilson et al. 1980). One unresolved issue is what range of doses is in the physiological, as opposed to pharmacological, range.

The effect of chronic glucagon administration for 2 – 3 days on nutrient transport has been studied in several laboratories and has been found, with some exceptions (Lorenz-Meyer et al. 1977) to result in increased glucose transport (Rudo and Rosenberg 1973; Debnam 1982a; Caspary and Lucke 1976; Thompson and Debnam 1986). Thompson and Debnam (1986) noted that one cannot exclude the possibility that the increase is a consequence of glucagon-induced alterations in insulin secretion. A possible underlying mechanism is an increased electrochemical gradient for Na^+-coupled nutrient transport, since Thompson and Debnam (1986) have observed a hyperpolarization of the jejunal brush border membrane of rats treated with glucagon for 2 days. They also found glucagon administration to influence the expression of nutrient absorption during the development of enterocytes.

Hormones might affect nutrient absorption through hypertrophic effects on intestinal mucosa, but with a probable time course exceeding 2 or 3 days (Karasov and Diamond 1983b). Gastrin has been much studied in this regard. A general correlation exists among decreased serum gastrin levels (Johnson 1976), mucosal atrophy (Johnson 1976), and decreased intestinal transport in fasted or parenterally fed animals (Karasov and Diamond 1983b). More direct evidence for a stimulatory effect of gastrin on nutrient absorption comes from recent studies by M. Z. Schwartz and Storozuk (1985, 1986). They found that chronic (2 weeks) administration of gastrin-17 either systemically or luminally in rats enhanced galactose and glycine absorption cm^{-2} nominal area small intestine.

Other hormones for which a humoral effect on intestinal structure and/or function has been suggested include cholecystokinin (Hughes et al. 1981), secretin (Weser et al. 1981), insulin (Caspary 1973; Olsen and Rosenberg 1970), prolactin (Campbell and Fell 1964), epinephrine (Kinzie et al. 1976), prostaglandin E_1 (Nassar et al. 1982), and androgenic and estrogenic steroids (Habibi et al. 1983; Habibi and Ince 1984). For most of these the evidence for a direct effect is equivocal. There are no well-studied examples of effects of nerves on sugar or amino acid transport.

4 Patterns of Adaptation

Having examined the pathways for sugar and amino acid absorption and some of the mechanisms and signals involved in alteration of absorption rates, I now review some of the patterns of adaptation within and between species of vertebrates. With regards to variation within a species, the emphasis of this section will be on changes which might occur in the natural life of animals. Effects of experimental manipulations such as intestinal resection or drug-induced diabetes are reviewed elsewhere (Karasov and Diamond 1983 b, 1986). Here, I shall evaluate transport adaptations within the context of the feeding and nutritional ecology of animals and their environmental physiology. In the former case I consider transport adaptations to the nutrient composition of the diet and how much food is eaten per day, and in the latter case transport adaptations to changes in temperature and (for aquatic organisms) salinity.

4.1 Adaptation to Dietary Composition

4.1.1 Carnivores vs Omnivores/Herbivores

That diet composition can affect digestive tract function has been known at least since the time of Pavlov, who demonstrated induction of digestive enzymes by dietary changes in their substrate levels. Analogously, differences between species in some digestive enzymes may also be related to diet. In fish, for example, at least six comparative studies have found that amylase activity mg^{-1} pancreas is higher in species which are herbivorous or omnivorous compared with those which are carnivorous (reviewed in Kapoor et al. 1975).

A parallel adaptation apparently occurs for nutrient transport. Karasov et al. (1985 a) tested for intestinal transporter specialization in vertebrates with different dietary habits by comparing ratios of glucose uptake to proline uptake measured at concentrations which saturated the carriers. A virtue of this approach is that it bypasses entirely the issue as to what measure of intestinal tissue nutrient uptake should be normalized (e.g., to cm length, cm^2 nominal area, mass, or protein content) when making comparisons between species. This ratio was independent of body mass, but averaged more than three times higher in omnivores/herbivores than carnivores. Figure 1 presents a similar analysis based on many more species using the inverse ratio. Results of several other studies are also indicative of lower glucose transport rates in carnivores: Ferraris and Ahearn (1983) for carnivorous and herbivorous fish, Karasov and Diamond (1981) for carnivorous and herbivorous lizards.

Most of the variation in glucose/proline ratios arises from species differences in glucose uptake (Karasov et al. 1985 a). This probably relates to the fact that

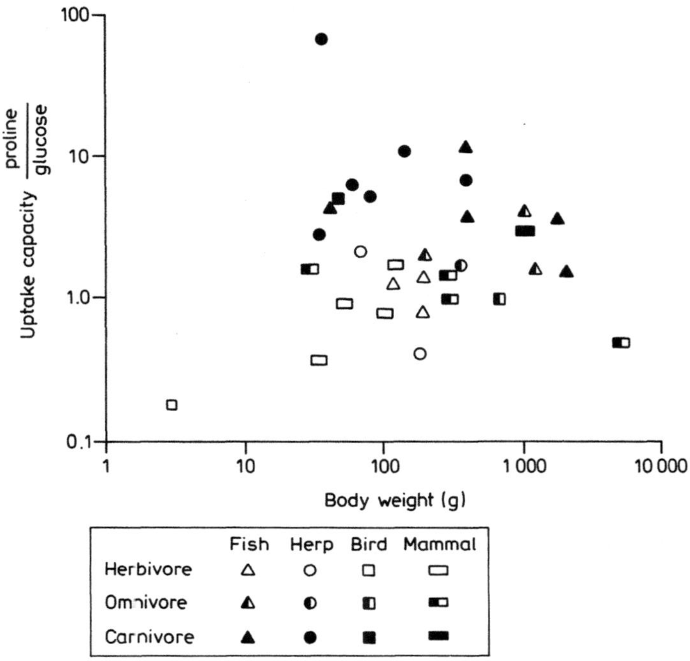

Fig. 1. Relative rates of sugar and amino acid transport in herbivorous, omnivorous, and carnivorous species of all five higher vertebrate classes. Each species cate its natural diet or else one of similar nutrient composition to the natural diet. The ***ordinate*** is the ratio of the intestine's total uptake capacity for proline to its uptake capacity for glucose. Uptake capacity was calculated by measuring solute uptake cm^{-1} at carrier-saturating concentrations at several positions along the small intestine and then interpolating uptake rates linearly between successive positions and summing over the length of the small intestine. The proline/glucose ratio is highest for carnivores, and this trend holds true in all five higher vertebrate classes. *Herp*, amphibia plus reptiles (Diamond and Buddington 1987)

since carbohydrates are nonessential nutrients only herbivores and omnivores which normally eat them maintain a substantial capacity to absorb them. Amino acids, in contrast, are required by all animals and all maintain substantial capacity to absorb them.

Most of the animals represented in Fig. 1 were fed their natural diets, or ones of similar nutrient composition. Do these differences in relative glucose and amino acid uptake rates reflect a genotypic influence or are they simply phenotypic adaptations to the differing amounts of carbohydrate in the diet? Even if animals with different natural feeding habits were maintained on the same diet in the laboratory, could the diets they ate early in life somehow "fix" for their lifetimes their rates of nutrient absorption? In the following sections I review these three types of adaptation.

4.1.2 Phenotypic Adaptation to Diet Composition

Adaptation of Sugar Transport to Dietary Carbohydrate. Phenotypic adaptation to dietary composition has been studied most extensively in laboratory rodents.

In rats and mice intestinal monosaccharide absorption, whether measured in vivo (Hahn and Koldovsky 1966; Ginsburg and Heggeness 1968) or in vitro (Bode et al. 1981; Diamond and Karasov 1984; Wolffram and Scharrer 1984) increased monotonically with dietary carbohydrate content (Diamond and Karasov 1987). Evidence that this enhancement is specific for carrier-mediated monosaccharide uptake at the brush border comes from studies with mice (Diamond and Karasov 1984). In adjacent tissues from the same mice passive influx of L-glucose was unchanged and uptake of L-proline was either unchanged or decreased depending on the particular diet. The increased glucose uptake was due to induction of glucose transporters (Ferraris and Diamond 1986) and occurred within 24 h of a switch in diet (Diamond and Karasov 1984). As discussed above, the signal for dietary adaptation to carbohydrate might be the concentration of the transported monosaccharides themselves, measured in the gut lumen, enterocyte, or blood.

Adaptation of Amino Acid Transport to Dietary Protein. Amino acid transport is responsive to dietary protein content. In mice and rats fed high protein rations absorption of amino acids was enhanced whether measured in vivo (Lis et al. 1972; Sugiyama et al. 1983) or in vitro (Scharrer et al. 1981; Karasov et al. 1983, 1985c, 1987). In mice the response was specific for amino acid transport and occurred within 24 h of a switch in diet (Karasov et al. 1983). The brush border transporters for imino and acidic amino acids appeared to be enhanced to a greater extent than the transporters for neutral and basic amino acids (Karasov et al. 1987).

There have been at least a dozen studies of amino acid transport in animals fed low protein or protein-free rations (reviewed in Karasov et al. 1987) and these rations have variously been reported to increase, decrease, or not to change amino acid absorption. The inconsistency is due partly to differences among studies in methods used for measuring absorption, partly to differences in the amino acids selected, and in the use of only one or two amino acids and no other solutes in most studies, and perhaps largely to the complications introduced by rations deficient in total nitrogen (N) and essential amino acids. These rations generally lead to mobilization of intestinal protein to satisfy amino acid requirements and thus also to changes in intestinal mass and structure (Syme 1982). Thus, much of the apparent conflict may reflect different authors' choices of different bases of normalization, as is the case for the apparent conflict regarding the intestinal response to fasting or starvation (Karasov and Diamond 1983b). When rations deficient in total N but not in any essential amino acid were fed to mice for up to 2 weeks, the mice did not undergo any significant decrease in total intestinal mass, mucosa mass, or nominal surface area (Karasov et al. 1987). In this study there was no change in brush border uptake of three amino acids after 4 days of N deficiency, but after 14 days the uptake of nonessential imino and acidic amino acids was depressed, whereas uptakes were maintained or even slightly enhanced for essential neutral and basic amino acids and a neutral amino acid sharing the neutral carrier.

Functional Significance of Dietary Adaptation of Transport. Diamond and Karasov (1987) have reviewed the contrasting patterns for effects of diet on in-

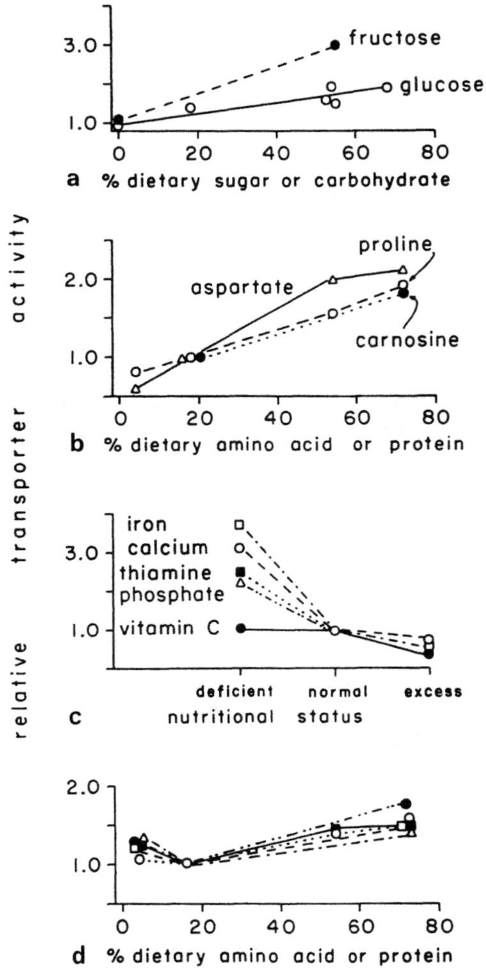

Fig. 2a–d. Summary of how activities of intestinal brush border nutrient transporters vary with dietary substrate levels or nutritional status. *a* Fructose or glucose transport as a function of dietary fructose or carbohydrate level. *b* Aspartate, proline, or carnosine transport as a function of dietary level of casein or of an amino acid mixture corresponding to a complete hydrolysate of casein. *c* Brush border receptors for iron in guinea pig proximal intestine, calcium uptake in chick duodenum, thiamine uptake in rat jejunum, phosphate transport in rat jejunum, and vitamin C uptake in guinea pig ileum, on rations with the given nutrient at inadequate levels, normal levels, or in excess. *d* As in *b* but for histidine (● –·–·–), lysine (○ ...), leucine (■———), alanine (□ – – –), and methionine (▽ –·–·–·–). Sources for all of these data are given in Diamond and Karasov (1987). All panels except *c* are based on transport rates measured at carrier-saturating concentrations. Ordinate values are *relative* transporter activity: i.e., transport in mice eating a ration with the substrate level shown on the *abscissa,* divived by transport in mice eating a ration with some standardized substrate level. The standardized level was a fructose- or carbohydrate-free ration in *a*, a 18% casein or 18% amino acid ration in *b* and *d*. Panel *c* gives transport in animals on a ration with the given nutrient in excess or at inadequate levels, divided by transport in animals with the nutrient at adequate levels. The measure of transport used in each case is total uptake, except that only the active component of D-glucose uptake is shown. Most transport was carrier-mediated, as shown by Na^+ dependence or a saturable uptake/concentration relation.
The depicted trends in transporter activity with dietary substrate level or nutritional status are all significant at the $P < 0.05$ level or better, except that the differences between vitamin C transport on a deficient and normal diet, and between transport of the five amino acids shown in *d* on a 4 and 18% amino acid diet, are not significant (Diamond and Karasov 1987)

testinal uptake of sugars, dipeptide and amino acids, minerals, and vitamins. The pattern of variation varies greatly among solutes (Fig. 2). With increasing dietary substrate level, uptake may increase (Fig. 2a, b), decrease (Fig. 2c), go through a minimum (Fig. 2d), or initially remain constant and then increase (Fig. 2d). Diamond and Karasov (1987) identified trade-offs among four factors which may set these patterns (in an evolutionary sense): (1) caloric pay-off from uptake of energy-yielding nutrients; (2) fixed daily requirements for some essential nutrients; (3) toxicity of some nutrients at high levels; and (4) the costs (biosynthetic and other) of synthesizing and maintaining nutrient transporters. In combination with this last factor, factor 1 should tend to result in a substrate up-regulating its transporter, while factors 2 and 3 would tend to result in down-regulation.

Adaptation of sugar transport to dietary carbohydrate fits one of these cases. Since carbohydrate is nonessential and nontoxic but is a source of energy, its uptake should vary directly with dietary substrate level (cf. Fig. 2a and above discussion). The reasoning is that the costs of adding transporters on a high carbohydrate ration would be more than met by the resulting additional calorie absorption, while decreased absorption of nonessential carbohydrate on a low carbohydrate ration would be safe and would save transporter costs.

Adaptation of essential yet potentially toxic nutrients not used for energy (e.g., vitamins and minerals) fits another of these cases. Their transport should vary inversely with dietary substrate level (cf. Fig. 2c). The reasoning is that the transporters would be most needed when the essential nutrients were in short supply, but would be less needed at high dietary levels at which absorption by passive diffusion could satisfy nutritional requirements (or at which the same amount could be extracted with lower efficiency by a lower number of transporters). Down-regulation at high dietary levels would in addition protect the animal against risk of intoxication.

For amino acid and peptide transporters the predictions from some factors conflict with predictions from other factors. A positive relation between amino acid or peptide transporter activity and dietary substrate level is expected because amino acids and peptides yield calories; more absorption yields more calories. An inverse relation is expected because all amino acids and peptides are collectively essential as a nitrogen source and because some amino acids are specifically essential. Interestingly, regulatory patterns for amino acid absorption appear to combine elements of the patterns seen for both sugar transport and for mineral and vitamin transport, especially in the case of essential amino acids or those nonessential amino acids sharing their carriers (e.g., nonessential alanine in Fig. 2d).

4.1.3 Genotypic Adaptation to Diet Composition

If all species included in Fig. 2 were capable of regulation of transporter activity according to dietary substrate level, as in the above examples, phenotypic regulation would account for most of the observed pattern. As a test for genotypic regulation Buddington et al. (1987) studied eight fish species including herbivores, omnivores, and carnivores which were fed the same ration containing

45% protein and ca. 25% digestible carbohydrate. The ratio of glucose uptake to proline uptake was highest in the herbivores and lowest in the carnivores. Thus, in these fish species there has apparently been a natural selection for relative glucose and amino acid transport rates that correspond to the relative amounts of carbohydrate and protein in each species' diet in nature.

One might also expect that the genetic capacity for phenotypic adaptation would be a possible outcome of natural selection. So far, there have been published accounts of phenotypic regulation of intestinal glucose transport for omnivorous mammals (mice and rats) and fish (Buddington 1987). It will be interesting to learn in future studies whether most omnivores share this ability, and whether obligate carnivores have the ability to regulate glucose transporter activity.

4.1.4 Developmental Adaptation to Diet

Many species undergo developmental changes in diet and it is possible that nutrient transport changes in parallel, either as a result of dietary induction or according to a genetically fixed schedule. Around the time of weaning in ruminants, for example, intestinal glucose transport activity becomes rudimentary (Scharrer 1976). The regression of glucose transport activity from higher levels during nursing is specific (i.e., does not affect amino acid transport; Scharrer 1975), and parallels the transition from a milk-based diet high in monosaccharides to a plant-based diet which is fermented in the rumen and thus releases little monosaccharide to the intestine. Thus, adult ruminants likely are exceptions to the rule that herbivores and omnivores have relatively high glucose absorption rates (Wolffram et al. 1986), but they represent an exception which "proves the rule".

Buddington et al. (1987) tested whether a developmental diet change in a fish was associated with a genetically programmed change in nutrient transport. The fish was the monkeyface prickleback (*Cebidichthys violaceus*) which during development undergoes a transition from carnivory to herbivory in the wild (Montgomery 1977). In the laboratory when they were fed the same carbohydrate-containing ration throughout development, small fish had relatively low glucose uptake rates typical of carnivores, whereas large fish had relatively high glucose uptake rates typical of herbivores. Thus, the shifts in glucose uptake in the fish demonstrate genetically programmed changes.

Karasov et al. (1985c) tested whether in mice there might be any irreversible effect of diet during an early critical period of development on nutrient transport activity in the adult, but found no effect.

4.2 Adaptation to Level of Intake

4.2.1 Phenotypic Adaptation to Intake Level

Hyperphagia. Hyperphagia (increased food intake) occurs under a variety of conditions. One of the best studied situations with respect to nutrient absorption is the increased food intake during lactation. In laboratory mice, for example, intake increased as much as four times during lactation (Studier 1979). In spite of the much greater nutrient flux through the digestive tract lactating mice were able to extract a similar proportion of energy from the food as nonreproductives (Studier 1979). I have already discussed how mice and other rodents display a nonspecific increase in the absorption rate cm^{-1} intestine for many nutrients during lactation, with an increase in absorptive surface area (villus are per cm length; Fell 1972) as the proximate mechanism.

Another natural situation in which hyperphagia occurs and for which the same kind of nonspecific increase in absorption rate may occur is the increased food intake in homeothermic animals exposed to cold temperatures. Rats and voles maintained at 5 °C approximately doubled their food intake and exhibited intestinal hyperplasia (Jacobs et al. 1975; Gross et al. 1985), and enhanced glucose absorption was noted in the rats.

In these examples hyperphagia occurs in response to increases in the demand for energy and possibly certain nutrients. Many animals can also be made hyperphagic through dilution of the metabolizable calories in their ration with nonnutritive fillers such as fiber, kaolin, or nonmetabolizable substrates. These experimental manipulations might be considered analogous to the more natural situation for plant-eaters of calorie dilution by increasing levels of plant cell wall content, or for omnivores of switches between "rich" and "poor" diets (e.g., Sibly 1981). But the situation is complicated by the possibility of direct effects of the nonnutritive fillers on intestine structure and/or function. For example, fiber, even at low dietary levels (5 – 10% of diet dry mass), has been reported to reduce absorption of glucose (S. E. Schwartz and Levine 1980) and leucine (Freeman 1984).

There is no clear picture, yet, of what effect hyperphagia due to calorie dilution has on intestinal structure and function. In rats, calorie dilution with kaolin in amounts causing increases in food intake of three to four times has been variously reported to have no effect (Dowling et al. 1967) or to cause enlargement (Kennedy and McCance 1958) of the small intestine. In voles calorie dilution with cellulose almost doubled food intake but had no effect on small intestine length or dry mass (Gross et al. 1985). In mice diet calorie dilution with the poorly or nonmetabolized substrates galactose and 3-O-methylglucose doubled food intake and caused increases in intestinal length and mucosa mass (Solberg and Diamond 1987). Several species of omnivorous birds exhibit increases in food intake and intestine length when switched from rich to poor diets (Sibly 1981).

Increases in sugar absorption rate in parallel with the hyperplasia occurred in the rats (Dowling et al. 1967) and mice (Solberg and Diamond 1987). However, since retention time of digesta in the small intestine does not necessarily decrease when the ration is diluted (e.g., Dowling et al. 1967; Savory and Gentle 1976), the nutritional significance of these increases in absorption rate are not apparent (Karasov and Diamond 1983 b).

Fasting and Semistarvation. The literature on effects of fasting (total removal of food) and semistarvation (feeding submaintenance amounts of foods) is voluminous and has yielded conflicting results. Several detailed reviews of this literature are available (Karasov and Diamond 1983b; Levin 1984). Most studies of intestinal structure concur that apparent absorptive surface area is reduced due to decreased enterocyte number and villous height and this is reflected in lower intestinal weight per cm intestine. As might be expected, most well-designed studies measuring absorption cm^{-1} or over the entire intestine have shown decreased monosaccharide absorption after a fast of several days (Karasov and Diamond 1983b).

This may not, however, reflect what is happening in individual enterocytes. There is evidence that the migration rate of enterocytes up the villous is decreased (Brown et al. 1963; Dauncey et al. 1983) and this might be expected to make individual enterocytes hypermature on the villus (Williamson 1982). Support for this idea comes from the finding of Debnam and Thompson (1984) that fasting for 1 or 3 days resulted in an increased V_m of enterocytes from both jejunum and ileum of rat intestine, and that the depolarization induced by the addition of either D-galactose of L-valine to the mucosal fluid was enhanced by fasting. The depolarization might be an indirect measure of Na^+-coupled transport across the brush border membrane of the individual enterocyte. Dauncey et al. (1983) measured by autoradiography an increased Na^+-dependent alanine uptake by enterocytes from pigs maintained on a restricted diet.

Thus, there may be enhanced transport at the level of the individual enterocyte in fasted or semistarved animals. This may be why when corrections are made for the overall decrease in surface area, and uptake is thereby normalized to villous surface area as done by Levin (1984) for chicken intestine, transport appears enhanced in fasted animals. But whatever changes are occurring at the level of the individual enterocyte, they usually are not quite large enough to compensate for the overall reduction in number of enterocytes since absorption considered from the perspective of the whole animal is usually found to be decreased. Still other phenomena adding to the complexity of the fasting/semistarvation model include apparent differential responses of different regions of the intestine and of the individual transporters for monosaccharides and amino acids (Debnam 1982a; Levin 1984).

4.2.2 Genotypic Adaptation to Level of Intake

If we were to predict the daily energy or nutrient requirements of an animal, the two most important things to know would be its taxa and its body size. Within each class of vertebrates requirements are generally an increasing function of body size, and endothermic mammals and birds exceed ectothermic fish, amphibians, and reptiles by about an order of magnitude.

There are several ways larger animals or endotherms could achieve higher rates of nutrient absorption. They could ingest food at the same rate as smaller ones and ectotherms but digest it more efficiently. But whole-animal feeding trials fail to support this idea (cf. Karasov et al. 1986b). Similarly, the extraction efficiency of the small intestine when measured in the intact animal is fairly similar among animals of differing body sizes from the various vertebrate classes. In fact, in man and domestic animals practically all of the dietary soluble carbohydrate and protein is digested and absorbed by the time the digesta contents reach the ileum or the distal portion of the gut. Extraction efficiencies for amino acids from soybeans averaged 90% in rainbow trout (Dabrowski and Dabrowska 1981) and 93% in chickens (Achinewhu and Hewitt 1979); for glucose and sucrose >95% in fish (NRC, 1983), 97.6% in pigs (Smabrook 1979), and ≥97% in nectarivorous birds (Karasov et al. 1986a). Thus, higher nutrient absorption rates are not achieved by higher extraction efficiencies and must be achieved by ingestion of more food.

Are higher ingestion rates met by a higher absorptive capacity of the intestine? A priori we can list several ways in which the intestine could achieve higher absorptive capacity and thereby make possible increased feeding rates: (1) higher passive permeability to nutrients; (2) higher affinity carriers (lower K_m^*); (3) more active transport sites (as reflected by higher V_{max}); (4) greater intestinal surface area at the macroscopic and microscopic level; (5) the advantage of operating the intestine at high temperatures all day (as in most mammals and birds) or part of the day (as in reptiles, amphibians, and some fish) rather than at low temperatures (as in most fish). Inasmuch as I have already found no major trends in glucose passive permeability and K_m^* for glucose and proline among ectotherms and endotherms of differing sizes (Sect. 3.1.3), and consider it difficult anyway to establish unambiguously differences between species in K_m^*, I shall focus on the latter three factors affecting absorptive capacity.

Relationships Between Nutrient Transport, Body Size, and Taxa. Karasov et al. (1985b) viewed the apportioning of intestinal nutrient uptake capacity between physiological, environmental, and anatomic causes as follows:

$$J_{s,T} = (J_T/X)(X) \quad , \qquad (2)$$

where $J_{s,T}$ is the uptake capacity of the small intestine at a given temperature, T, J_T/X is the uptake rate per unit intestine (e.g., cm length or cm^2 nominal area) at the temperature, and X is an anatomic measure of the small intestine. $J_{s,T}$ was measured using the everted sleeve method (Karasov and Diamond 1983a) by measuring solute uptake cm^{-1} at saturating concentrations at several positions along the small intestine and then interpolating uptake rates linearly between successive positions and summing over the length of the small intestine.

An index of the total absorptive capability of the intestine for sugars and amino acids was calculated by adding together for each species the summed uptake rates for glucose and proline (Fig. 3a). When comparing mammals, reptiles, and fish (Karasov et al. 1985a, b; Karasov 1987) summed uptake rates for glucose and proline were found to be an increasing function of body size. The relations are usefully described by allometric equations of the general form $Y = aX^b$ and are linear when plotted on logarithmic coordinates. The small differences in slope were not found to be significantly different and so data for each taxa were fitted to a common slope. Comparison of the proportionality coefficients (a, the intercept at unity) showed that mammals exceeded both reptiles and fish in summed uptake by 4 and 13 times, respectively.

What is the basis for differences in summed uptake within and between taxa within the context of Eq. (2)? To answer this question Karasov (1987) compared

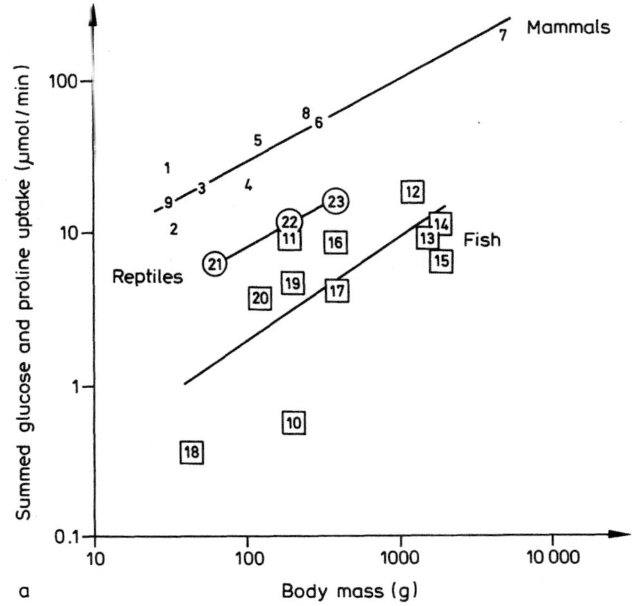

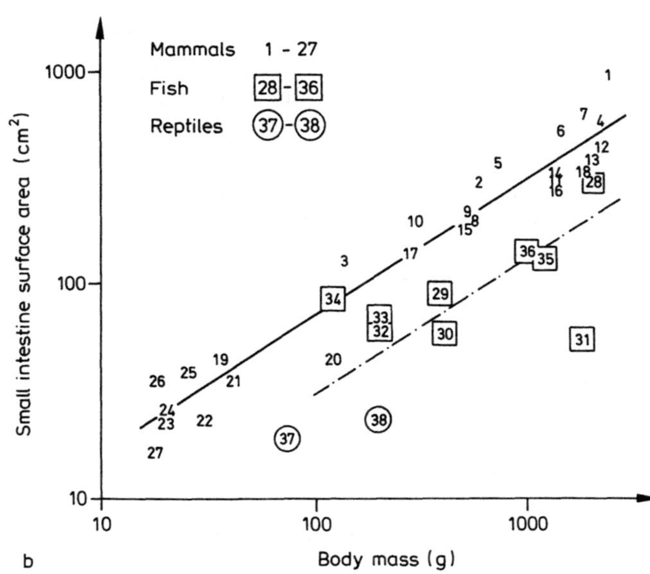

Fig. 3 a, b. Relationships between body mass and the sum of glucose and proline uptake capacity (*a*) and nominal surface area of the small intestine (*b*) in fish, reptiles, and mammals. *Numbers* in both figures are species designations given in Karasov (1987). In *a* for each species uptake capacity for both solutes was determined at either 25 or 50 mM. The relationships do not differ significantly in slope. When fitted to a common slope (0.57) the calculated proportionality coefficients are: fish, 0.17; reptiles, 0.58; mammals, 2.15. In *b* surface areas for fish include pyloric caeca when present. Data for each taxa are fitted to a common slope (0.63) and the calculated proportionality coefficients are: fish, 1.68; mammals, 3.94. (Both figures from Karasov 1987)

among the species brush border uptake rates for glucose and proline for the section of the intestine where uptake was greatest. Plots of uptake cm^{-2} nominal area as a function of body size indicated that there was no dependence of tissue-specific transport activity on body size for either glucose or proline. Uptake rates of reptiles did not differ significantly from those of mammals for either glucose or proline, but uptake rates for proline and glucose by fish intestine were significantly below those in mammals and reptiles. But the factorial difference between mammals and fish in uptake cm^{-2} (ca. $3-5\times$) did not account for the 13-fold difference in summed uptake between mammals and fish. Uptake in fish had been measured at 20 °C, whereas in mammals and reptiles at 37 °C. When uptakes were measured at the same temperature, there was little difference between the groups.

If there are no differences in uptake cm^{-2} (i.e., J_T/X) among different sized species within a taxa, and only small differences in uptake cm^{-2} between these taxa, presumably the differences in summed uptake can be attributed to differences in surface area [i.e., X in Eq. (2)]. Indeed, nominal intestinal surface area does increase with increasing body size within a taxa, and mammals of a given size do significantly exceed fish and reptiles of the same size in nominal surface area (Fig. 3b).

Thus, at least three differences in intestinal transport between endothermic mammals and ectothermic reptiles and fish contribute to higher summed uptake rates in mammals. First, mammals may exceed fish in uptake cm^{-2} at 20 °C, but only by a factor of about two. Second, fish intestine operates at a lower temperature than mammal intestine all day, and reptile intestine usually for part of the day (during inactivity when they don't bask). Third, mammals appear to have about four times more intestinal nominal surface area than fish and reptiles, though this conclusion was based on relatively few data (Karasov 1987). Increased surface area also accounts for the increase in summed uptake with body size within a taxa. While differences in surface area between the groups can be further partitioned into differences at the villus and microvillus level (cf. Table 3), those differences do not affect this analysis because their effect on uptake rate was taken into account in the measures of uptake cm^{-2} nominal area.

The increased nutrient requirements associated phylogenetically with increasing body size or the evolution of endothermy involve increased requirements for all nutrients. This increase is met by the evolution of more intestinal tissue, the simplest solution to the problem of absorbing more of everything.

4.3 Adaptation to Environmental Conditions

4.3.1 Temperature

The response of the intestine to changing temperature can be studied by altering incubation temperature during experimental measurements of transport, and by altering the rearing (i.e., acclimation) temperature of animals. One might think that for homeothermic mammals and birds the former manipulation has little physiological significance. But, quite a few species in these taxa display significant changes in body temperature on a daily basis (e.g., nocturnal hypothermia; hyperthermia in desert animals). In the case of ectothermic fish, amphibians, and reptiles, daily change in body temperature can be routine. Of course, long-term changes in temperature can be a normal part of life for certain species from any of the vertebrate classes.

The transport response of intestinal tissue to change in incubation temperature can be expressed as a Q_{10} value (rate change with a 10 °C increase). As the following examples show, Q_{10} values for nutrient transport in intestine determined in vitro are close to expected values for processes such as diffusion (Q_{10} ~1.4) and for metabolic processes (Q_{10} ~2 – 3). Thomson and Dietschy (1980b) estimated the Q_{10} for passive glucose uptake across rabbit jejunal brush border membrane to be 1.2. They also pointed out that the USL can affect the Q_{10} determination for active transport when it is measured in the presence of passive flux. Q_{10} estimates based on direct measures of carrier-mediated D-glucose uptake (i.e., in the absence of passive uptake) range between 1.8 – 3.4 in fish, amphibians, reptiles, and mammals (Thomson and Dietschy 1980b; Groot et al. 1983; Karasov et al. 1985a, b). Q_{10} values of hibernating and normothermic Belding's ground squirrels were not significantly different (Karasov unpubl data).

Carrier-mediated D-glucose uptake in reptiles and mammals increased with increasing incubation temperature to a point (the optimum) and then fell (Karasov et al. 1985b). In woodrats, box turtles, and desert iguanas the temperature optimum was 45° – 50 °C, exceeding the normal body temperature or preferred body temperature. Above this temperature the integrity of the intestine as a barrier begins to break down (e.g., Shapiro et al. 1986).

Acclimation temperature has been found to result in structural changes in mammal and fish intestine with resulting nonspecific and sometimes specific effects on nutrient transport. The simplest situation occurs for mammals. Chronic exposure of rodents to heat reduces intestinal weight and cold exposure increases it, the changes occurring in about 2 weeks (Carpenter and Musacchia 1978; Chu et al. 1979; Gross et al. 1985). In parallel with these changes, mucosal uptake of glucose by everted sacs is inversely related to exposure temperature, unless the uptake is normalized to tissue weight in which case there is no effect of temperature, or uptake per gram is higher in animals exposed to chronic heat. This latter confusing result is due to the normalization of transport to the decreased tissue weight. All of these changes probably result from the inverse relation between temperature and feeding rate in mammals and therefore represent another example of the hyperphagia phenomena discussed above.

The effect of acclimation temperature on the structure and function of goldfish intestine has been studied in several laboratories. Goldfish acclimated 12 days or more at 8° – 10 °C have longer intestines which weigh more cm^{-2} nominal area and which have greater mucosal area than goldfish acclimated at 25° – 30 °C (Kitchin and Morris 1971; Groot et al. 1983). This change is not consistent with the hypothesis of hyperphagia induced hypertrophy because in fish feeding rate is probably positively related to environmental temperature.

The change in morphology makes a comparison between intestinal transport in low and high temperature fish complicated. Glucose evoked potential changes across intestinal segments were higher from fish acclimated to low temperature at every incubation temperature (Smith 1966; Groot et al. 1983). Cold-adapted fish also had higher glucose net flux and glucose-induced current cm^{-2} nominal area (Groot et al. 1983). Groot et al. (1983) considered the basis for these changes to be an approximately threefold increase in mucosal area in cold-adapted fish, since when fluxes and currents were normalized to estimated mucosal area they were slightly higher in the warm-adapated fish. However, Smith (1966) found changes in the glucose-evoked potential 20 h after change in acclimation temperature and this seems much too rapid to be explained by macroscopic changes in gut morphology which generally occur over days (Karasov and Diamond 1983 b). This suggests other kinds of changes in intestine structure and function.

These other changes are further suggested by studies of the effect of acclimation temperature on amino acid transport. In Smith's (1970) studies of the effects of acclimation temperature on amino acid tissue/medium ratios in sacs of goldfish intestine, acclimation effects seemed specific to hydrophilic amino acids. The time course for change in serosal transfer of amino acids after change in environmental temperature (~48 h) matched that for change in percentage composition of fatty acids in mucosal phospholipids (Smith and Kemp 1971). Because acclimation temperature did not seem to affect brush border uptake of amino acids (unpubl. data cited in Smith and Kemp 1971) Smith (1983) concluded that acclimation temperature probably affected amino acid efflux across the basolateral membrane through its effect on the lipid nature of that membrane.

In summary, temperature affects nutrient transport in vertebrates over the short term through its expected effects on rate processes which are passive and active, and in mammals and fish (one species) over the long term through effects on intestinal structure at the macroscopic level and possibly at the biochemical level. I discuss below the issue of whether these changes in nutrient transport in response to acclimation temperature are necessarily "adaptive".

4.3.2 Salinity

The intestine plays a major role in water and electrolyte balance in fish. In hypertonic media, such as seawater (SW), drinking and water absorption must occur to compensate for osmotic losses through the gills. In hypotonic media, such as fresh water (FW), there is no need to absorb water through the gut. The functional differences in FW and SW are reflected in differences in electrical properties of intestine of FW fish and SW fish (Lahlou 1983). Also, there may be differences in the degree of unsaturation of membrane fatty acids affecting membrane fluidity (Leray et al. 1984). A relevant question with respect to nutrient transport is whether there are differences in the transport characteristics of the intestine.

Ferraris and Ahearn (1984) reviewed data on nutrient transport in fish and suggested that there were differences between SW and FW fish in several measures of transport including passive permeability, tissue/medium ratios, and net fluxes. In general, the evidence for differences was equivocal.

With regards to passive permeability, for example, K_d^*'s for different solutes measured by different techniques were compared for only two species of SW and two species of FW fish. Transepithelial electrical resistance was not significantly different between FW and SW species but does decrease within a species when transferred from FW to SW (Ferraris and Ahearn 1984; Collie 1985). Suggestive of a permeability difference is Collie's (1985) finding that serosa to mucosa transepithelial fluxes in short-circuited intestine of salmon were greater for both proline and glucose in SW fish. But passive proline uptake at the brush border was not significantly higher in normal SW smolts. Perhaps the difference in transepithelial permeability is due to a difference at the basolateral membrane. Reshkin and Ahearn (1987) found that passive permeability to glucose across the brush border was enhanced in SW-adapted African tilapia.

Tissue/medium ratios in FW and SW fish have rarely been measured for the same solute at the same concentration, making rigorous comparison impossible. Nonetheless, most studies have found in FW fish intestine tissue/medium ratios exceeding 1.0 for both sugars and amino acids, whereas in SW fish these ratios rarely exceed 1.0 (Ferraris and Ahearn 1984). Studies with non- or poorly metabolizable sugars and amino acids would be useful for determining whether these apparent differences are due to differences in the metabolism of absorbed solute or in backflux from cell to solution.

There is no set of measurements of nutrient flux or uptake made under similar conditions with similar techniques with which to compare different species of SW and FW fish. Using brush border membrane vesicles, Reshkin and Ahearn (1987) measured in SW-acclimated tilapia a higher Na^+-dependent glucose uptake than in FW tilapia. Collie (1985) used intact tissue but observed a similar pattern in salmon; higher unidirectional uptake of proline at the brush border in SW compared with FW salmon. Curiously, Collie found that the mucosa to serosa flux and net flux were higher in the FW salmon, and not in the SW fish which had higher rates of brush border uptake. A difference such as this might be explained by a large change at the basolateral membrane opposite in direction to that at the brush border. There was no significant difference in glucose net flux. Recall that Collie's results for serosa to mucosa flux in combination with those for passive uptake at the brush border were also suggestive of a change at the basolateral membrane. If the transition from FW to SW does alter net proline absorption in salmon, perhaps it is through the same kind of partly specific effect on basolateral efflux as suggested by Smith (1983) for the case of temperature acclimation.

In summary, there is suggestive evidence for differences in the nutrient transport characteristics of intestine from FW and SW fish. Studies on effects of salinity should be extended but comparisons need to be made systematically using uniform techniques which can discriminate effects at the brush border and basolateral membranes.

4.4 Is Intestinal Adaptation "Adaptive"?

It is clear that rates of nutrient absorption by the small intestine vary in response to many factors. Do these changes, however, really help an animal "cope" better with its new physiological state, diet, or environmental conditions? Are absorption rates nutritionally important?

Some of the aforementioned examples of variation in nutrient transport may not have an adaptive basis. It is not clear, for example, how a change in acclimation temperature or salinity alters a fish's nutritional status thereby requiring a compensatory adaptation by the intestine. In fact, poorer amino acid transport in SW fish might be maladaptive as suggested by Dabrowski et al.'s (1986) conclusion that SW trout were less efficient at the whole-animal level than FW trout at extracting amino acids from a test meal. Perhaps the apparent changes in nutrient transport characteristics of fish intestine in response to acclimation temperature or salinity are simply consequential side effects of regulated changes in membrane structure which enable cells to maintain ionic gradients in the face of changing salinity or temperature. This view recognizes that the intestine serves several functions at the whole animal level and that requisite changes in processes serving one function might affect processes serving other functions.

Other objections to the view that adaptations as discussed in this section are physiologically significant include the following (Snook 1974): (1) There may be plenty of spare capacity to absorb nutrients under normal conditions. (2) A change in transport rate might be physiologically insignificant if transport is not the rate-limiting step in the digestive process. Let us consider these objections.

As discussed above, natural selection would probably select against maintenance of excessive capacity due to the costs (biosynthetic and other) of intestinal tissue and transporters. It is cheaper to maintain the level of transport activity necessary for normal functioning, and at the same time maintain the ability for adaptive response in transport when the normal absorptive capacity is inadequate. By this reasoning, the increased absorption rates in mammals during lactation or exposure to low temperatures, or on diets with high levels of carbohydrate or protein, reflect the fact that the body's normal absorptive capacity does not constitute an infinite reserve and would be inadequate in these cases if it were not for the increase in absorption rate. More direct evidence that the intestine's absorptive capacity is not infinite and that adaptation does enable an animal to cope better following a perturbation comes from a review of nutritional studies extending over the period following intestinal resection (Karasov and Diamond 1986).

If the gut's normal capacity to digest and absorb food is exceeded, one might expect to see adaptation first or only in that step of digestion which is rate-limiting. As I have already discussed (Sect. 2.2), there are some examples where transport has been shown to be the limiting step. Also, there can be no generalizations for vertebrates as to what step of overall digestion is limiting since this probably varies with at least diet. The reality of the situation is that many steps of the digestive process (e.g., hydrolytic enzymes, transporters, motility) probably adapt in concert (Karasov and Diamond 1986). Consider the previously discussed examples of enhanced amylase activity and glucose transport activity in herbivorous compared with carnivorous fish (Sect. 4.1.1).

Finally, it is worth noting that even if the intestine normally absorbs virtually all the ingested soluble sugar and amino acid as appears to be the case for domestic animals (Sect. 4.2.2), the rate of absorption itself might be important. Absorption rates affect the postprandial circulating concentration of nutrients, which in turn can affect the subsequent metabolism of those nutrients or their incorporation into other molecules. Certain amino acids and choline, for example, are known to be precursors for neurotransmitters whose rates of formation are precursor-controlled. It is possible, therefore, that transport rates could ultimately affect nervous system function. Thus, adaptive regulation of nutrient transport rates is significant for its effects on the rate as well as completeness of absorption.

Acknowledgements. I thank Jared Diamond, Ronald Ferraris, Randy Buddington, David Solberg, Peter Lee, Steve Chang, and Eric Stein for sharing with me their unpublished manuscripts. Sue Boshers helped in preparation of the manuscript. Portions of this work were funded by the National Science Foundation (BSR-8452089).

References

Achinewhu SC, Hewitt D (1979) Assessment of the nutritional quality of proteins: the use of "ileal" digestibilities of amino acids as measures of the availabilities. Br J Nutr 41:559 – 571

Adibi SA, Kim YS (1981) Peptide absorption and hydrolysis. In: Johnson LR (ed) Physiology of the gastrointestinal tract. Raven Press, New York, pp 1073 – 1095

Adibi SA, Mercer DW (1973) Protein digestion in human intestine as reflected in luminal, mucosal and plasma amino acid concentrations after meals. J Clin Invest 52:1586 – 1594

Albus H, Lippens F, Siegenbeek van Heukelom J (1983) Sodium-dependent sugar and amino acid transport in isolated goldfish intestinal epithelium; electrophysiological evidence against direct interactions at the carrier level. Pfluegers Arch 398:10 – 17

Alvarado F, Monreal J (1967) Na^+-dependent active transport of phenylglucosides in the chicken small intestine. Comp Biochem Physiol 20:471 – 488

Anderson JH, Taylor AB (1973) Scanning and transmission electron microscopic studies of jejunal microvilli of the rat, hamster, and dog. J Morphol 141:281 – 291

Barry PH, Diamond JM (1984) Effects of unstirred layers on membrane phenomena. Physiol Rev 64:763 – 872

Barry RE (1976) Mucosal surface areas and villous morphology of the small intestine of small mammals: functional interpretations. J Mammal 57:273 – 289

Bode Ch, Eisenhardt JM, Haberich FJ, Bode J Ch (1981) Influence of feeding fructose on fructose and glucose absorption in rat jejunum and ileum. Res Exp Med 179:163 – 168

Brot-Laroche E, Alvarado F (1983) Mechanisms of sugar transport across the intestinal brush border membrane. In: Gilles-Baillien M, Gilles R (ed) Intestinal transport. Springer, Berlin Heidelberg New York, pp 147 – 169

Brown HO, Levine ML, Lipkin M (1963) Inhibition of intestinal epithelial cell renewal and migration induced by starvation. Am J Physiol 205:868 – 872

Buddington RK (1987) Does the natural diet influence the intestine's ability to regulate glucose absorption? J Comp Physiol B, 157:677 – 688

Buddington RK, Diamond JM (1987) Pyloric caeca of fish, a "new" absorptive organ. Am J Physiol 252:G65 – G76

Buddington RK, Chen JW, Diamond JM (1987) Genetic and phenotypic adaptation of intestinal nutrient transport to diet in fish. J Physiol (London) 3:261 – 281

Campbell RM, Fell BF (1964) Gastrointestinal hypertrophy in the lactating rat and its relation to food intake. J Physiol (London) 171:90 – 97

Carpenter M, Musacchia XJ (1978) Chronic heat exposure and intestinal absorption of sugars in hamster. Comp Biochem Physiol 60A:411 – 416

Carter-Su C, Kimmich GA (1980) Effect of membrane potential on Na^+-dependent sugar transport by ATP-depleted intestinal cells. Am J Physiol 238:C73 – C80

Caspary WF (1973) Effect of insulin and experimental diabetes mellitus on the digestive absorptive function of the small intestine. Digestion 9:248 – 263

Caspary WF, Lucke H (1976) Intestinal adaptation in response to chronic glucagon administration. In: Robinson JWL (ed) Intestinal ion transport. MTP Press, Lancaster, pp 345–352
Cheeseman CI (1986) Expression of amino acid and peptide transport systems in rat small intestine. Am J Physiol 251:G636–G641
Chivers DJ, Hladik CM (1980) Morphology of the gastrointestinal tract in primates: comparisons with other mammals in relation to diet. J Morphol 166:337–386
Chu DT, Yousef MK, Haghani Z, Hillyard SD (1979) Intestinal absorption in heat and cold acclimated desert woodrats. Proc Soc Exp Biol Med 160:483–485
Chung YC, Kim YS, Shadchehr A, Garrido A, Macgreggor IL, Sleisenger MH (1979) Protein digestion and absorption in human small intestine. Gastroenterology 76:1415–1421
Collie NL (1985) Intestinal nutrient transport in coho salmon (*Oncorhynchus kisutch*) and the effects of development, starvation, and seawater adaptation. J Comp Physiol B 156:163–174
Coulson RA, Hernandez T (1970) Protein digestion and amino acid absorption in the Cayman. J Nutr 100:810–826
Crane RK (1960) Intestinal absorption of sugars. Physiol Rev 40:785–825
Cremaschi D, James PS, Meyer G, Peacock MA, Smith MW (1982) Membrane potentials of differentiating enterocytes. Biochim Biophys Acta 688:271–274
Cripps AW, Williams VJ (1975) The effect of pregnancy and lactation on food intake, gastrointestinal anatomy, and the absorptive capacity of the small intestine in the albino rat. Br J Nutr 33:17–32
Crompton DWT, Nasheim MC (1969) Amino acid patterns during digestion in the small intestine of ducks. J Nutr 99:43–50
Csaky TZ, Fischer E (1981) Intestinal sugar transport in experimental diabetes. Diabetes 30:568–574
Csaky TZ, Fischer E (1984) Effects of ketohexosemia on the ketohexose transport in the small intestine of rats. Biochim Biophys Acta 772:259–263
Dabrowski K (1983) Digestion of protein and amino acid absorption in stomachless fish, common carp (*Cyprinus carpio*). Comp Biochem Physiol 74A:409–415
Dabrowski K, Dabrowska H (1981) Digestion of protein by rainbow trout (*Salmo gairdneri*, Rich.) and absorption of amino acids within the alimentary tract. Comp Biochem Physiol 69A: 99–111
Dabrowski K, Leray C, Nonnotte G, Colin DA (1986) Protein digestion and ion concentrations in rainbow trout (*Salmo gairdneri* Rich.) digestive tract in sea- and fresh water. Comp Biochem Physiol 83A:27–39
Daumerie C, Henquin JC (1982) Somatostatin and the intestinal transport of glucose and other nutrients in the anaesthetized rat. Gut 23:140–145
Dauncey MJ, Ingram DL, James PS, Smith MW (1983) Modification by diet and environmental temperature of enterocyte function in piglet intestine. J Physiol (London) 341:441–452
Davies NT, Williams RB (1976) The effects of pregnancy on uptake and distribution of copper in the rat. Abstr Proc Nutr Soc 35:4A
Davies NT, Williams RB (1977) The effect of pregnancy and lactation on the absorption of zinc and lysine by the rat duodenum in situ. Br J Nutr 38:417–423
Debnam ES (1982a) Effects of fasting and glucagon on the kinetics of active hexose absorption across the rat distal ileum in vivo. Q J Exp Physiol 67:587–594
Debnam ES (1982b) Effect of sodium concentration and plasma sugar concentration on hexose absorption by the rat jejunum in vivo. Pfluegers Arch 393:104–108
Debnam ES (1985) Adaptation of hexose uptake by the rat jejunum induced by the perfusion of sugars into the distal ileum. Digestion 31:25–30
Debnam ES, Levin RJ (1975) Effects of fasting and semistarvation on the kinetics of active and passive sugar absorption across the small intestine in vivo. J Physiol (London) 252:681–700
Debnam ES, Levin RJ (1976) Influence of specific dietary sugars on the jejunal mechanisms for glucose, galactose, and alpha-methyl glucoside absorption: evidence for multiple sugar carriers. Gut 17:92–99
Debnam ES, Thompson CS (1984) The effects of fasting on the potential difference across the brush border membrane of enterocytes in rat small intestine. J Physiol (London) 355:449–456
Debnam ES, Karasov WH, Thompson CS (1988) Nutrient uptake by rat enterocytes during diabetes mellitus: evidence for an increased Na-electrochemical gradient. J Physiol (London) 397: 503–512

Diamond JM, Buddington RK (1987) Intestinal nutrient absorption in herbivores and carnivores. In: Dejours P, Bolis L, Taylor CR, Weibel ER (eds) Comparative physiology: life in water and on land. Liviana Press and Springer, Berlin Heidelberg NewYork, pp 193–203
Diamond JM, Karasov WH (1984) Effect of dietary carbohydrate on monosaccharide uptake by mouse small intestine in vitro. J Physiol (London) 349:419–440
Diamond JM, Karasov WH (1987) Adaptive regulation of intestinal nutrient transporters. Proc Natl Acad Sci USA 84:2242–2245
DiCostanzo G, Duportail G, Floretz A, Leray C (1983) The brush border membrane of trout intestine: influence of its lipid composition on ion permeability, enzyme activity and membrane fluidity. Mol Physiol 4:270–290
Dowling RH, Riecken EO, Laws JW, Booth CC (1967) The intestinal response to high bulk feeding in the rat. Clin Sci 32:1–9
Elias E, Dowling RH (1973) The effect of lactation on intestinal structure and function following by-pass or transposition of the jejunum in the rat – evidence for hormonal stimulation of intestinal mucosal hyperplasia. In: Dowling RH, Riecken EO (eds) Intestinal adaptation. Schattauer, Stuttgart, pp 93–102
Fell BF (1972) Adaptations of the digestive tract during reproduction in the mammal. World Rev Nutr Diet 14:180–256
Ferraris RP, Ahearn GA (1983) Sugar and amino acid transport in fish intestine: a review. Comp Biochem Physiol 77A:397–413
Ferraris RP, Ahearn GA (1984) Intestinal glucose transport in carnivorous and herbivorous fishes. J Comp Physiol 152:79–90
Ferraris RP, Diamond JM (1986) Use of phlorizin binding to demonstrate induction of intestinal glucose transporters. J Membr Biol 94:77–82
Fisher RB, Parsons DS (1950) The gradient of mucosal surface area in the small intestine of the rat. J Anat 84:272–282
Freeman HJ (1984) Amino acid and dipeptide absorption in rats fed chemically defined diets of differing fiber composition. Can J Pharmacol 62:1097–1101
Garg VK (1978) Site and rate of sugar and amino acid transport in the intestine and pyloric caeca of teleost fishes. Indian J Exp Biol 16:1095–1097
Ginsburg JM, Heggeness FW (1968) Adaptation in monosaccharide absorption in infant and adult rats. J Nutr 96:494–498
Gray GM, Santiago NA (1966) Disaccharide absorption in normal and diseased human intestine. Gastroenterology 51:489–498
Groot JA, Albus H, Bakker R, Dekker K (1983) Changes in sugar transport and in electro-physiological characteristics of intestinal preparations of temperature-acclimated goldfish (*Carrassius auratus* L.). J Comp Physiol 151:163–170
Gross JE, Wang Z, Wunder BA (1985) Effects of food quality and energy needs: changes in gut morphology and capacity of *Microtus ochrogaster*. J Mammal 66:661–667
Gunter-Smith P, Grasset E, Schultz SG (1982) Sodium-coupled amino acid and sugar transport by *Necturus* small intestine. An equivalent electrical circuit analysis of a rheogenic co-transport system. J Membr Biol 71:89–94
Habibi H, Ince BW (1984) Intestinal transport of leucine in intact and gonadectomized underyearling rainbow trout, *Salmo gairdneri* Rich. Comp Biochem Physiol A 79:349–352
Habibi H, Ince BW, Matty AJ (1983) Effects of 17-alpha-methyltestosterone and 17-beta-oestradiol on intestinal transport and absorption of L-(^{14}C)-leucine in vitro in rainbow trout (*Salmo gairdneri*). J Comp Physiol B 151:247–252
Hahn P, Koldovsky O (1966) Utilization of nutrients during postnatal development. Pergamon Press, Oxford New York
Hanson WR, Rijke RPC, Plaisier HM, Ewijk W van, Osborne JW (1977) The effect of intestinal resection on Thiry-Vella fistulae of jejunal and ileal origin in the rat: evidence for a systemic control mechanism of cell renewal. Cell Tissue Kinet 10:543–555
Henriques de Jesus C, Da Graca Emilio M, Santos MA (1979) Electrical characterizations of sodium-glucose carriers in in vivo intestinal preparations. Gastroenterol Clin Biol 3:172–173
Hokazono S, Tanaka Y, Katayama T (1979) Intestinal transport of L-lysine in rainbow trout (*Salmo gairdneri*). Bull Jpn Soc Sci Fish 45:845–848

Holdsworth GD, Wilson TH (1967) Development of active sugar and amino acid transport in the yolk sac and intestine of the chicken. Am J Physiol 212:233 – 240
Honegger P, Semenza G (1973) Multiplicity of carriers for free glucalogues in hamster small intestine. Biochim Biophys Acta 318:390 – 410
Hoover WH, Heitmann RN (1975) Cecal nitrogen metabolism and amino acid absorption in the rabbit. J Nutr 105:245 – 252
Hopfer U (1975) Diabetes mellitus: changes in the transport properties of isolated intestinal microvillus membranes. Proc Natl Acad Sci USA 72:2027 – 2031
Hopfer U (1987) Membrane transport mechanisms for hexoses and amino acids in the small intestine. In: Johnson LR (ed) Physiology of the gastrointestinal tract, 2nd edn. Raven Press, New York, pp 1499 – 1526
Hornicke H, Bjornhag G (1980) Coprophagy and related strategies for digesta utilization. In: Ruckebusch Y, Thirend P (eds) Digestive physiology and metabolism in ruminants. AVI, Westport, Connecticut, pp 707 – 730
Hughes CA, Breuer RS, Ducker DA, Hatoff DE, Dowling RH (1981) The effect of cholecystokinin and secretin on intestinal and pancreatic structure and function. In: Robinson JWL, Dowling RH, Riecken EO (eds) Mechanisms of intestinal adaptation. MTP Press, Lancaster, pp 435 – 450
Ilundain A, Naftalin RJ (1981) Na^+-dependent co-transport of alpha-methyl-D-glucose across the mucosal border of rabbit descending colon. Biochim Biophys Acta 644:316 – 322
Jacobs LR, Bloom S, Harsoulis P, Dowling RH (1975) Intestinal adaptation to hypothermic hyperphagia. Abstr Clin Sci Mol Med 48:14p
Johnson LR (1976) The trophic action of gastrointestinal hormones. Gastroenterology 70:278 – 288
Kapoor BG, Smit H, Verighina IA (1975) The alimentary canal and digestion in teleosts. Adv Mar Biol 13:109 – 239
Karasov WH (1987) Nutrient requirements and the design and function of guts in fish, reptiles, and mammals. In: Dejours P, Bolis L, Taylor CR, Weibel ER (eds) Comparative physiology: life in water and on land. Liviana Press and Springer, Berlin Heidelberg New York Tokyo, pp 181 – 191
Karasov WH (1988) The nutritional significance of regulation of intestinal nutrient transport. In: Morley J, Sterman MB, Walsh J (eds) Nutritional modulation of neural function. Academic Press, London New York, pp 51 – 63
Karasov WH, Debnam ES (1987) Rapid adaptation of intestinal glucose transport: a brush-border or basolateral phenomenon? Am J Physiol 253:G54 – G61
Karasov WH, Diamond JM (1981) Intestinal glucose transport varies with taxa and diet. Abstr Am Zool 21:1031
Karasov WH, Diamond JM (1983a) A simple method for measuring intestinal solute uptake in vitro. J Comp Physiol 152:105 – 116
Karasov WH, Diamond JM (1983b) Adaptive regulation of sugar and amino acid transport by vertebrate intestine. Am J Physiol 245:G443 – G462
Karasov WH, Diamond JM (1986) Adaptation in intestinal nutrient transport. In: Johnson LR (ed) Physiology of the gastrointestinal tract, vol 2. Raven Press, New York, pp 1489 – 1497
Karasov WH, Pond RS III, Solberg DH, Diamond JM (1983) Regulation of proline and glucose transport in mouse intestine by dietary substrate levels. Proc Natl Acad Sci USA 80:7674 – 7677
Karasov WH, Buddington RK, Diamond JM (1985a) Adaptation of intestinal sugar and amino acid transport in vertebrate evolution. In: Gilles R, Gilles-Baillien MG (eds) Transport processes, iono- and osmoregulation. Springer, Berlin Heidelberg New York Tokyo, pp 227 – 239
Karasov WH, Solberg DH, Diamond JM (1985b) What transport adaptations enable mammals to absorb sugars and amino acids faster than reptiles? Am J Physiol 249:G271 – G283
Karasov WH, Solberg DH, Chang SD, Stein ED, Hughes M, Diamond JM (1985c) Is intestinal transport of sugars and amino acids subject to critical-period programming? Am J Physiol 249:G770 – G785
Karasov WH, Phan D, Diamond JM, Carpenter FL (1986a) Food passage and intestinal nutrient absorption in hummingbirds. Auk 103:453 – 464
Karasov WH, Petrossian E, Rosenberg L, Diamond JM (1986b) How do food passage rate and assimilation differ between herbivorous lizards and nonruminant mammals? J Comp Physiol 156:599 – 609
Karasov WH, Solberg D, Carter S, Hughes M, Phan D, Zollman F, Diamond JM (1986c) Uptake pathways for amino acids in mouse intestine. Am J Physiol 251:G501 – G508

Karasov WH, Solberg DH, Diamond JM (1987) Dependence of intestinal amino acid uptake on dietary protein or amino acid levels. Am J Physiol 252:G614 – G625

Kaunitz JD, Wright EM (1984) Kinetics of sodium D-glucose cotransport in bovine intestinal brush border vesicles. J Membr Biol 79:41 – 51

Kaunitz JD, Gunther R, Wright EM (1982) Involvement of multiple sodium ions in intestinal D-glucose transport. Proc Natl Acad Sci USA 79:2315 – 2318

Keegan DJ, Modinger R (1979) Microvilli of the intestinal mucosal cells of *Rousettus aegyptiacus*. S Afr J Zool 14:220 – 223

Kennedy GC, McCance RA (1958) Endocrine aspects of overnutrition. In: Gardner-Hill H (ed) Modern trends in endocrinology. Butterworth, London, p 79

Kilberg MS, Han H, Barber EF, Chiles TC (1985) Adaptive regulation of neutral amino acid transport system A in rat H4 hepatoma cell. J Cell Physiol 122:290 – 298

Kimmich GA (1981) Intestinal absorption of sugar. In: Johnson LR (ed) Physiology of the gastrointestinal tract. Raven Press, New York, pp 1035 – 1061

Kimmich GA, Randles J (1984) Sodium-sugar coupling stoichiometry in chick intestinal epithelial cells. Am J Physiol 247:C74 – C82

King IS, Sepulveda FV, Smith MW (1981) Cellular distribution of neutral and basic amino acid transport system in rabbit ileal mucosa. J Physiol (London) 319:355 – 368

Kinter WB, Wilson TH (1965) Autoradiographic study of sugar and amino acid absorption by everted sacs of hamster intestine. J Cell Biol 25:19 – 39

Kinzie JL, Grimme NL, Alpers DH (1976) Cyclic AMP-dependent amino acid uptake in intestine – the importance of beta-andrenergic agonists. Biochem Pharmacol 25:2727 – 2731

Kitchin SE, Morris D (1971) The effect of acclimation temperature on amino acid transport in the goldfish intestine. Comp Biochem Physiol A 40:431 – 443

Kotler DP, Levine GM, Shiau YF (1981) Effects of luminal nutrition and metabolic status on in vivo glucose absorption. Am J Physiol 240:G432 – G436

Krejs GJ, Browne R, Raskin P (1980) Effect of intravenous somatostatin on jejunal absorption of glucose, amino acids, water, and electrolytes. Gastroenterology 78:26 – 31

Lahlou B (1983) Intestinal transport and osmoregulation in fishes. In: Gilles-Baillien M, Gilles R (eds) Intestinal transport. Springer, Berlin Heidelberg New York, pp 341 – 353

Leray C, Chapelle S, Duprotail G, Florentz A (1984) Changes in fluidity and 22:6(n-3) content in phospholipids of trout intestinal brush border membrane as related to environmental salinity. Biochim Biophys Acta 778:233 – 238

Lerner J (1984) Cell membrane amino acid transport processes in the domestic fowl (*Gallus domesticus*). Comp Biochem Physiol 78A:205 – 215

Lerner J, Sattelmeyer P, Rush R (1975) Kinetics of methionine influx into various regions of chicken intestine. Comp Biochem Physiol 50A:113 – 120

Levin RJ (1984) Intestinal adaptation to dietary change as exemplified by dietary restriction studies. In: Batt RM, Lawrence TLJ (eds) Function and dysfunction of the small intestine. Liverpool Univ Press, Liverpool, pp 77 – 93

Levine GM (1986) Nonspecific adaptation of jejunal amino acid uptake in the rat. Gastroenterology 91:49 – 55

Lherminier M, Alvarado F (1981) Virtual elimination of the interference of unstirred water layers on intestinal sugar transport kinetics by use of the tissue accumulation method at appropriate shaking rates. Pfluegers Arch 389:155 – 158

Linder MC, Munro HN (1973) Iron and copper metabolism during development. Enzyme 15:111 – 138

Linder MC, Munro HC (1977) The mechanism of iron absorption and its regulation. Fed Proc 36:2017 – 2023

Lis MT, Crampton RF, Matthews DM (1972) Effect of dietary changes on intestinal absorption of L-methionine and L-methionyl-L-methionine in the rat. Br J Nutr 27:159 – 167

Lorenz-Meyer H, Menge H, Riecken EO (1977) Functional and morphological studies on intestinal mucosa of the rat under chronic glucagon application. Res Exp Med 170:181 – 192

Low AG (1979) Studies on digestion and absorption in the intestines of growing pigs 6. Measurements of flow of amino acids. Br J Nutr 41:147 – 156

Lukie BE, Westergaard H, Dietschy J (1974) Validation of a chamber that allows measurement of both tissue uptake rates and unstirred layer thickness in the intestine under conditions of controlled stirring. Gastroenterology 67:652 – 661

Maenz DD, Cheeseman CI (1986) Effect of hyperglycaemia on D-glucose transport across the brush-border and basolateral membrane of rat small intestine. Biochim Biophys Acta 860:277–285
Marsh MN, Swift JA (1969) A study of the mucosal structure of the small intestine using the scanning electron microscope. Q J Med 38:515–516
Mavrias DA, Mayer RJ (1973) Metabolism of fructose in the small intestine. I. The effect of fructose feeding on fructose transport and metabolism in the rat small intestine. Biochim Biophys Acta 291:531–537
Mayhew TM, Middleton C (1985) Crypts, villi and microvilli in the small intestine of the rat. A stereological study of their variability within and between animals. J Anat 141:1–17
Menge H, Sepulveda FV, Smith MW (1983) Cellular adaptation of amino acid transport following intestinal resection in the rat. J Physiol (London) 334:213–223
Mepham TB, Smith MW (1966) Amino acid transport in the goldfish intestine. J Physiol (London) 184:673–684
Moin-Un-Nisa Q, Zain BK, Zain-Ul-Abedin M (1970) Intestinal transport of amino acids in a lizard during hibernation and activity. Comp Biochem Physiol 36:569–577
Montgomery WL (1977) Diet and gut morphology in fishes, with special reference to the monkeyface prickleback, *Cebidichthys violaceus* (Stichaeidae: Blennioidei). Copeia 1977:178–182
Moran A, Turner RJ, Handler JS (1983) Regulation of sodium coupled glucose transport by glucose in a cultured epithelium. J Biol Chem 258:15087–15090
Moran A, Turner RJ, Handler JS (1984) Hexose regulation of sodium-hexose transport in LLC-PK, epithelia: the nature of the signal. J Membr Biol 82:59–65
Mortensen A, Tindall A (1981) On caecal synthesis and absorption of amino acids and their importance for nitrogen recycling in willow ptarmigan (*Lagopus lagopus lagopus*). Acta Physiol Scand 113:465–469
Munck BG (1981) Intestinal absorption of amino acids. In: Johnson LR (ed) Physiology of the gastrointestinal tract. Raven Press, New York, pp 1097–1122
Munck BG (1983) Comparative aspects of amino acid transport in guinea pig, rabbit and rat small intestine. In: Gilles-Baillien M, Gilles R (eds) Intestinal transport. Springer, Berlin Heidelberg New York, pp 260–283
Musacchia XJ, Hartner AM (1970) Intestinal absorption of glucose and blood glucose and hematocrit in pregnant and nonpregnant hamsters. Proc Soc Exp Biol Med 135:307–310
Nassar CF, Haddad ME, Habbal ZM (1982) Prostaglandin E_1 inhibition of lysine transport across the rat, rabbit, and turtle small intestine. Comp Biochem Physiol 72:483–487
National Research Council (1983) Nutrient requirements of coldwater fishes. Natl Acad Sci USA
Nixon SE, Mawer GE (1970) The digestion and absorption of protein in man. 2. The form in which digested protein is absorbed. Br J Nutr 24:241–258
Olsen WA, Rosenberg IH (1970) Intestinal transport of sugars and amino acids in diabetic rats. J Clin Invest 49:96–105
Palay SL, Karlin LJ (1959) An electron microscopic study of the intestinal villus. II. The pathway of fat absorption. J Biophys Biochem Cytol 5:373–383
Penzes L, Simon G (1968) Intestinal absorption and turnover of L-methionine during reproduction in the rat. Jpn J Physiol 18:288–296
Pfeiffer CJ (1968) Intestinal villous morphology. Postgrad Med 43:215–221
Pilch PF, Thomson PA, Czech MP (1980) Coordinate modulation of D-glucose transport activity and bilayer fluidity in plasma membranes derived from control and insulin-treated adipocytes. Proc Natl Acad Sci USA 77:915–918
Reshkin SJ, Ahearn GA (1987) Intestinal glucose transport and salinity adaptation in a euryhaline teleost. Am J Physiol 252:R567–R578
Richter GC, Levine GM, Shiau YF (1983) Effects of luminal glucose versus nonnutritive infusates on jejunal mass and absorption in the rat. Gastroenterology 85:1105–1112
Robinson JWL, Lusier AL, Mircovitch V (1963) Transport of amino acids and sugars by the dog colon's mucosa. Pfluegers Arch 345:315–326
Ross SA, Shaffer EA (1983) Influence of GI hormones on glucose absorption. Clin Res (Abstr) 31:60
Roy CC, Dubois RS (1972) Monosaccharide induction of 3-O-methylglucose transport through rat jejunum. Proc Soc Exp Biol Med 139:883–886
Rudo ND, Rosenberg IH (1973) Chronic glucagon administration enhances intestinal transport in the rat. Proc Soc Exp Biol Med 142:521–525

Sambrook IE (1979) Studies on digestion and absorption in the intestines of growing pigs 7. Measurements of the flow of total carbohydrate, total reducing substances and glucose. Br J Nutr 42:267 – 277

Savory CJ, Gentle MJ (1976) Effects of dietary dilution with fibre on the food intake and gut dimensions of Japanese quail. B. Poult Sci 17:561 – 570

Scharrer E (1975) Developmental changes of sugar and amino acid transport in different tissues of ruminants. In: McDonald IW (ed) Digestion and metabolism in the ruminant. Univ New England Publ Unit, Armidale, Australia, pp 49 – 59

Scharrer E (1976) Developmental changes of sugar transport in the ovine small intestine. Pfluegers Arch 366:147 – 151

Scharrer E, Wolffram S, Raab W, Amman B, Agne N (1981) Adaptive changes of amino acid and sugar transport across the brush border of rat jejunum. In: Robinson JWL, Dowling RH, Riecken EO (eds) Mechanisms of intestinal adaptation. MTP Press, Lancaster, pp 123 – 137

Schultz SG, Strecker CK (1970) Fructose influx across the brush-border of rabbit ileum. Biochim Biophys Acta 211:586 – 588

Schwartz MZ, Storozuk RB (1985) Enhancement of small intestine function by gastrin. J Surg Res 38:613

Schwartz MZ, Storozuk RB (1986) Enhancement of small intestine absorption by intraluminal gastrin. J Surg Res 40:421 – 425

Schwartz SE, Levine GD (1980) Effects of dietary fiber on intestinal glucose absorption and glucose tolerance in rats. Gastroenterology 79:833 – 836

Shapiro Y, Alkan M, Epstein Y, Newman F, Magazunika A (1986) Increase in rat intestinal permeability to endotoxin during hyperthermia. Eur J Appl Physiol Occup Phys 55:410 – 412

Shimada T, Hoshi T (1986) A comparative study of specificity of the intestinal Na^+/sugar cotransport among vertebrates. Comp Biochem Physiol 84A:365 – 370

Shotwell MA, Kilberg MS, Oxender DL (1983) The regulation of neutral amino acid transport in mammalian cells. Biochim Biophys Acta 737:267 – 284

Sibly RM (1981) Strategies of digestion and defecation. In: Townsend CR, Calow P (eds) Physiological ecology. Sinauer, Sunderland, Mass, pp 109 – 139

Smith MW (1966) Time course and nature of temperature-induced changes in sodium-glucose interactions of the goldfish intestine. J Physiol (London) 183:649 – 657

Smith MW (1970) Selective regulation of amino acid transport by the intestine of goldfish (*Carassius auratus* L.). Comp Biochem Physiol 35:387 – 401

Smith MW (1983) Membrane transport in fish intestine. Comp Biochem Physiol 75A:325 – 335

Smith MW (1985) Comparative aspects of enterocyte differentiation. In: Gilles R, Gilles-Baillien M (eds) Transport processes, iono- and osmoregulation. Springer, Berlin Heidelberg New York Tokyo, pp 218 – 226

Smith MW, Kemp P (1971) Parallel temperature-induced changes in membrane fatty acids and in the transport of amino acids by the intestine of goldfish (*Carassius auratus*). Comp Biochem Physiol 39B:357 – 365

Smith MW, Paterson JYF, Peacock MA (1984) A comprehensive description of brush border membrane development applying to enterocytes taken from a wide variety of mammalian species. Comp Biochem Physiol 77A:655 – 662

Snook JT (1974) Adaptive and nonadaptive changes in digestive enzyme capacity influencing digestive function. Fed Proc 33:88 – 93

Solberg DH, Diamond JM (1987) Comparison of different dietary sugars as inducers of intestinal sugar transporters. Am J Physiol 252:G574 – G584

Spector MH, Levine GM, Deren JJ (1977) Direct and indirect effects of dextrose and amino acids on gut mass. Gastroenterology 72:706 – 710

Spector MH, Traylor J, Young EA, Weser E (1981) Stimulation of mucosal growth by gastric and ileal infusion of single amino acids in parenterally nourished rats. Digestion 21:33 – 40

Stein ED, Chang SD, Diamond JM (1987) Comparison of different dietary amino acids as inducers of intestinal amino acid transport. Am J Physiol 252:G626 – G635

Steiner M, Gray SJ (1969) Effect of starvation on intestinal amino acid absorption. Am J Physiol 217:747 – 752

Stevens BR, Kaunitz JD, Wright EM (1984) Intestinal transport of amino acids and sugars: advances using membrane vesicles. Annu Rev Physiol 46:417 – 433

Storelli C, Vilella S, Cassano G (1986) Na-dependent D-glucose and L-alanine transport in eel intestinal brush border membrane vesicles. Am J Physiol 251:R463–R469
Studier EH (1979) Bioenergetics of growth, pregnancy, and lactation in the laboratory mouse, *Mus musculus*. Comp Biochem Physiol 64A:473–481
Sugiyama K, Ohkubo T, Muramatsu K (1983) Effects of quantity and quality of dietary protein on the jejunal absorption for L-leucine and glycylglycine in rats. Agric Biol Chem 47:309–314
Syme G (1982) The effect of protein-deficient isoenergetic diets on the growth of rat jejunum to protein deficiency. Br J Nutr 48:25–36
Tedman RA, Hall LS (1985) The absorptive surface area of the small intestine of *Pteropus poliocephalus* (Megachiroptera: Pteropodidae): an important factor in rapid food transit? Aust Mammal 8:271–278
Thompson CS, Debnam ES (1986) Hyperglucagonaemia: effects on active nutrient uptake by the rat jejunum. J Endocrinol 111:37–42
Thomson ABR, Dietschy JM (1980a) Intestinal kinetic parameters: effects of unstirred layers and transport preparation. Am J Physiol 239:G372–G377
Thomson ABR, Dietschy JM (1980b) Experimental demonstration of the effect of the unstirred water layer on the kinetic constants of the membrane transport of D-glucose in rabbit jejunum. J Membr Biol 54:221–229
Thomson ABR, Dietschy JM (1984) The role of the unstirred water layer in intestinal permeation. In: Csaky TZ (ed) Handbook of experimental pharmacology, vol 70/II. Springer, Berlin Heidelberg New York Tokyo, pp 165–269
Thomson ABR, Hotke CA, Weinstein WM (1982) Comparison of kinetic constants of hexose uptake in four animal species and man. Comp Biochem Physiol 72A:225–236
Tucker AM, Kimmich GA (1973) Characteristics of amino acid accumulation by isolated intestinal epithelial cells. J Membr Biol 12:1–22
Verzar F, McDougall EJ (1936) Absorption from the intestine. Longmans and Green, London
Vinardell MP, Lopera MT, Moreto M (1986) Absorption of 3-oxy-methyl-D-glucose by chicken cecum and jejunum in vivo. Comp Biochem Physiol 85A:171–173
Weser E, Bell D, Tawil T (1981) Effects of octapeptide-cholecystokinin, secretin, and glucagon on intestinal mucosal growth in parenterally nourished rats. Dig Dis Sci 26:409–416
Westergaard H, Dietschy JM (1974) Delineation of the dimensions and permeability characteristics of the two major diffusion barriers to passive mucosal uptake in the rabbit intestine. J Clin Invest 54:718–732
Williamson RCN (1982) Intestinal adaptation: factors that influence morphology. Scand J Gastroentol 17 (Suppl 74):21–29
Williamson RCN, Malt RA (1981) Humoral modulation of compensatory intestinal hyperplasis. In: Dowling RH, Riecken EO (eds) Mechanisms of intestinal adaptations. MTP, Lancaster, pp 215–224
Williamson RCN, Buchholtz TW, Malt RA (1978) Humoral stimulation of cell proliferation in small bowel after transection and resection in rats. Gastroenterology 75:249–254
Wilson FA, Antonson DL, Hart BL, Warr TA, Cherrington AD, Lifenquist JE (1980) The effect of somatostatin on the intestinal transport of glucose in vivo and in vitro in the rat. Endocrinology 106:1562–1567
Wilson TH (1957) In vitro studies on intestinal absorption in fish. Biol Bull 113:362
Wilson TH (1962) Intestinal absorption. Saunders, Philadelphia
Wilson TH, Landau BR (1960) Specificity of sugar transport by the intestine of hamster. Am J Physiol 198:99–102
Winne D (1973) Unstirred layer, source of biased Michaelis constant in membrane transport. Biochim Biophys Acta 298:27–31
Winne D (1976) Unstirred layer thickness in perfused rat jejunum in vivo. Experientia 32:1278–1279
Winne D (1977) Correction of the apparent michaelis constant, biased by an unstirred layer, if a passive transport component is present. Biochim Biophys Acta 464:118–126
Wolffram S, Scharrer E (1984) Effect of feeding a high protein diet on amino acid uptake into rat intestinal brush border membrane vesicles. Pfluegers Arch 400:34–39
Wolffram S, Giering H, Scharrer E (1984) Na^+-gradient dependence of basic amino acid transport into rat intestinal brush border membrane vesicles. Comp Biochem Physiol 78A:475–480
Wolffram S, Eggengerger E, Scharrer E (1986) Kinetics of D-glucose and L-leucine transport into sheep and pig intestinal brush border membrane vesicles. Comp Biochem Physiol 84A:589–593
Wood HO (1944) The surface area of the intestinal mucosa in the rat and in the cat. J Anat 78:103–105

Chapter 5

Nutrient Transport Across the Integument of Marine Invertebrates

S.H. Wright

Contents

1 Introduction

Since the turn of this century there has been recurrent interest in the ability of aquatic organisms to make direct nutritional use of dissolved organic material (DOM). Upon examination, it is a particularly compelling idea, i.e., routine exposure of an animal's integument to its surroundings results in an accumulation of exogenous material to support nutritional and/or other physiological needs. Indeed, some animals, for example, pogonophorans, and others lacking a gut, would seem dependent on integumental uptake. In fact, all animals should

Department of Physiology, College of Medicine, University of Arizona, Tucson, AZ 85724, USA

Advances in Comparative and Environmental Physiology. Vol. 2

(it would seem) benefit from such a capacity. Evidence to support this contention, however, has only recently been generally accepted. We are currently in a period of renewed interest in integumental transport of DOM, stemming from studies using several new technical innovations and experimental systems. It is increasingly apparent that the integument of marine invertebrates not only serves as an effective barrier that protects the animal from intrusion of the sea, but also permits the regulated exchange of organic materials between organism and environment. Furthermore, it is now clear that this exchange can play a significant role in animal nutrition.

As mentioned above, this idea is not new. Pütter (1909) is generally credited for formalizing the first hypothesis that DOM represents an important (and in his mind, essential) source of nutrition for aquatic animals. His ideas were influenced to a large extent by the inaccuracy of contemporary methods to measure particulate and dissolved organic material. Still, the debate that this idea initiated continued until the early 1930s, when Krogh (1934) reported the results of his studies on this issue (Jørgensen 1976, provides an excellent review of the work of Pütter, Krogh, and their contemporaries). Krogh had turned his attention first to the measurement of DOM in natural waters. As I shall emphasize several times during the course of this review, such information remains crucial to an understanding of the role of integumental transport processes. Because of the technical difficulties involved with the analysis of marine samples, Krogh focused on measurements of DOM in freshwater. He found that levels were rather lower than those believed to exist by Pütter, and found little evidence to support the contention that aquatic organisms can make significant use of this resource (Krogh 1934). This conclusion influenced the field for 25 years.

Interest in this subject was minimal until the late 1950s and early 1960s when G.C. Stephens and his colleagues reported results of work employing improved analytical procedures, including the use of radioactively labeled substrates to study integumental transport. While this work confirmed Krogh's conclusion that DOM was essentially unavailable to freshwater invertebrates, he also found that amino acids and sugars were accumulated by the integument of many marine invertebrates (Stephens and Schinske 1961). This observation led to a burst of study on the nature and distribution of such uptake processes. Work through the mid-1970s produced a body of information clearly supporting the idea that carrier-mediated transport is a general feature of the integument of marine invertebrates. This information has been reviewed extensively (Jørgensen 1976; Stephens 1968, 1972, 1982; Stewart 1979).

As it became clear that integumental uptake of DOM is a general phenomenon, interest turned from efforts to catalog the existence of such processes toward the study of their role in animal physiology. Again, significant advances have arisen from the introduction of new analytical procedures, and the results of recent studies indicate that uptake of DOM can represent a significant supplemental source of nutrition for (at least) some marine invertebrates.

Interest in the physiological role of integumental transport has also led to the study of the physiology of the process itself. There is a growing awareness that the integument is not an inert barrier, but instead is designed to permit a controlled exchange of material, including nutrients, between the animal and its environment (see Bereiter-Hahn et al. 1984). Recent work suggests that many of the transport mechanisms found in mammalian and other vertebrate systems have analogs in the integument of marine invertebrates. Study of the specific "design" of these processes is leading to explanations of their extraordinary ability to accumulate organic material from extemely dilute solutions, without sacrificing the ability of the integument to sustain the large concentrations of small organic molecules that are characteristic of marine invertebrate tissues.

This review includes a brief summary of the experimental observations that led to the general conclusion that DOM uptake is a widespread phenomenon in marine invertebrates (see also Jørgenson 1976; Stewart 1979). It then examines in more detail the observations, many of them relatively recent, that firmly establish the nutritional potential of integumental transport. The focus of this discussion is on studies of representatives from three phyla about which our information is most complete: the mollusks, echinoderms, and annelids. Each of these groups has a unique feature or set of features making the study of integumental processes of particular interest. Though these features may differentiate between the groups, it should become apparent that the general characteristics of integumental transport are quite similar, justifying the extrapolation (albeit, with care) of this information to other marine phyla. The review closes with a discussion of recent information on the mechanism(s) of DOM accumulation, emphasizing some elements that may prove to be common to integumental transport in general.

Throughout the review I will attempt to point out areas about which crucial information is still lacking, or which remain controversial. In this way I hope to stimulate interest on subjects which may influence future interpretation of the physiology of integumental uptake.

Finally, though I have taken this opportunity to discuss a considerable fraction of the previous work on integumental transport processes in marine invertebrates, this is not intended to be an exhaustive review of the extensive literature on this subject. Besides the earlier reviews already cited, the interested reader is directed to several recent surveys of integumental transport phenomena in specific groups of marine invertebrates (Annelida: Gomme 1984; Coelenterata: Schlichter 1984; Echinodermata: Bamford 1984; Mollusca: Stewart 1984; Pogonophora: Southward 1984).

2 Background Information

2.1 Amount and Nature of DOM in the World's Oceans

DOM in the oceans represents an enormous pool of reduced carbon. It is comparable in mass to all other sources of organic carbon on earth, excluding fossil fuels (Woodwell et al. 1978). Despite its absolute mass, DOM concentrations are quite low, typically on the order of a few milligrams per liter (Williams 1975), and less than 10% of this material has been identified chemically. Furthermore, the bulk of the DOM appears to be relatively refractory, with a turnover time of the entire DOM pool measured in thousands of years (Williams et al. 1969).

A small percentage of the DOM, however, consists of well-defined classes of organic material, and has been shown to be readily accessible to marine organisms. Dissolved free amino acids (DFAAs) represent approximately 1% of the DOM (10 μgC l^{-1}, or approx. 0.5 μM total DFAA), and the turnover time of DFAAs in surface waters has been estimated to be between $1-2$ days (see Carlucci et al. 1985). Less is known amount the carbohydrate and fatty acid content of sea water but levels of these compounds are estimated to be on the order of 200 μgC l^{-1} and 5 μgC l^{-1}, respectively (Williams 1975).

The concentration of these different classes of DOM in marine waters is extremely variable. In broad terms, two categories of habitat can be defined: the water column and the sediment. Reported variations in DFAA levels will serve to document this point. DFAA levels in near shore waters have been found to range from as high as $1-3$ μM (e.g., Braven et al. 1984; Siebers and Winkler 1984) to as low as 10 to 100 nM (e.g., Henrichs and Williams 1985). Rather than representing technical differences in measurement, this range probably represents the actual variability of this parameter in the marine environment. Environmental DFAA concentrations vary with season (Braven et al. 1984; N.O.G. Jørgensen 1982), time of day (Mopper and Lindroth 1982), and depth (Henrichs and Williams 1985; Mopper and Lindroth 1982). Moreover, striking spatial heterogeneity of DFAA concentrations in surface waters have also been noted on a horizontal scale ranging from hundreds of meters (Poulet et al. 1985) down, perhaps to <2 m (Manahan and Arnold 1983; Smith 1986), and on a vertical scale of several centimeters (the sea surface "microlayer" vs subsurface water; Henrichs and Williams 1985). It is worthwhile to note that while these levels of DFAA are low, they are in the same range as the amount of organic material available as filterable phytoplankton in near shore waters (Parsons 1975).

The concentration of DFAA in the interstitial water of near shore sediments is 10 to 100 times higher than that of the water column, ranging from 10 to 100 μM. Stephens et al. (1978) measured a mean value for total primary amines (as fluorescamine-positive material; FPM) of 115 μM in cores of a sandy sediment along the coast of southern California, and used thin layer chromatography to identify the major constituents as amino acids. Jørgensen (1979) also measured FPM in sediment from a small estuary, and found it to range from 16 to 56 μM. In a subsequent study, Jørgensen et al. (1981) used high performance liquid chromatography (HPLC) to measure directly DFAAs in pore water extracted from sediments, and reported values ranging from approximately 1 to 20 μM. Considerable variation was noted as a function of both depth and physical quality of the sediment (i.e., porosity, grain size, etc.). Gas chromatography on interstitial water from a shallow bay bottom and from deep water samples revealed DFAA concentrations of $\simeq 8$ to 50 μM (Henrichs and Farrington 1979).

Though the concentration of DFAAs (and other organic compounds) in sediment waters are greater than those in the overlying water column, it is not clear to what extent these compounds are available to sediment dwelling organisms. Many burrowing animals actively ventilate their burrows. Consequently, the concentration of DOM to which the integument of these animals is exposed probably lies between the extremes of concentration noted above for these diverse habitats.

The variation in availability of DOM obviously influences any estimates of the nutritional role of DOM uptake. Through the course of this review it should become apparent that the potential input from integumental transport is large; in many organisms uptake can represent a significant source of nutrition, given exposure to adequate levels of exogenous substrate. Furthermore, "adequate levels" quite often lie within the range of concentrations reported to occur in the marine environment. A thorough survey of the distribution of DOM in the environment remains crucial to our understanding of the nutritional role of integumental transport processes.

It may seem that this discussion, as well as those to come, dwells inordinately on the integumental uptake of amino acids, to the exclusion of the "remaining" 99% of the DOM pool. However, with the exception of a comparatively few studies on uptake of sugars or fatty acids, the overwhelming majority of the work in this area has dealt with the accumulation of DFAA. This emphasis on amino acid transport stems primarily from three factors. First, amino acids are (and have been) amongst the easiest components of the DOM to measure. Second, though they comprise a relatively small percentage of the total DOM pool, that nevertheless represents, in absolute terms, a very large amount of reduced carbon (and organic nitrogen). Finally, amino acid metabolism, including the control of tissue levels of amino acids, is of major physiological importance to marine organisms. Nevertheless, the comparative lack of understanding of the possible accumulation of other elements of the DOM pool is a major omission in the study of integumental transport. Breakthroughs in our understanding of such processes will probably require development of the kind of simple technologies for the measurement of dissolved organic carbon that currently exist for measurement of amino acids.

2.2 Evidence for Integumental Transport of DOM

Stephens (1960) reported that ^{14}C-glucose can be removed from sea water by the solitary coral, *Fungia scutaria*. This uptake was later (Stephens 1962) shown to be a saturable process that did not involve binding to mucus. Furthermore, it was not influenced by plugging the mouth of the animal with paraffin, indicating that the accumulation of glucose occurred directly into the general integument of the animal. Following this report came a series of papers from Stephens' laboratory, as well as from other groups, documenting the presence of integumental transport of sugars, amino acids, and fatty acids in marine invertebrates. The results of these studies, through 1978, have been extensively reviewed (Stewart 1979), and can be summarized as follows.

2.2.1 Distribution of Integumental Transport in Marine Invertebrates

The ability to accumulate radioactively labeled, low molecular weight, dissolved material has been reported to occur in members of 18 classes, representing 13 phyla, of marine organisms (Table 1). This uptake appears to occur directly into

Table 1. Integumental transport of dissolved organic material in marine invertebrates. Taxonomic groups for which at least one representative has been shown to accumulate DOM from sea water

List of taxa	
Phylum / *Class*	*Phylum* / *Class*
Porifera	Annelida
Demospongiae	Polychaeta
Coelenterata	Oligochaeta
Anthozoa	Mollusca
Scyphozoa	Bivalvia
Hydrozoa	Gastropoda
Platyhelminthes	Hemichordata
Turbellaria	Enteropneusta
Nemertea	Pogonophora
Anopla	Echnidermata
Brachipoda	Echinoidea
Articulata	Asteroidea
Ectoprocta	Ophiuroidea
Gymnolaemata	Holothuroidea
Sipunculoidea	Chordata
	Ascidiacae

body surfaces exposed to sea water, i.e., into the integument. Ligation of the mouth and anus does not influence rates of accumulation into marine worms (Fisher and Oaks 1978; Stephens 1963; Taylor 1969), indicating that uptake does does not involve ingestion of sea water and a subsequent transport by digestive epithelia. Reports of uptake into isolated body wall of marine worms (e.g., Chien et al. 1972) and isolated gill tissue of marine bivalves (e.g., Bamford and Gingles 1974; DuPaul and Webb 1971; Stewart and Bamford 1975; Wright et al. 1975) and the isolated echinoderm test (de Burgh 1978; de Burgh et al. 1977; Stephens et al. 1978) further implicate the general integument as the site of DOM uptake. Furthermore, autoradiograms of animals exposed to exogenously applied radiolabeled organic molecules show that accumulated materials appear first in integumental tissues (e.g., bivalves: Péquignat 1973; brachipods: McCammon and Reynolds 1976; echinoderms: Ferguson 1967a; Pearse and Pearse 1973; pogonophores: Little and Gupta 1968; Southward and Southward 1968). The term "integumental transport" can be applied to this general class of transport processes.

Marine arthropods appear to be the only taxonomic group of marine invertebrates whose members cannot accumulate DOM. Though there were early reports that some crustaceans could make direct use of dissolved sugars and fatty acids (McWhinnie and Johanneck 1966), later studies suggested that the apparent uptake of DOM into arthropods was probably the result of uptake into epiphytic microorganisms (Anderson and Stephens 1969). Continuing controversy over the possible role of DOM in arthropod nutrition (e.g., McWhinnie and Urbanski

1971) led to a reexamination of the issue (Castille and Lawrence 1979), with a final conclusion that no evidence could be found to support a direct utilization of dissolved substrates by marine arthropods. Consequently, the term "soft-bodied" marine invertebrates is often used to describe collectively the animals possessing integumental transport pathways.

Uptake of DOM by freshwater animals has also been of continuing interest. As mentioned earlier, Krogh (1934) concluded that freshwater animals were incapable of accumulating dissolved materials. With the introduction of radiotracer methodologies, this question was reexamined. In a study of freshwater animals from 12 genera representing 6 phyla, Stephens (1964) found no evidence of uptake of amino acids or sugars that could not be dismissed as the result of uptake into contaminating microorganisms. He also reported that glycine accumulation into a brackish water polychaete, *Nereis limnicola,* was strongly influenced by salinity, and that uptake dropped sharply at external chlorosities below approximately 200 mEq l^{-1}. Later studies did indicate, however, the presence of carrier-mediated uptake of DOM in at least some freshwater animals (e.g., glucose and glycine uptake into a freshwater bivalve, Efford and Tsumura 1973; uptake of amino acids, carbohydrates, and short-chain fatty acids into freshwater oligochaetes, Richards and Arme 1979, 1980a, b, respectively). However, the rates of uptake reported in these studies are generally very low compared to those measured in soft-bodied marine invertebrates, leading to suggestions that these processes are not capable of a net accumulation of substrate in freshwater animals (e.g., Richards and Arme 1980b).

The apparent lack in freshwater animals of the type of integumental transport processes found commonly in marine invertebrates is consistent with current ideas on the importance (necessity?) of extracellular monovalent cations (particularly Na^+) in the concentrative transport of organic solutes (see Sect. 6.1). The very low ionic strength of "freshwater" (i.e., total salt concentrations typically $\leqslant$10 mM) precludes exposure of the body surface to solutions having the high concentrations of, for example, Na^+ that are required for carrier-mediated uptake in many systems (see Crane 1977). Therefore, practically speaking, the "burden of proof" must be on those who claim a concentrative uptake of DOM by a freshwater animal. With this in mind there does appear to be at least one freshwater candidate for integumental transport. Thomas et al. (1984) reported a net accumulation of short-chain carboxylic acids into the freshwater pulmonate snail, *Biomphalaria glabrata* (the invertebrate host of *Schistosoma mansoni*), at rates which suggest that this process could play a role in animal nutrition. Thus, the distribution and role of integumental transport in freshwater organisms must still be considered an open question.

2.2.2 Evidence for Carrier-Mediation of Integumental Transport

The "diagnostic" criteria commonly applied to studies of carrier-mediated transport processes include (1) saturability, and (2) structural specificity. The early literature on integumental transport processes focused on demonstrations that each of these criteria applied to uptake of DOM.

Saturability: The Kinetics of Integumental Transport. Stephens (1962) first showed that the integumental accumulation of an organic molecule (glucose) reaches a maximal rate (i.e., becomes saturated) upon addition of high concentrations of that substrate to sea water. Since then, a number of studies (summarized by Preston and Stevens 1982) have measured the concentration dependency of integumental transport. The kinetics of uptake have usually been adequately described by the Michaelis-Menten equation:

$$J = \frac{J_{max} \cdot [S]}{K_t + [S]}, \tag{1}$$

where J is the rate of uptake from an external substrate concentration of [S], J_{max} is the maximal rate of uptake occurring when the transport process is saturated with substrate, and K_t (the Michaelis constant) is the substrate concentration resulting in half-maximal uptake.

As is evident from Eq. (1), the rate of uptake of DOM is defined by the kinetic constants, J_{max} and K_t. Though these parameters can be used to assist the study of mechanisms of membrane transport, their use in studies of integumental uptake has generally been phenomenological. Mechanistic interpretation of kinetic data relies on experimental evidence that is usually lacking, including assurances that initial rates of uptake are being measured, that transmembranous driving forces are well defined and are not changing as a consequence of transport, and that uptake of the molecule of interest is not occurring by more than one carrier-mediated pathway.

Nevertheless, measurement of the concentration dependency of integumental transporters has provided valuable insights into their possible role in animal physiology. The J_{max} is a measure of the total transport capacity of the uptake pathway in question. Comparison of the maximal rate at which substrate can be provided to the tissue, albeit under saturating conditions unlikely to occur in the environment, to measured rates of metabolism is a simple method to gauge the potential of the pathway to provide substrate at rates capable of influencing tissue or animal physiology. The Michaelis constant provides a means of estimating to what extent the transport capacity is actually realized under physiologically or environmentally relevant conditions. A large K_t, even when coupled with a large J_{max}, may mean that an inconsequential fraction of the transport capacity is realized when the tissue exposed to the micromolar or submicromolar concentrations of the environment. Likewise, what might appear to be a relatively low J_{max} might result in a large input of substrate when coupled with a Michaelis constant in the micromolar range.

Structural Specificity of Integumental Transport. The ability to discriminate between different structural classes of substrate is a hallmark of carrier-mediated transport processes. Integumental amino acid uptake, for example, ordinarily involves multiple transporters, each with a degree of specificity for a separate class of these compounds (e.g., bivalves: Bamford and McCrea 1975; Stewart 1977; Wright et al. 1975; coelenterates: Schlichter 1978; echinoderms: de Burgh 1978; nemerteans: Roe et al. 1981; polychaetes: Siebers 1976; Siebers and Bulnheim

1977; Taylor 1969; rhyncocoeleans: Fisher and Oaks 1978). Specific, carrier-mediated pathways for sugars have also been found to be broadly distributed (e.g., bivalves: Bamford and Gingles 1974; coelenterates: Schlichter 1975; polychaetes: Ahearn and Gomme 1975; Albrechtsen and Gomme 1984). Other compounds for which integumental uptake has been reported include fatty acids (Bunde and Fried 1978; Testerman 1972) and creatinine (van Pilsum et al. 1975).

2.2.3 Integumental Transport: Does it Result in a Net Uptake of Substrate?

The accumulation of a radioactively labeled substrate provides no information on the direction of net substrate flux, unless the system has achieved a steady state with respect to the movement of label. Unfortunately, the rate of integumental transport is low enough that, when combined with the large concentrations of solutes, such as amino acids, found in integumental tissues, it becomes virtually impossible to achieve such a steady state in experiments with marine invertebrates (Stevens and Preston 1980a). With tissue levels of endogenous, nonlabeled substrates that are often from 10^4 to 10^7 times that in sea water, it is not unreasonable to suggest that rates of efflux may be significant, possibly exceeding carrier-mediated rates of influx. This suggestion, first outlined by Johannes et al. (1969), became a critical test of whether integumental uptake of DOM could play a nutritional role in marine invertebrates.

In fact, Stephens and Schinske (1961) used the colorometric reagent, ninhydrin, to show that soft-bodied marine invertebrates could effect a net uptake of glycine from sea water; though the concentration of substrate used in this study was generally very large (2 mM), in some tests net uptake occurred from concentrations as low as 20 μM. Ferguson (1971) used gas-liquid chromatography to demonstrate net uptake of amino acids into several sea stars from ambient concentrations >1 μM. But the widespread recognition that integumental transport results in a net accumulation of amino acid followed the introduction of the fluorometric reagent, fluorescamine, which permitted the rapid determination of total primary amines in sea water (North 1975). Stephens' (1975) demonstration that the polychaetes, *Capitella capitata* and *Nereis diversicolor,* could accumulate naturally occurring primary amines was subsequently supported by a series of studies using fluorescamine to show the net uptake of amino acids from micromolar concentrations into a variety of animals (bivalves: Crowe et al. 1977; Rice et al. 1980; Wright and Stephens 1977, 1978; echinoderms: Ferguson 1980a, 1982; Stephens et al. 1978; polychaetes: Costopulos et al. 1979; Jørgensen 1980; Siebers and Bulnheim 1977).

Still, wide variation in the specific fluorescence of reaction products of fluorescamine with different amino acids (North 1975), and this reagent's small but significant reaction with ammonia (Jørgensen 1980), made it difficult to conclude with any degree of certainty the ambient concentration from which carrier-mediated uptake equaled the passive loss of these compounds from the integument. This is not an academic question, as many organisms are probably exposed, at least transiently, to concentrations of less than 0.1 μM. For many organisms, clear knowledge of whether net uptake can still occur from such *very* low concentrations is critical for understanding the nutritional role of integumental uptake, and is not readily obtained from studies with fluorescamine. The recent introduction of sensitive, and comparatively simple, chromatographic procedures for amino acids have enabled this important question to be addressed directly (e.g., Manahan et al. 1982) and has shown that integumental uptake is capable of a net accumulation of amino acids from sea water concentration of less than 10 nmol l^{-1}.

Thus, studies through the late 1970s established quite convincingly that integumental transport of a wide variety of small organic molecules is a ubiquitous characteristic of soft-bodied marine invertebrates. The work of the last 4 to 6 years has centered primarily on efforts to determine the role(s) of such processes in animal physiology, and the mechanism by which the integument controls its permeability to small molecules. The following sections consist of a more detailed survey of current information on the nutritional role of integumental transport in three different phyla having features of particular interest to this general area: mollusks, echinoderms, and annelids.

3 Integumental Transport in Marine Mollusks

Mollusks are often used in studies of environmental physiology, due in large part to their predominance in a number of marine habitats (e.g., the rocky intertidal). In the study of integumental transport, mollusks offer experimental systems that are easy to manipulate, as well as ones that pose certain unique questions concerning the physiology of the integument. For example, the habit of filter feeding, developed to an extraordinary degree in a number of bivalve mollusks, causes the tissues of the mantle cavity to be exposed to a passing current of sea water. This current is both a continuously renewed source of dissolved foodstuffs to cells capable of accumulating it, and an "infinite sink" into which small molecules can diffuse from the animal to the environment. With respect to this latter point, it is important to emphasize that DOM concentrations in sea water are at least several orders of magnitude less than the levels of small organic molecules found in animal cells. The study of DOM uptake in marine mollusks has often focused on these opposing characteristics of the integument: the need to provide a barrier to the diffusive loss of endogenous organic molecules vs the advantages associated with having carrier-mediated pathways for uptake of these compounds.

3.1 General Characteristics of DOM Uptake in Marine Mollusks

The gill of animals such as *Mytilus edulis* is the primary source of uptake of small organic molecules (Péquignat 1973), and is responsible for between 70 and 80% of the uptake of amino acids (Jørgensen 1983; Wright and Secomb 1984). As a consequence, and because this tissue is particularly amenable to in vitro manipulation, most of what is known about DOM uptake in marine bivalves has come from studies of transport in isolated gill tissue.

3.1.1 Kinetics of Integumental Influx of Organic Material

The rate of DOM uptake into gill tissue is a function of the substrate concentration in the external solution. The kinetics of uptake have been adequately described by the Michealis-Menten equation [see Eq. (1)]. For example, Fig. 1 shows the effect of increasing alanine concentration on its uptake into gill tissue from *Mytilus californianus.* Though there were early suggestions that uptake into gills included a significant diffusive component (e.g., Stewart and Bamford 1975), this confusion probably stemmed from the presence of external tissue compartments from which it is difficult to rinse adherent radiolabeled substrate following exposure to a test solution (Wright 1979; Wright and Stephens 1977). Upon reflection, it is apparent that diffusion is unlikely to play a significant role in integumental uptake of DOM. Diffusion is a symmetrical process, with *net* substrate movement occurring in response to the orientation of a substrate's electrochemical gradient. The gradient for diffusive amino acid movement across integumental surfaces lies strongly in favor of a net efflux from the animal. Accordingly, if amino acids or other small molecules can diffuse *into* integumental cells at appreciable rates from the concentrations found in sea water, then their rate of diffusion *from* the integument would be such that it would be impossible for these cells to maintain their characteristically high cytoplasmic concentrations of small metabolites. As described in a later section, recent measurements (Wright et al. 1987) of passive permeability coefficients for amino acids in gill tissue are very low, supporting the conclusion that diffusion plays a minimal role in integumental uptake.

The kinetics of amino acid uptake into bivale gills indicate that these process are of relatively "low capacity" (i.e., have low J_{max}'s) but have a "high affinity" (i.e., have low K_t's). By low capacity, I refer to comparisons with values for J_{max} typically found in studies of uptake in the intestine of vertebrates or invertebrates (see Chapters 3 and 4, this Vol.). Table 2 shows a representative summary of J_{max} values for integumental transport processes in the gills of several marine bivalves. Values for J_{max} for amino acids in *Mytilus* gill range from between approximately 10 and 40 μmol g^{-1}-h^{-1}), or 0.3 and 1.2 × 10^{-9} mol cm^{-2} min^{-1}). In contrast,

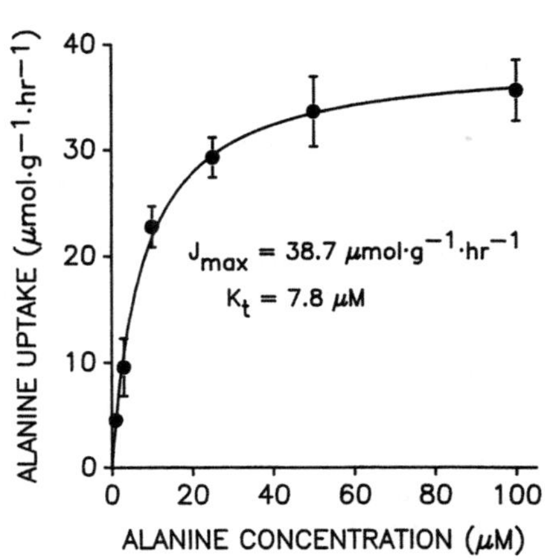

Fig. 1. Concentration dependency of ^{14}C-alanine uptake into gill tissue isolated from the mussel, *Mytilus californianus.* Lateral-ciliary activity in the isolated gills was reestablished by treating the tissue with 10 μM 5-HT (see text). Each point is a mean value determined from four different experiments (±1 SE). Uptakes were based on 5-min incubations and were expressed in terms of wet tissue mass. The kinetic constants indicated in the figure were determined from a nonlinear regression fit of the data to the Michaelis-Menten equation. The line was calculated by inserting these constants into the Michaelis-Menten equation. (Data from Wright 1987)

Table 2. Kinetics of amino acid uptake into bivalve gill tissue

Animal	SUB	K_t (μM)	J_{max} (μmol g^{-1} h^{-1})	Temp.	Reference
Bankia gouldii	GLY	8.2	13.4[a]	23 °C	Stewart and Dean (1980)
Modiolus demissus	GLY	21	21.4[a]	RT	Crowe et al. (1977)
Mya arenaria	ALA	94	6.9	10	Stewart and Bamford (1975)
Mytilus	ALA	7.7	39	20	Wright (1987)
californianus	TAU	4.0	11.4	20	Wright (1987)
Mytilus edulis[b]	GLY	4.2	12.8[c]	14	Jørgensen (1983)

[a] Original units expressed per gram dry mass; conversion made assuming dry mass to be 20% of wet mass.
[b] Uptake into intact, actively pumping animals.
[c] Original units expressed per gram dry mass of soft body parts; conversion made assuming the gill to be 15% of total tissue mass, and dry mass to be 20% of wet mass.

those in invertebrate intestine are on the order of 100×10^{-9} mol cm^{-2} min^{-1}) (Ahearn, this Vol.), and in mammalian intestine, values of between 60 and 600×10^{-9} mol cm^{-2} min^{-1}) (Karazov, this Vol.) are not unusual. Thus, while integumental transport must be considered to be of low capacity on an absolute scale, it will be shown below that it is of adequate capacity to influence invertebrate nutrition.

Measurement of K_t's for transport in gill tissue has been more equivocal. The apparent K_t values listed in Table 2 probably overestimate, in some cases by factors of more than ten, the effective half-saturation constants of the integumental transporters. Early reports contained K_t values for uptake into isolated gills ranging from 0.1 mM (Stewart 1978) to 0.25 mM (Wright et al. 1975). Considering the very low concentration of amino acid normally found in the environment (≤ 1 μM), these estimates for half-saturation constants led to suggestions that integumental pathways could play a comparatively small role in whole animal physiology (e.g., Bishop 1976). Studies with intact animals, however, resulted in much smaller apparent K_t's, i.e., 1 to 10 μM (Jørgensen 1983; Wright and Stephens 1978). This discrepancy stemmed from the effect of ciliary activity on the movement of water through the gill (Wright 1979; Wright et al. 1980). Lateral cilia are responsible for the normal, unidirectional flow of water through the gill (see Jorgensen 1975), and isolation of gill tissue results in an inhibition or complete cessation of lateral ciliary activity (Aiello 1970). The result is the development of "unstirred layers" above the gill surface. Membrane transport reduces substrate concentration in this stagnant layer, generating a concentration gradient between the "bulk" medium and the integument. If uptake is particularly vigorous, the rate of delivery of substrate to the transporter can be diffusion-limited. The extent of the discrepancy between substrate concentration in the medium and at the transporter is directly proportional to the transport capacity of the membrane (i.e., the J_{max}) and the thickness of the unstirred layer (see Winne 1973).

Unstirred layers have comparatively little effect on measurement of J_{max}'s; concentrations in the bulk medium can be raised to a point where diffusion through the unstirred layer provides a saturating level of substrate to the transporter. Michaelis constants, however, can be profoundly overestimated. Under subsaturating conditions, the substrate concentration in the medium is always greater than that at the membrane. In a tissue as morphologically complex as the bivalve gill, unstirred layers will vary in thickness over different regions of the organ. However, the complete cessation of lateral ciliary activity that occurs upon isolation of *Mytilus* gill results in the formation of unstirred layers of approximately 400 μm in "effective" thickness, and imparts upon the integumental uptake of glycine an "apparent K_t" as high as 130 μM, though the actual K_t is less than 5 μM (Wright et al. 1980). When ciliary activity in isolated gills is reestablished by addition of 5-hydroxytryptamine (5-HT) to the medium, uptake of amino acids is stimulated through a reduction in the apparent K_t (Wright 1979). It is now apparent that the uptake of amino acids into mussel gills, measured either in intact animals (Jørgensen 1983; Wright and Stephens 1978) or in vitro (Wright 1987; Wright and Secomb 1984) is characterized by apparent Michaelis constants ranging from 1 to 10 μM.

These comparatively low estimates for half-saturation constants still represent overestimates of the "true" K_t's of integumental transporters in the gill. Even with normal ciliary activity, the action of "upstream" transport sites results in the exposure of "downstream" sites to substrate concentrations that are 30–50% lower than that in the bulk medium. Gomme (1982a) has calculated that measurements of apparent K_t's in intact gill tissue overestimate the actual Michaelis constant of an integumental transporter by two- to tenfold, and this prediction has been confirmed in recent studies of uptake into gills of *Mytilus* (Wright 1985; Wright and Secomb 1984). Nevertheless, as emphasized by Jørgensen (1983), it is the concentration of substrate in sea water that results in a half-maximal activation of integumental uptake that applies to estimates of the role of DOM uptake to animal physiology.

This extended discussion of the influence of unstirred layers on transport kinetics has a relevance to our understanding of integumental transport that goes beyond the determination of the "true" kinetics of membrane transport. The integumental transporters of most organisms must ordinarily contend with unstirred layers; the effect of gill cilia in eliminating unstirred layers is a coincidence of the geometry of this specific tissue. An appreciation of the physiological role of an integumental transport process should be based upon an understanding of the workings of the transporter, uninfluenced by artifacts such as those introduced by unstirred layers, and on an understanding of the extent to which phenomena such as unstirred layers become an effective part of the system when viewed at a tissue or organismic level.

This latter point was emphasized in work by Gomme (1981a), who suggested that integumental transport, when at work in a system influenced by an unstirred layer, can result in a significant recovery of endogenous material lost from surface cells via passive diffusion. Gomme's studies were with the marine polychaete, *N. diversicolor,* and will be considered in more detail in a later section. His idea is, however, applicable to other integumental systems, including those

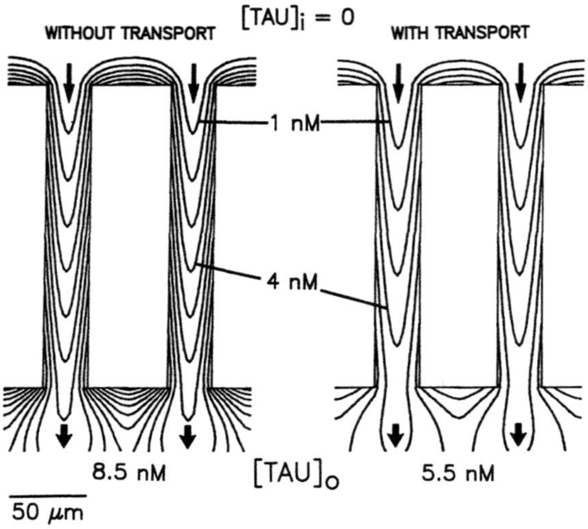

Fig. 2. Computer-generated contour plots showing the steady-state concentration of taurine in the water stream flowing between adjacent gill filaments as a result of the normal efflux of taurine from the gill epithelium in the absence (*left plot*) and presence (*right plot*) of normal integumental taurine transport activity. Incoming (i.e., incurrent) sea water was assumed to have no taurine. *Heavy arrows* show the direction of water flow, from the frontal toward the abfrontal aspect of the gill filaments. See Wright and Secomb (1984) for a discussion of the mathematical model used to generate these plots. The following values were assumed in the model: peak water flow between filaments of 0.12 cm s^{-1}: a taurine leak of 4×10^{-14} mol cm^{-2} s^{-1}; and for the case when taurine influx was considered, a K_t of 1.5×10^{-9} mol cm^{-3} with a J_{max} of 4×10^{-12} mol cm^{-2} s^{-1}. Integumental transport was estimated to recover approximately 35% of the taurine otherwise lost to the excurrent water. (After Wright and Secomb 1986)

of the bivalve gill. Materials lost from upstream regions of the gill become substrate for downstream regions. A model that takes into account the geometry and fluid-flow characteristics of gill tissue was recently introduced (the "convection-diffusion" model; Wright and Secomb 1984). Calculations made using this model indicate that integumental uptake of substrates such as taurine can result in a recovery of from 30 to 50% of the material lost from the gill by passive efflux (Fig. 2). This idea is supported by the experimental observation of increased taurine loss from bivalves when the integumental taurine transporter is blocked by addition of the competitive inhibitor, γ-aminoisobutyric acid (Wright and Secomb 1986). By sparing the loss of osmotically active substrates such as taurine, integumental transport pathways not only influence the nutritional status of the organism, but also play a role in the maintenance of normal cell volume.

3.1.2 Kinetics of Net Integumental Flux

Marine bivalves were the first organisms used to study the relationship between net substrate flux and the rate of influx of radiolabeled compounds (Crowe et al. 1977; Wright and Stephens 1977). Clearance from the medium of a radioactively

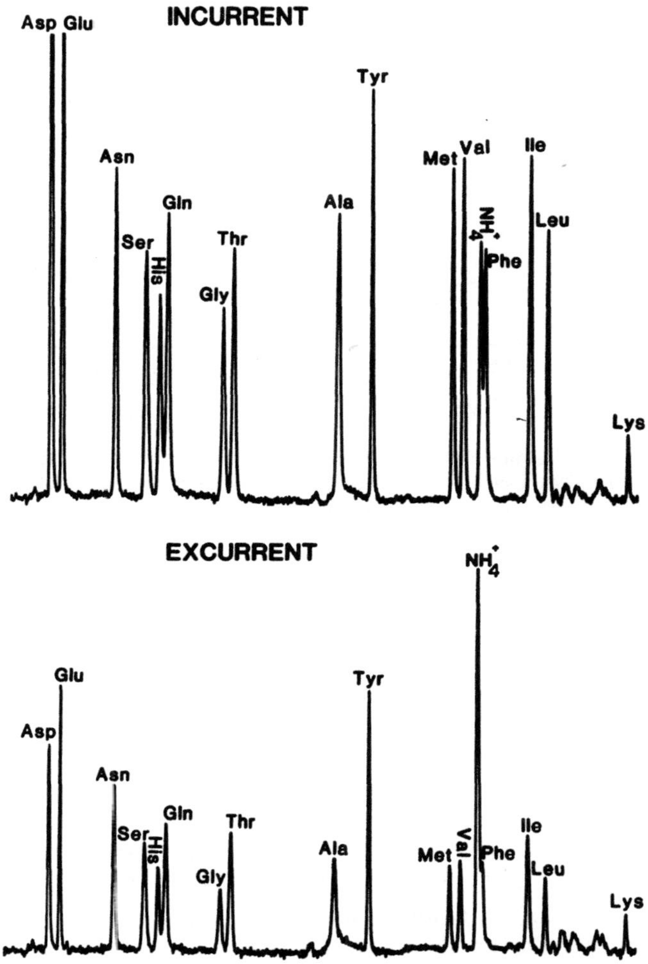

Fig. 3. HPLC chromatograms showing uptake of 16 free amino acids and release of ammonia by an intact *Mytilus edulis* during a single passage of water through the mantle cavity. Each amino acid was initially present in the artificial sea water medium at a concentration of 100 nM. *Incurrent* refers to a sample from the bulk medium; *excurrent* refers to a sample collected from the exhalant water stream following a circuit through the mantle cavity (Manahan et al. 1983a)

labeled amino acid was compared to clearance of total primary amines (TPA) as determined using fluorescamine. The former provided a relative measure of influx, the latter a measure of the net flux of amino acid. These studies showed that uptake of labeled substrate resulted in a net clearance of TPA, indicating that carrier-mediated influx did not involve an obligatory exchange for endogenous material. Fluorescamine has been used to demonstrate net accumulation of amino acids into bivalves from external concentrations as low as 0.1 µM (Ferguson 1982; Jørgensen 1983). Though, as mentioned earlier, these studies have not permitted an unambiguous measure of the lowest concentrations from which net amino acid uptake can occur.

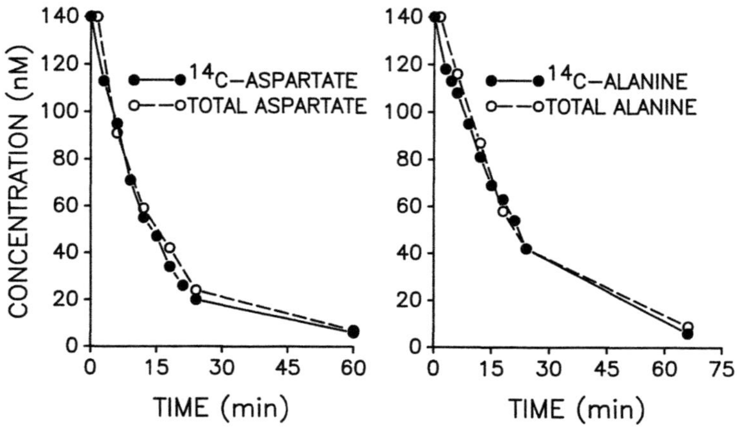

Fig. 4. Clearance from sea water of aspartate (*left plot*) or alanine (*right plot*) by intact *Mytilus edulis*. Clearance of ^{14}C-labeled substrate was as measure of influx into integumental tissues; clearance of total amino acid, chemically determined using HPLC, was a measure of the net substrate flux (Manahan et al. 1983 a)

The introduction of high performance liquid chromatographic (HPLC) techniques for the assay of amino acids in sea water (Lindroth and Mopper 1979) made it possible to clearly document that integumental transport in bivalves is capable of a net accumulation of substrate from environmentally relevant concentrations. Manahan et al. (1982) used HPLC to show a net uptake of amino acids into intact specimens of *M. edulis* from concentrations of less than 1 μM in natural sea water. Figure 3, taken from Manahan et al. (1983 a), compares the amino acid content of samples of exhalant water to that of samples of the surrounding medium, and reveals that 40 – 60% of each of the components of a mixture of 16 amino acids can be removed during a single circuit of water through the mantle cavity. The sensitivity of the HPLC technique also permitted a more accurate comparison of the relative rates of influx and net flux into intact mussels. As shown in Fig. 4, the clearance of radioactively labeled aspartate and alanine was virtually identical to the net clearance of these compounds, down to ambient concentrations that approached the limits of resolution of the HPLC technique ($\simeq$1 nM), indicating that, ordinarily, efflux of amino acids from mussels is virtually unmeasurable.

The results of experiments such as those shown in Fig. 4 contrast sharply with those of Bayne and his colleagues (see Bayne and Scullard 1977) who have reported that excretion of amino acids from *M. edulis* can represent a caloric drain of as much as 60% of these animals' energy requirements. I should emphasize that these earlier studies did not use analytical procedures as sensitive or specific as the HPLC techniques used in the more recent studies of Manahan et al. (1982, 1983 a) and others (Wright and Secomb 1986; Wright et al. 1987; Wright, unpubl. observations). Still, it is possible that these published discrepancies in apparent rates of amino acid loss from mussels are the product of differences in either animal handling procedures or in the populations of mussels used in the different studies. It would be worthwhile to consider these latter possibilities in a future, comparative study.

3.1.3 Multiplicity of Transport Pathways for Amino Acids

Amino acids, though often lumped together into a single generic class of organic substrate, are more accurately viewed as consisting of several structural categories which have in common an amino group ($-NH_3$) and an acidic group (either $-CO_2H$ or $-SO_3H$). The common "neutral" (i.e., zwitterionic) amino acids include alanine, glycine, and serine. The acidic (anionic) amino acids, including aspartic acid and glutamic acid, and the basic (cationic) amino acids, including lysine and arginine, comprise two other structural classes of these compounds. Finally, the beta-amino acid, taurine (2-aminoethane sulfonic acid), represents a unique structural class of amino acid. It is also the primary constituent of the free amino acid pool of the integumental tissues of many bivalves, including *Mytilus* (Zurburg and de Zwaan 1981).

Studies of the inhibitory interactions of different amino acids on integumental uptake in bivalves indicate that there are several, separate transport pathways for these compounds in the gill. There appear to be at least four separate amino acid transporters in *Mytilus* gill (Wright 1985), including two pathways for α-neutral amino acids, one of which will interact with lysine, while the other shows a specificity for the imino amino acid, proline; a third pathway specific for the anionic amino acids, aspartate and glutamate; and a fourth pathway for β-amino acids, including taurine, which is insensitive to α-amino acids. Of interest, taurine transport into *M. edulis* mantle tissues is inhibited by α-neutral amino acids (Rice and Stephens, in press), suggesting the presence of two different transporters for β-amino acids in this animal. In *Modiolus demissus*, transport of neutral amino acids, including glycine, occurs by a transport pathway (or set of pathways) that is distinct from those for anionic and cationic substrates (Crowe et al. 1977). In a series of studies on the kinetics of amino acid uptake into gills from the soft-shelled clam, *Mya arenaria*, (summarized in his 1979 review) Stewart found evidence of at least four separate pathways for (1) zwitterionic, (2) imino, (3) basic, and (4) acidic amino acids. Amino acid transporters in gill tissue show a limited degree of stereospecificity, selecting for L-isomers (e.g., Stewart and Bamford 1975; Wright et al. 1975), though D-isomers show a significant degree of competitive interaction and can be accumulated as well (Preston 1986; Stewart and Bamford 1975).

3.2 Role of Amino Acid Uptake in Bivalve Nutrition

Two approaches have been used to estimate the nutritional effect of DOM uptake by mollusks, and by other marine organisms, as well. The "direct" approach involves holding animals in a particulate-free medium for extended periods and monitoring growth as a function of exposure to various diets composed of dissolved foodstuffs. While evidence gained from such tests would appear to be the most convincing in terms of establishing a nutritional role for integumental transport, interpretation of the results of direct studies has been difficult. Animals used in these experiments have usually been (unavoidably) contaminated with microorganisms. Microorganisms can certainly make use of dissolved substrates and are expected to subsist on such material. Thus, it can be a problem to discern to what extent observed growth is a result of a direct utilization of DOM via integumental uptake, or is indirect, via ingestion of microorganisms. Still, through the use of flow systems and filtered water, studies have been made in an effort to test directly the nutritional role of integumental DOM uptake (e.g., Ferguson 1980b). The caveats associated with the use of growth studies have, however, made the direct approach of limited use.

One significant exception to this general statement has been the recent development of procedures for the rearing of cultures of axenic and/or nonfeeding invertebrate larvae. Use of these preparations represents one of the most exciting breakthroughs in the study of integumental transport, and will be discussed in more detail below.

A more commonly used approach to assess the nutritional potential of DOM uptake has been "indirect", and involves comparing the rate of utilization of foodstuffs to the rate at which nutrients can be supplied by integumental transport. Such calculations are based upon the stoichiometric relationship between oxidation of a nutrient and, for example, the consumption of O_2 (Q_{O_2}). The logic is as follows. Consider an uptake process that provides substrate at a rate of five units per hour. Further, assume that the oxidation of each unit of substrate requires 1 ml O_2. If the whole animal Q_{O_2} is 10 ml h^{-1}, then uptake has the *potential* to support 50% of the animal's needs for oxidizable substrate.

Through such calculations, integumental DFAA uptake can be shown to have a very large nutritional potential in organisms such as *Mytilus*. The maximal rate of amino acid uptake, with all the integumental pathways for amino acid taken into consideration, is approximately 9 μmol g^{-1} dry mass of soft body parts per hour, with an effective half-saturation constant of approximately 4 μM (estimated from Jørgensen 1983; Wright 1985). The complete oxidation of a mixture of amino acids reflecting that found in the marine environment (see Braven et al. 1984) requires approximately 1 ml O_2 mg^{-1} amino acid. It is useful to convert rates of amino acid uptake into units which reflect the oxidative potential of accumulated substrate. Maximal rates of integumental substrate input can thus be seen to support 0.9 ml O_2 consumption g^{-1} h^{-1}. The Q_{O_2} for both *M. edulis* and *M. californianus* ranges from "standard" rates (i.e., nonfeeding) of approximately 0.25 ml O_2 (g^{-1} dry mass h^{-1}) to "routine" rates (i.e., actively feeding) of 0.5 ml O_2 (g^{-1} h^{-1}) (Bayne and Newell 1983). Uptake of amino acids may then provide substrate to account for between 200 and 400% of the total needs of these animals for oxidizable substrate.

Of course, it is unlikely that integumental transport pathways will be exposed to saturating levels of substrate, and the above calculation represents the upper limit of the nutritional input of integumental amino acid uptake. Of critical importance, then, in assessing the effective nutritional role of this process is knowledge of the concentration of amino acids in the environment. Recent studies employing sensitive HPLC procedures have reported total DFAA concentrations of near shore waters ranging from as low as 10 – 200 nM (Mopper and Lindroth 1982) to as high as 1 – 3 μM (Braven et al. 1984; Siebers and Winkler 1984). This disparity quite probably reflects a real variation of DFAA levels in the marine environment. As mentioned previously, amino acid concentrations can vary with season, depth, and time of day, and can vary to a remarkable extent (1000-fold) over horizontal distances of several meters. For the purposes of calculating the input of amino acids from integumental transport, a range of 200 to 700 nM is one most mussels will probably encounter, either daily or seasonally. With the kinetic parameters for uptake noted above, integumental input of amino acids could sustain from between 0.04 and 0.13 ml O_2 consumption $g^{-1}\ h^{-1}$, or from between 8 and 55% of a mussel's total needs for substrate to support oxidative metabolism. The "significance" of a nutritional input of this size is difficult to estimate. Still, it is apparent that integumental amino acid uptake can provide substrates capable of supporting a large percentage of these animals' nutritional requirements from concentrations that are well within the range to which they are normally exposed.

3.2.1 Assumptions Inherent in Estimates of the Nutritional Role of Integumental Transport

The calculations used to make an indirect estimate of the nutritional potential of DOM uptake in animal nutrition are based upon several important assumptions. The first concerns the substrate concentration to which the animals are exposed. In the case for bivalves noted above, the range of 0.2 to 0.7 μM appears to be reasonable, based upon the current literature, but these values and the techniques used to measure them are not uncontroversial (Fuhrman and Bell 1985). Furthermore, there is no clear indication of how often animals are exposed to the extremes of substrate concentration; i.e., when is uptake providing a major input of substrate, when is it minor? I think that the major outstanding issue in attempts to quantify the nutritional role of integumental transport is the question of environmental substrate availability. Until a more detailed survey of environmental concentrations of DFAA (and other constituents of the DOM pool) is made, this field will be limited to rather vague estimates of "nutritional potential".

A second assumption that is pivotal to the use of indirect assessments ot the physiological role of DOM uptake is that O_2 consumption is a valid measure of an organism's energy metabolism. Many marine invertebrates, including most marine bivalves, are facultative anaerobes. To the extent that anaerobic pathways are used to support an organism, the Q_{O_2} will underestimate the needs of energy metabolism. Until recently, there was speculation that a number of marine bivalves, including *M. edulis*, make extensive use of anaerobic energy-producing pathways, even when exposed to fully oxygenated sea water (Hammen 1979, 1980). It is now quite clear, however, that O_2 consumption can provide a quantitative measure of energy metabolism, based upon direct comparisons of rates of Q_{O_2} with rates of total heat production (Hammen 1983). It should be stressed, however, that rates of respiration underestimate the nutrient needs of growth and reproduction. In the case of bivalve filter feeders, these may exceed by 50% the energy required to support maintenance metabolism (Bayne and Newell 1983).

A third assumption is that accumulated nutrients are available to the general metabolic pool of the cell. Though comparison of "input to needs" does not necessitate that accumulated materials be immediately funneled into oxidative pathways, it does imply that they be available, eventually, to the general cellular metabolic pool. Outside of the specialized literature on the influence of osmotic concentration on the metabolism of accumulated amino acids by bivalve tissues (see Bishop et al. 1983), there is little information on the metabolic "fate" of accumulated DOM. On a gross level, increased times of exposure to labeled amino acids does result in an increase in the amount of label recovered in "alcohol-insoluble" fractions of bivalve tissue (e.g., Anderson 1975; Péquinat 1973; Stewart 1977; Wright et al. 1984), and in the labeling of autoradiograms of nonglutaraldehyde fixed tissues (Péquignat 1973; see Manahan and Crisp 1983), which suggests that the labeled carbon skeletons are gradually incorporated into large molecular weight compounds. These observations, however, provide no insight into such questions as to whether the metabolic state of the organism influences the utilization of accumulated material, or if transported substrates are funneled into metabolic compartments that are distinct from those of the general cellular pool. This latter phenomenon has been suggested to occur in some tissue culture systems (Wheatley and Inglis 1984), and the very large free pools of amino acid in integumental cells of marine invertebrates pose some interesting questions for the regulation of metabolic processes (see also Gomme 1984). The integration of integumental uptake with cellular metabolism represents an important area of future research.

Related to this discussion is the question of whether materials that enter the integument are distributed to the "deeper" tissues of the body. In the case of bivalves, autoradiography has been used to document the translocation of radioactive label from integumental tissues of adult *M. edulis* (Péquignat 1973) and larval *M. edulis, Crassostrea gigas,* and *Ostrea edulis* (Manahan and Crisp 1983) to deeper tissues of the viscera. Whether all integumental surfaces participate equally in "exporting" accumulated materials is not clear, however. Jørgensen (1983) reported that there was no significant loss of taurine from gills of *M. edulis* 24 h following exposure to labeled material, suggesting that there had been no significant translocation of this material to deeper tissues. Rice and Stephens (in press) found that surgical removal of the gill from *M. edulis* has no effect on the rate of appearance of radiolabeled amino acids into the hemolymph. The implication of these studies is that the translocation evident from autoradiographic studies is limited to exchange between the mantle (probably) and the deeper viscera. Thus, it may not be valid to compare uptake into gill tissue to the metabolic needs of the whole animal, but, instead, to view such uptake as providing for the needs of the gill itself (see C.B. Jørgensen 1982). Questions about the extent of "transintegumental" DOM exchange have arisen as a critical subject for study.

Finally, the estimation of integumental input is based upon rates of substrate accumulation measured under laboratory conditions. Great pains are often taken in studies of energy metabolism to control environmental parameters, such as temperature and feeding ration. Studies of integumental transport processes have not been characterized by such attention to the metabolic status of the test organism. Typically, animals are collected and held in the laboratory, often in "artificial sea water", for days or weeks prior to use in experiments measuring rates of uptake. Holding temperatures seldom mimic those of the environment, and the animals are usually kept starved. Both of these latter conditions have been shown to influence energy metabolism of marine bivalves (Bayne et al. 1976), and probably influence other types of organisms as well. It is, therefore, appropriate to view with caution comparisons of metabolic rates to rates of uptake. To help resolve this issue, it would represent a major advance in our understanding of the nutritional role of integumental uptake to couple the direct calorimetric measurement of energy metabolism in freshly collected animals with studies of net flux and influx of amino acids (e.g., is there a "specific dynamic action" associated with integumental exposure to amino acids?). The techniques for such a study are currently available.

C. B. Jørgensen (1982) offered a further caveat regarding quantitative interpretation of laboratory measurements of integumental uptake. He suggested that animals may "up-regulate" the integumental transport process following exposure to amino acid-free sea water, i.e., normal laboratory holding conditions. This idea arose from his conclusion that measured rates of uptake into gill tissue are so large that they could profoundly perturb the intracellular composition of the tissue-free amino acid pool. He proposed that exposure to the normal levels of amino acid in sea water might result in a adaptative "down-regulation" of integumental transport. Evidence for this type of regulatory response in mollusks is, however, very limited. Prolonged (24 h) exposure to large (2 mM) glycine concentrations does reduce the rate of net glycine uptake into intact *M. edulis* (Jørgensen 1983), though the physiological relevance of this observation is unclear given the high concentration of amino acid used to elicit the response. Furthermore, rates of net amino acid uptake from natural sea water into freshly collected *M. edulis* (Siebers and Winkler 1984) are comparable to rates measured in laboratory studies (Wright and Stephens 1978; Manahan et al. 1983a; Jørgensen 1983). Nevertheless, the issue of regulation of integumental transport in bivalves remains essentially unexplored, and warrants further study.

3.3 Uptake of DOM in Larval Bivalves

Though examination of integumental transport has been largely confined to work with adult organisms, there has been a recent surge of interest in uptake of DOM by larval invertebrates (see Manahan and Crisp 1982). These preparations not only permit the study of developmental questions pertaining to nutrient transport, but also serve as model systems for the study of issues of general importance to the field of integumental transport.

In a series of studies, Manahan and his colleagues have characterized the integumental uptake of free amino acids into several larval bivalves. Activation of glycine uptake into *C. gigas* eggs occurs upon fertilization, with rates of uptake increasing for at least 3 h thereafter (Manahan 1983 a). Autoradiograms of larval oysters (*C. gigas* and *O. edulis*) and mussels (*M. edulis*) exposed briefly to ^{3}H-glycine showed that the velum is the principal site of uptake, while in newly settled oysters and scallops (*Pecten maximums*), the developing gill buds were found to be the major site of transport (Manahan and Crisp 1983). Accumulated amino acids are quickly metabolized in larval oysters and mussels. Following a 100-min exposure to ^{14}C-glycine, 47% of the radioactive label was found in protein (i.e., cold TCA-insoluble material), while less than 2% was found lipid (Manahan 1983 b). By 100 min, the rate of production of $^{14}CO_2$ was 38% of the rate of accumulation of labeled material in the larvae. The apparent Michaelis constants for glycine uptake into *C. gigas* and for alanine uptake into *M. edulis* larvae were 3.7 and 3.5 μM, respectively (Manahan 1983 b). The latter value is comparable to the K_t for alanine uptake in adult *Mytilus* (Wright 1985). The J_{max}'s for alanine uptake ranged from 1 – 2 ng per larva h^{-1}, which after appropriate conversion of units represents a maximal transport capacity of approximately ten times that noted in studies with adult bivalves, a probable reflection of the difference in surface to volume ratios of the larval and adult organism (Manahan 1983 b). These rates of uptake could supply an estimated 2 to 10% of the material required to account for larval protein synthesis (Manahan 1983 b).

Bivalve larvae were also used to study competition for dissolved substrates between bacteria and animal tissue. The K_t's of bacterial transport processes are often in the nanomolar range, while those for integumental transport are on the order of 1 to 10 μM or more. Such comparisons lead to the suggestion (Sepers 1977) that bacteria can utilize more efficiently than invertebrates the low concentrations of DOM in the environment. This hypothesis was tested (Manahan and Richardson 1983) by directly comparing the uptake of glycine into larvae (*M. edulis*) and bacteria. At substrate concentrations above 0.3 μM, uptake into the larvae exceeded that into an equivalent biomass of bacteria, reflecting that though the K_t for glycine uptake into bacteria was less than that of the larval transporter, the larval J_{max} for uptake was also much greater than the bacterial J_{max}. Manahan and Richardson (1983) concluded that over the concentration range larval (and adult) bivalves are exposed to, integumental transport processes are capable of effectively competing with bacteria for dissolved substrates.

Larval systems are particularly attractive to studies of the nutritional potential of integumental DOM uptake. Langdon (1983) reported that a particulate-free diet can support growth of bacteria-free cultures of larval oysters (*C. gigas*). Shell growth of 6-day-old larvae was significantly greater when exposed to a semidefined diet of dissolved material, compared to starved controls. Unfortunately, this study did not include data on changes in dry organic mass of the larvae, so it is difficult to conclude that uptake of DOM supported net growth. Such measurements were made, however, in a recent study from Manahan's laboratory. In a preliminary report (Manahan and Jaeckle 1986), exposure to naturally occurring levels of DOM in particulate-free sea water was found to result in a doubling of the organic biomass of nonfeeding stages of larval *C. gigas*. Taken together, these studies represent the best evidence to date that marine invertebrates can make direct nutritional use of exogenous DOM.

3.4 Regulation of Integumental Transport in Marine Mollusks

The means by which integumental DOM uptake is regulated has received comparatively little attention. C.B. Jørgensen's (1982) observation, mentioned earlier, that prolonged exposure to large (2 mM) concentrations of amino acid inhibits net integumental uptake warrants further study. In particular, it would be of interest to measure changes in the composition of the tissue and hemolymph pools of free amino acid, as well as changes in rates of integumental uptake, following the controlled, prolonged exposure of intact animals to large but environmentally relevant levels (e.g., 2 to 5 μM) of these compounds.

Control of integumental permeability, including mediated transport pathways, has been examined in the context of volume regulation in marine bivalves. The large free amino acid concentrations characteristic of bivalve tissues are believed to be involved in isosmotic volume regulation (e.g., Gilles 1979). Exposure to dilute sea water results in a rapid loss of free amino acids from bivalve tissues to the surrounding medium (e.g., Henry and Mangum 1980; Livingston et al. 1979; Pierce and Greenberg 1972; Wright et al. 1987). The mechanism responsible for controlling membrane permeability to amino acids is not known, but efflux from isolated ventricles from the mussel, *M. demissus*, has been shown to be controlled by the ambient concentration of the divalent cations, Ca^{2+} and Mg^{2+} (Pierce and Greenberg 1973). Furthermore, removal of Ca^{2+} and Mg^{2+} from sea water has been shown to inhibit integumental uptake of glycine into gill tissue from *M. californianus*, while also stimulating efflux of amino acid from this tissue (Swinehart et al. 1980). Whether divalent cations have a direct effect on integumental transport processes, or instead exert their primary influence on general membrane permeability to amino acids is not clear. The role of such effectors in the regulation of integumental permeability remains an extremely exciting area of future research.

3.5 Integumental Transport of Non-Amino Acid Constituents of the DOM

In addition to the multiple pathways for uptake of amino acids, bivalve gill has separate pathways for the accumulation of a wide variety of other small organic compounds. Dipeptides, such as glycyl-L-phenylalanine may be accumulated by *M. edulis* gill (Stewart 1981) by a process separate from that for individual amino acids. Péquignat's (1973) autoradiographic study of integumental uptake in *M. edulis* provided evidence for accumulation of glucose into a marine bivalve. The characteristics of carrier-mediated sugar transport have subsequently been described in gill tissue from *C. gigas* (glucose, Bamford and Gingles 1974), the oysters, *O. edulis* (α-methylglucoside, Riley 1981), and *Saccostrea commercialis* (glucose, Nell et al. 1983). Short-chain carboxylic acid uptake has been demonstrated in gills of *Mercenaria* sp. (DiDomenico and Iverson 1977), while long-chain fatty acid uptake was found in intact specimens of *Crassostrea virginica* (Bunde and Fried 1978). Finally, concentrative transport of several prostaglandins has been found to occur in tissues, especially the gill, from *Spisula solidissima* and *Placopecten magellanicus* (Bito 1972).

4 Integumental DOM Transport in Echinoderms

Initial interest in DOM uptake in echinoderms focused primarily on suggestions (Ferguson 1967a, b) that there is limited exchange of materials between the integument and the viscera, and that the integument, having an "aggressive digestive" capacity (Péquignat 1966), may largely rely on materials absorbed from sea water. This idea received support from autoradiographic studies showing little exchange of materials between the epidermis and deeper tissues including the gut (e.g., Ferguson 1967a, 1970). Later studies, however, cast doubt on the presence of a barrier to exchange between the integument and deeper tissues of echinoderms (Pearse and Pearse 1973; Péquinat 1972), and there is recent evidence that accumulation of DOM is not essential to the maintenance of the integument of asteroids (Ferguson 1980b). Nevertheless, measured rates of integumental DOM transport indicate that it can serve to supplement to the nutrition of some echinoderms. Furthermore, there is an increasing interest in the possible role of integumental accumulation of DOM in the support of tissue regeneration by the nonfeeding forms of animals that use radical autotomy or evisceration as a means to escape predation (Smith et al. 1981).

4.1 General Characteristics of Integumental Transport in Echinoderms

Amino acid uptake is generally saturable and adequately described by Michaelis-Menten kinetics, though at high substrate concentrations ($>10-50$ μM), a first-order process is sometimes evident which may represent diffusion (e.g., Ahearn and Townsley 1975; de Burgh et al. 1977). Zwitterionic amino acids are transported by a pathway or set of pathways with a broad structural specificity for this structural class of amino acid (Ahearn and Townsley 1975; de Burgh 1978); cationic and anionic amino acids do not share this pathway (de Burgh 1978). Stephens et al. (1978) reported that, compared to uptake of neutral amino acids, epidermal tissues from the sand dollar *Dendraster excentricus* accumulated aspartate and lysine slowly, and glutamate not at all. Davis and Stephens (1984a) confirmed these observations in a recent study of net amino acid uptake into bacteria-free larvae of *D. excentricus.* They also noted low rates of taurine uptake in this animal. In contrast, intact specimens of the brittle star *Ophionereis annulata* accumulate anionic amino acids and taurine at rates comparable to those for α-neutral substrates, though lysine accumulation is comparatively low (Davis et al. 1985b).

It is difficult to compare J_{max}'s of transport in echinoderms expressed per gram of tissue with those measured in other groups of organisms because of the presence of skeletal plates in the tissues of the former. Even after taking this consideration into account, the transport capacity of the integument has often been found to be relatively low. The J_{max} for alanine uptake into epidermis of spines from the sea urchin, *Paracentrotus lividus,* expressed per gram dry wt. of decalcified tissue, is approximately 1.6 μmol h^{-1} (de Burgh et al. 1977). The J_{max} for uptake of α-aminoisobutyric acid into the body wall of the holothuroid,

Chirodota rigida (which does not have calcarous plates), is 9 nmol g^{-1} wet wt. h^{-1}. While these systems are of low capacity, compared even to other integumental processes (see Table 2), they have a comparatively high affinity for substrate, with K_t's ranging from 6 to 8 μM (Ahearn and Townsley 1975; de Burgh et al. 1977). Thus, under naturally occurring conditions, uptake into these echinoderms may operate at a considerable fraction of their maximal capacity. Uptake of glycine into *D. excentricus,* however, has a capacity on a par with transport in bivalves (J_{max} = 3.6 nmol cm^{-2} min^{-1}), though it is also characterized by a rather low affinity for substrate (apparent K_t of 74 μM; Stephens et al. 1978).

Siebers (1982) suggested that uptake of DOM into echinoderms and other marine invertebrates may be complicated by uptake into epiphytic bacteria. His subsequent work (Siebers and Winkler 1984) has not, however, supported this hypothesis. Furthermore, Davis and Stephens (1984a) compared uptake of 14 different amino acids into axenic and nonaxenic cultures of *D. excentricus* larvae; over time courses of 4 h or less, there was no significant difference in rates of uptake into the two cultures. Though the role of the subcuticular bacteria common to many echinoderms (Holland and Nealson 1978; see Sect. 4.2) in integumental transport is not clear, it is increasingly apparent that epiphytic bacteria play no significant role in integumental transport by soft-bodied marine invertebrates.

Integumental uptake of amino acids in echinoderms is usually representative of a net accumulation of substrate. This was first shown by Ferguson (1971), and was later confirmed in studies of uptake of total primary amines (Ferguson 1980a, 1982; Stephens et al. 1978). These general observations have been extended in more recent studies using HPLC techniques (Davis et al. 1985b; Davis and Stephens 1984a, b; Manahan et al. 1983b) to measure net uptake of 14–18 different amino acids. It is clear that integumental transporters in echinoderms are capable of a net accumulation of DFAA from ambient concentrations of less than 0.1 μM.

4.2 Nutritional Role of Integumental Uptake in Adult Echinoderms

Stephens et al. (1978) estimated the potential contribution of integumental amino acid transport to the nutrition of *D. excentricus.* This animal is found partially buried in sandy sediments and therefore is likely to be exposed to concentrations of substrate greater than those in the water column. They measured uptake of naturally occurring primary amines collected from the sediments in which the animals were collected. Rates of uptake from the high concentrations found in sediment water (115 ± 60 μM) suggested that such uptake could account for between 60% to >100% of the animals' oxidative requirements. More recently, Davis et al. (1985b) used HPLC to measure the amino acid concentrations in samples from the benthic habitat of the brittle star, *O. annulata.* The total concentration of identified free amino acids was 35 ± 8.7 μM. Net uptake of substrate from such concentrations could account for approximately 75% of the measured O_2 consumption of these animals. Finally, Ferguson (1982) reported rates of net uptake of primary amines from concentrations representative of those he measured in Tampa Bay. He found that between 0.2 and 37.2% of the oxidative needs of five different species of echinoderm could be met by net epidermal uptake.

It should be stressed that, as with marine mollusks, the impact of integumental uptake on echinoderm nutrition will be a direct function of the substrate concentration that the animals are exposed to in the environment. Little is known about the concentration of DOM at the integumental surface of an organism living in or on the sediment. Ferguson's (1982) comparatively low estimates of nutritional potential reflected his measurement of environmental amino acid concentrations of approximately 0.8 μM, i.e., the primary amine concentration several centimeters above the benthos. In light of evidence that there is a steady release of amino acids from sediments to overlying waters (N. O. G. Jørgensen 1982), this may have underestimated the actual substrate concentration of surfaces in direct contact with the sediments. On the other hand, the amino acid concentrations of interstitial water may overestimate the concentration of substrate that reaches epidermal surfaces of burrowing animals; mixing of media overlying the integument is probably minimal, introducing the likelihood of a substrate-depleted unstirred layer next to transporting surfaces. Thus, while it is evident that DOM uptake can play an important supplementary role in the nutrition of benthic echinoderms, further work needs to be done on the dynamics of substrate availability in the environment of these animals.

Integumental uptake of DOM has also been suggested to play a role in supporting the growth and development of regenerating tissues in those echinoderms that undergo a nonfeeding stage following the partial or complete evisceration or autotomization used as a means to avoid predation. Smith et al. (1981) reported an increased production of $^{14}CO_2$ by newly eviscerated specimens of the crinoid, *Cenometra bella,* exposed to a mixture of radiolabeled amino acid. Similarly, in a recent preliminary report, Clements (1986) noted that the brittle star, *Microphiopholus gracillima*, which goes through a nonfeeding stage following loss of the oral disk, increases the rate of net accumulation of amino acids following autotomy. Interpretation of these results is complicated by the presence of endosymbiotic bacteria between the cuticle and the integumental epithelium of many echinoderms (e.g., Holland and Nealson 1978). Indeed, there are recent suggestions that the presence of such bacteria may play a role in uptake of DOM (Lesser 1986). Still, these studies represent the introduction of valuable new experimental systems whith which to study the nutritional role of integumental uptake of DOM.

4.3 Transport in Larval Echinoderms

There is a considerable literature on this topic, stemming from a long-standing interest in the membrane physiology of newly fertilized sea urchin eggs, and a more recent interest in DOM transport by axenic cultures of echinoderm larvae.

Early observations (e.g., Berg 1968) showed that developing sea urchin embryos could accumulate radiolabeled amino acids from sea water. Epel (1972) reported that a Na^+-dependent transport of amino acid by eggs of *Strongylocentrotus purpuratus* is activated within minutes of fertilization. This has been confirmed and expanded upon in several recent studies (Allemand et al. 1984, 1985, 1986; Schneider 1985). These studies have shown the presence of a single, major transport pathway with broad specificity for neutral amino acids. However, unlike the situation in adult and larval echinoderms noted above, influx of radiolabeled amino acid by fertilized eggs is not necessarily indicative of a net accumulation of amino acid. In a preliminary report, Manahan (1986) observed that unfertilized eggs of *S. purpuratus* lose glycine to sea water; following fertilization, this loss decreases, with a net influx of amino acid noted approximately 30 min after fertilization. He suggested that activation of influx pathways may serve to reduce losses of amino acids, as well as to acquire metabolic substrates from sea water.

The question of the possible nutritional role of developing integumental transport processes has been addressed more fully in studies of uptake in axenic cultures of echinoid larvae (5 – 7 days postfertilization). Manahan et al. (1983 b) reported rates of net uptake of neutral amino acids (total conc. = 1.63 μM) into *S. purpuratus* larvae that could account for 79% their oxidative requirements. Also of interest was their calculation that the accumulation of nitrogen represented by this uptake was equivalent to that lost as ammonia. Thus, amino acid uptake may play an important role in maintaining the nitrogen balance of growing larvae. A similar study with axenic *D. excentricus* larvae (Davis and Stephens 1984 a) indicated that net uptake from an ambient concentration of 3.5 μM mixed amino acid (250 nM each of 14 different amino acids) could account for 68% of the larvae's oxidative needs, and 73% of the nitrogen excreted as ammonia. Furthermore, this uptake accounted for approximately 12% of the amino acid required for the synthesis of new protein.

Again, the quantitative impact of integumental uptake hinges on the environmental concentration of substrate. As pointed out earlier, amino acid concentrations in near and offshore waters are low, with levels of 20 to 200 nM being common in offshore waters (e.g., Mopper and Lindroth 1982). However, the patchy distribution of dissolved organic nitrogen (Smith 1986) and free amino acids (Manahan and Arnold 1983) described previously suggests that free swimming larvae may be exposed, transiently or perhaps routinely, to higher substrate levels than "bulk water" samples would suggest and leeds to the exciting possibility that larval organisms may be attracted to micro "oases" of high DOM levels. In any event, measured rates of amino acid uptake from concentrations that have been measured in the water column indicate DOM may represent a significant source of supplemental nutrition during early stages of growth and development in echinoderms.

4.4 Regulation of Integumental Uptake in Echinoderms

Echinoderms, as well as other soft-bodied marine invertebrates, maintain intracellular free amino acid pools within tight limits. Davis and Stephens (1984 b), noting that this regulation occurs in the face of what is likely to be a highly variable input of free amino acids from the external environment, suggested that integumental transport is likely to be under some form of control responsive to the level of substrate in the ambient medium. They observed such a regulation following a 4-h exposure of *S. purpuratus* larvae to large (>50 – 1000 μM) concentrations of serine. This treatment, reminiscent of Jørgensen's (1983) study with adult *M. edulis* (see Sects. 3.2.1 and 3.4), resulted in a 16- to 23-fold increase in intracellular serine content. Though serine influx was not influenced by this exposure, a carrier-mediated efflux of serine did increase in parallel with increasing concentration of serine in the external solution. The result was a net loss of this compound from the enlarged intracellular pool. They pointed out the adaptive significance of a response that may maintain a "physiologically acceptable total concentration of the internal DFAA pool by selective efflux". Though no mechanism was proposed to account for their observation, this work points the way to studies on the regulation of membrane permeability as a mechanism to influence the intracellular pool of DFAA in integumental tissues.

4.5 Integumental Uptake of Non-Amino Acid Constituents of the DOM

Glucose has been shown to be accumulated by epidermal tissues of several echinoderms, including *P. lividus* (de Burgh et al. 1977), *Cucamaria lactea* and *Amphipholis squamata* (Fontaine and Chia 1968), and *A. forbesi* and *Henricia sanguinolenta* (Ferguson 1967b). Unfortunately, there is no information concerning the kinetics of these processes.

Nishioka and his colleagues (see Nishioka et al. 1985) have characterized the transport of nucleosides, including thymidine, in newly fertilized sea urchin eggs. This process is quite distinct from the transport pathways for amino acids in fertilized sea urchin eggs noted earlier.

5 Integumental Uptake of DOM in Annelids

The relatively simple, veriform structure of the typical worm has proved useful in the study of integumental uptake of DOM, resulting in abundant literature on transport in these animals (see Gomme 1984 for an extensive review of this material). Further, the sediment habitat of many burrowing marine annelids is particularly rich in organic material, so they are likely to be exposed to considerably higher concentrations of DOM than animals, such as most mollusks and echinoderms, that live on or above the surface of the benthos. So clear is the evidence that integumental transport can contribute to the nutrition of marine annelids, that reference to this feeding mechanism is now commonly found in basic texts of environmental physiology (e.g., Gordon 1982; Schmidt-Nielsen 1983).

5.1 General Characteristics of Integumental Transport in Annelids

Uptake of DOM into worms occurs directly into the general body surface, as shown by studies with animals whose mouth and anus are ligated (e.g., Ahearn and Gomme 1975; Stephens 1963), and by experiments using isolated pieces of body wall (Chien et al. 1972; Stevens and Preston 1980a–c). In some species, e.g., *G. dibranchiata,* higher rates of uptake have been noted in parapodial tissues (Chien and Rice 1985), and these surfaces may be the primary sites for integumental transport.

As in the cases cited previously, Michaelis-Menten kinetics provide an adequate description of the concentration dependency of integumental transport in annelids. Several studies have measured the concentration dependency of amino acid uptake into several species of annelids, including *G. dibranchiata* (Chien et al. 1972; Stevens and Preston 1980b), three species of *Nereis* (Jørgensen and Kristensen 1980), and the oligochaete, *Enchytraeus albidus* (Siebers and Bulnheim 1977). Maximal rates of uptake range from approximately 0.1 to 2.8 μmol g^{-1} h^{-1} (see also Preston and Stevens 1982). In a few studies, uptakes have been expressed on a surface area basis, permitting a more meaningful comparison to values determined in other systems. For uptake of alanine, Stevens and Preston (1980b) reported a J_{max} of 3.8×10^{-9} mol cm^{-2} min^{-1} (body wall of *G. dibranchiata*), and Gomme (1984) reported a preliminary value of 0.3×10^{-9} mol cm^{-2} min^{-1} (intact *N. diversicolor*). These numbers are quite comparable to those noted for amino acid uptake into gills from the mussel, *Mytilus* (Table 2), and into the integument of echinoderms noted earlier.

Estimates of the Michaelis constants for amino acid uptake in annelids range from values as low as 20 μM (Siebers 1976) to in excess of 2 mM (Chien et al. 1972), with the majority being within the range of 50 to 250 μM (Preston and Stevens 1982). This variability, as well as the comparatively large values of these apparent half-saturation constants, has certainly been influenced by the presence of unstirred layers in studies of uptake into worm tissues. However, the diffusional resistance introduced to integumental uptake by unstirred layers is a physiologically relevant consequence of the morphology of this transporting epithelium (in contrast to the case for bivalves, noted earlier) (see Gomme 1984). Consequently, these apparent K_t's probably provide a reasonable measure of the extent to which the transport capacity of annelid uptake pathways is realized by exposure to DOM levels in the environment of these animals. Furthermore, though these "effective" K_t's tend to be large compared to values noted in many bivalves and echinoderms, they are still within the range of the amino acid concentrations typical of the interstitial water of burrowing annelids.

This uptake of radiolabeled amino acids is indicative of a net uptake of substrate. Stephens (1975), in the first of such studies, used fluorescamine to show that *C. capitata* and *N. diversicolor* can accumulate naturally occurring primary amines collected from the sediments inhabited by these animals. The ability to effect a net uptake of amino acid from environmentally relevant concentrations was confirmed in several subsequent studies (Costopulos et al. 1979; Ferguson 1982; Jørgensen 1980; Siebers and Bulnheim 1977). Though, to my knowledge, these studies have not been extended through the use of HPLC techniques to examine net uptake of individual amino acids, in fact, such observations would not, arguably, represent a real advance in the understanding of annelid integumental transport. The fluorescamine studies cited above showed that marine worms can accumulate amino acids from external concentrations as low as 1 μM (e.g., Costopulos et al. 1979), which is much lower than the concentrations to which these animals are likely to be exposed. Furthermore, the latter study demonstrated that exposure to anoxic conditions, a condition likely to exist at low tide in many sediment habitats, does not result in an increase in the loss of endogenous amino acids from animals, but merely reduces the rate of uptake. Therefore, it seems clear that the integument of marine annelids can accumulate amino acids under the conditions ordinarily encountered in the sediment environment.

Though the evidence is limited, it appears likely that there are multiple pathways for amino acid uptake in marine annelids (see Jørgensen 1979; Siebers 1976). There is also evidence for the presence of transport processes specific for other structural classes of DOM, including short-chain (Holst and Zebe 1984) and long-chain (Testerman 1972) fatty acids. The short-chain fatty acid, proprionate, is accumulated by *Arenicola marina* (Holst and Zebe 1984) with a J_{max} of 0.6 μmol g^{-1} h^{-1} and an apparent K_t of 75 μM. Testerman (1972) reported that long-chain fatty acids are accumulated by *Stauronreis rudolphi* by saturable processes with J_{max}'s ranging from 0.02 to 0.18 μmol g^{-1} h^{-1}, and K_t's of from 1 to 20 μM. Creatine is transported by the body wall of *G. dibranchiata* with a J_{max} of 0.3 μmol g^{-1} h^{-1} and a K_t of 37 μM (van Pilsum et al. 1975), and though uptake is inhibited by several amino acids (e.g., glycine and glutamine) the interaction is noncompetitive in nature.

Finally, glucose uptake in *N. diversicolor* has received considerable attention from Gomme and his colleagues. The maximal rate of uptake into animals acclimated to 65% artificial sea water (22.4 ‰ salinity) is approximately 30×10^{-12} mol cm^{-2} min^{-1} (area as nominal frontal surface of the integument; Gomme 1981 b), and acclimation to 35% sea water results in a severalfold reduction in this transport. Detailed kinetic analysis has been complicated by a high degree of interanimal variability in rates of uptake, but examination of Gomme's data (his Fig. 5; 1981 b) suggests that the apparent K_t for glucose uptake is on the order of 25 μM. Net uptake of glucose from sea water was noted from ambient concentrations of less than 1 μM, and was completely blocked by phlorizdin, a specific inhibitor of concentrative glucose transport in a variety of epithelial cell systems. In a detailed study of the inhibitory effects of structural analogs on glucose uptake by this polychaete, Albrechtsen and Gomme (1984) found that the apparent specificity of the integumental transporter is similar to that of the apical membrane of vertebrate small intestine and proximal tubule.

5.2 Nutritional Role of Integumental Uptake in Marine Annelids

Comparisons of rates of DOM uptake to rates of oxidative metabolism have been made for a variety of annelid species. Stephens (1963) reported that uptake of amino acids by intact *Clymenella torquata* from the relatively large concentrations of ninhydrin-positive material which he collected from interstitial water (≃60 – 104 μM), could account for approximately 150% of the oxidative requirements of this animal. Similar calculations have shown that uptake of glycine and aspartate from the 1 to 3 μM concentrations measured in natural sediment water can sustain from 7 to 12% of the metabolism of *Nereis virens* (Jørgensen 1979). Testerman (1972) reported that uptake of free fatty acids could account for 16% of the oxidative needs of *S. rudolphi.*

Once again, the need for a better understanding of environmental DOM concentrations is apparent. With burrowing annelids, this is a particularly important question. For example, though it is evident that DFAA levels in sediment pore water are large in terms of producing a significant integumental input of nutrients, it is not clear whether these levels, measured in water samples expressed from sediments by pressure or centrifugation, are an accurate reflection of the effective substrate concentration at the surface of worm integument. Unstirred layers may play a significant modifying role in the kinetics of transport. But perhaps more important to the question of substrate availability is the extent to which active ventilation of worm burrows serves to reduce the high substrate levels of sediment water by introducing the comparatively low concentrations of DOM in the water overlying the benthos. Resolution of these issues will require a thorough study of the effects of convection and diffusion in sediments on DOM concentrations at integumental surfaces.

Finally, the issue of the "transintegumental" nature of DOM uptake has been raised recently with respect to the nutritional role of this process is annelids. Gomme (1984) has reviewed the literature on DOM uptake in worms and stressed that there is little information supporting the export of accumulated substrates to nonintegumental tissues. Chien and Rice (1985), however, recently provided evidence that radiolabeled amino acids are quickly transported across the thin parapodial tissues of the bloodworm, *G. dibranchiata*, and showed autoradiograms that revealed an extensive labeling of blood cells in this tissue. Taken with the recent evidence that transintegumental amino acid transport in *Mytilus* may also be confined to a specialized region of the body surface (i.e., the mantle; Rice and Stephens, in press), Chien and Rice's (1985) observations with *G. dibranchiata* suggest that in many species the nutritional impact of DOM uptake may vary with the specific integumental site of transport: over much of the surface of the integument, accumulated substrates may be effectively confined to the site of uptake and will support the nutritional needs of those cells; uptake into other integumental regions may result in a rapid transcellular movement of accumulated materials to the hemolymph for transport to deeper tissues. To the extent that this type of specialized "partitioning" of accumulated DOM occurs in the integument, the calculations of the nutritional potential of DOM uptake that have been described here will, of course, under- or overestimate the nutritional potential of uptake for a given organism or tissue. These speculations are related to the earlier discussion of the broader "fate" of accumulated materials, and offer some important areas for study.

5.2.1 Nutritional Role of Integumental Recycling of Endogenous DOM

Integumental transport processes can also influence the nutritional status of animals by preventing the loss to the environment of endogenous organic material. Gomme (1981 a) made the first quantitative assessment of the potential of integumental transporters in annelids to "recycle" organic substrates. The basis of this idea is that the cuticle overlying the epithelial cells of the integument serves to increase the diffusional resistance of the unstirred layers associated with this transporting surface. Molecules lost from integumental cells, or through the paracellular pathways between cells, must cross this cuticular layer, as well as the poorly stirred layer immediately above the surface of the integument, before being "lost" from the animal. While within this region of heightened resistance to diffusional resistance, endogenous organic molecules have an increased probability of being substrates for integumental transporters. Gomme's (1981a) calculations indicate that such recycling by *Nereis* may result in a recovery of as much as 40% of the glucose lost across the integument. Subsequent studies, mentioned earlier (Sect. 3.1.1), have confirmed and extended this hypothesis to include recycling of amino acids by *Mytilus* gill tissue (Wright and Secomb 1984, 1986; Wright et al. 1987). Solute recycling may represent an important general phenomenon of the invertebrate integument, and of epithelial tissues in general (Gomme 1984).

6 On the Mechanism of Integumental Transport

Net integumental uptake of DOM is an "active" process; solutes (e.g., amino acids) can be accumulated from external concentrations in the submicromolar range into cellular pools in excess of 0.1 M. Indeed, the chemical gradients against which integumental transporters can work are the most extreme reported in animal tissues. The maintenance of such large tissue concentrations of amino acid implies that the passive permeability of the integument be quite low. In fact, a recent study of net amino acid exchanges across the integument of *M. edulis* gill (Wright et al. 1987) showed that the passive permeability coefficients for several amino acids are less than 1×10^{-9} cm s^{-1}, which approaches the low permeability of artificial lipid bilayers (Young and Ellory 1977). However, the amino acid transporters of the gill result in a carrier-mediated permeability for these compounds of greater than 1×10^{-3} cm s^{-1} (i.e., J_{max}/K_t; see Wright et al. 1987). It might seem that the inclusion of potential carrier-mediated "leak pathways" in the integument would place the gill at risk of losing large amounts of amino acid, yet observations of net uptake from nanomolar substrate concentrations in sea water show that such losses do not occur. Studies of the mechanism of integumental transport have provided some insight into how these tissues serve simultaneously as barriers to loss of endogenous compounds and avenues for the rapid accumulation of exogenous materials.

6.1 Role of Na in Integumental Transport

The "active" transport of organic solutes by animal cells usually involves a coupling between the fluxes of an alkalai metal ion (most often Na^+) with the flux of the organic molecule. The potential energy in transmembranous electro-

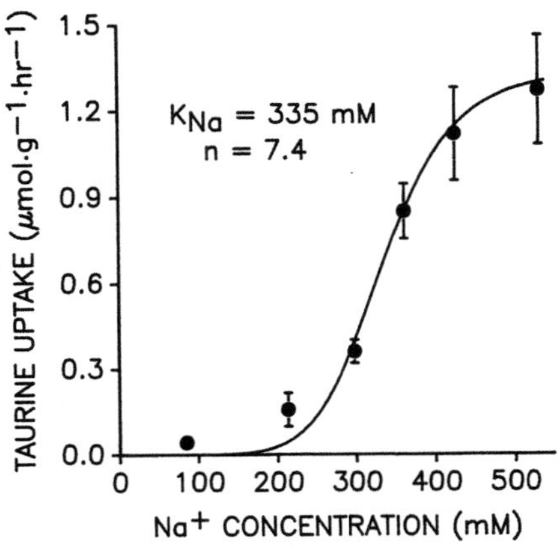

Fig. 5. Effect of increasing Na^+ concentration in sea water on the activation of taurine uptake into gill tissue isolated from *Mytilus californianus*. The K_{Na} (the Na concentration resulting in half-maximal uptake of 0.5 μM ^{14}C-taurine) and apparent Hill coefficient, n, were determined using a nonlinear regression fit of the data to the Hill equation (see Segel 1975). Use of the Hill equation to analyze these data should not be considered a suggestion that strong cooperativity necessarily exists between multiple Na^+ binding sites at the integumental taurine transporter; rather, it permits comparison of the characteristics of taurine uptake in bivalve gill tissue to similar processes observed in other tissues. (Data from Wright 1987)

chemical ion gradients, which are themselves the product of "primary" active, ATP-requiring transport processes, can thereby be used to "drive" the concentrative transport of other solutes. Because metabolic energy, i.e., ATP, is not used directly in this latter mode of transport, they are often referred to as being "secondary" active processes. This general model for transport, dubbed the "Na-gradient hypothesis", has proven an adequate explanation for concentrative solute uptake in a wide variety of tissues (see Crane 1977). Thus, it has not been surprising that replacement of Na^+ in sea water by other cations virtually eliminates integumental uptake of amino acids and sugars in bivalves (Anderson 1975; Bamford and Gingles 1974; Riley 1981; Stewart and Bamford 1975; Wright and Stephens 1977; Wright 1987), polychaetes (Ahearn and Gomme 1975; Albrechtsen and Gomme 1984; Gomme 1981 b; Stevens and Preston 1980 a, c), and echinoderms (Ahearn and Townsley 1975; Davis et al. 1985 a).

Studies of the kinetics of Na^+'s interaction with integumental DOM transporters have provided information concerning their mechanism. As shown in Fig. 5, increasing the external Na^+ concentration from 0 to 500 mM results in a sigmoid (rather than hyperbolic) activation of taurine uptake into gills from *M. californianus* (Wright 1987). This type of relationship indicates that more than one Na ion is interacting whith the integumental taurine transporter; Hill analysis of these data (see Segel 1975) suggests that seven to eight Na^+ ions may be involved with the activation of the taurine transporter. Similarly, Stevens and Preston (1980 c) reported that three Na^+ ions appear to be required for the activation of alanine uptake into body wall of *G. dibranchiata,* while Davis et al. (1985 a) concluded that alanine uptake into larval *S. purpuratus* behaves as if two

Na^+ ions are associated with the transporter. I should emphasize that this analytical approach does not provide the stoichiometry of "coupling" between a cotransport of Na^+ and a second substrate. In fact, it is difficult to establish or refute the presence of a cotransport process based on kinetic evidence alone (see Aronson 1981; Turner 1983). Furthermore, it is usually not possible, without independent information, to assign a definite value to the number of Na^+ ions involved in activation of transport; such conclusions are usually biased by selection of a particular "model" to describe the kinetics of sigmoid activation (see Wright et al. 1983). Nevertheless, these observations do clearly support the hypothesis, expanded upon below, that integumental transport requires a coupling of multiple Na^+ ions with each molecule of DOM.

6.1.1 Energetics of Transport

As emphasized earlier, one hallmark of the integumental transporters of marine invertebrates is their ability to accumulate organic substrates against extremely large chemical gradients. There is, of course, a limit to the chemical gradient which can be sustained by a coupling to the electrochemical gradient of a second solute. For example, assuming that integumental amino acid uptake is coupled to an influx of Na^+, the maximum substrate gradient against which net uptake can occur is given by the following relationship (Aronson 1981):

$$\frac{(A)_i}{(A)_0} \leq \frac{(Na)_0}{(Na)_i} \exp \frac{F}{RT} (\psi_0 - \psi_i)^{n_{Na}/n_A} , \tag{2}$$

in which $(A)_i$ and $(Na)_i$ refer to the cytoplasmic activities of amino acid and Na^+, respectively; $(A)_0$ and $(Na)_0$ are the activities of these solutes in sea water; ψ is the electrical potential of the sea water or cytoplasmic compartment; F, R, and T, are the Faraday constant, gas constant, and absolute temperature, respectively; and n refers to the number of Na ions or amino acid molecules transported in a single "translocation event." The ratio of coupled Na^+ and organic solute flows is also referred to as the "coupling coefficient" of the transport process.

The chemical gradient for Na^+ across the integumental cell membrane is probably less than 30 to 1, sea water to cytoplasm. From Eq. (2) it is evident that if the Na^+ electrochemical gradient is the sole source of energy used to drive integumental transport, then more than one Na ion must be coupled to the flux of each amino acid molecule to account for the observed amino acid gradients in excess of 10^6 to 1. This equation can be used to calculate the minimum coupling coefficient required to support a net flux of amino acid against these gradients. Such calculations have led to the conclusion that three Na ions must be coupled to the integumental transport of alanine in *Mytilus* (Manahan et al. 1983) and *G. dibranchiata* (Stevens and Preston 1980c). Davis et al. (1985a) suggested that a coupling coefficient of 2 may be adequate to account for serine transport in *S. purpuratus,* though this figure was based upon a very conservative estimate of the amino acid gradient against which net transport can occur.

Despite the observations of a profound Na dependency for integumental transport processes, and the results of calculations such as those just described, until recently there has been no evidence of an actual cotransport of Na^+ and DOM. Unfortunately, direct measurements of the stoichiometry of coupling between Na^+ and organic substrates have been difficult to perform in marine invertebrates, due to technical problems in resolving the coupled Na^+ flux from the background flux of this ion from all other transport pathways. Two recent studies of integumental amino acid uptake have, however, provided convincing evidence that this transport does involve a cotransport with Na^+. Pajor and Wright (1987) showed that a cell-free preparation of isolated integumental membranes from *M. edulis* gill can actively accumulate alanine when the only energy to sustain such uptake is in the form of an inwardly directed Na^+ gradient. The transduction of the energy in one solute gradient into the formation of a second gradient necessitates the physical coupling of the flows of the two solutes (Turner 1983). The second observation, by Davis et al. (1985a), was based upon the expected electrogenicity of transport arising from a coupling between a flux of Na^+ with an electroneutral amino acid. They found that exposure of fertilized sea urchin eggs to 25 µM leucine results in a rapid depolarization of the egg membrane, consistent with a coupling of Na^+ with leucine uptake. Therefore, in light of the observations described above, I think it is clear that integumental transport in marine invertebrates ordinarily involves a coupling between fluxes of Na^+ and DOM.

The nature of the energy source for Na-coupled cotransport results in several predictions. First, elimination of metabolic energy pools should not *directly* influence transport rates; instead, the fate of coupled transport processes should follow the course of the resulting degradation of the Na^+ electrochemical gradient. The second prediction is a corollary of the first, namely, poisoning of the Na,K-ATPase should influence transport only to the extent that this inhibition results in an increase in intracellular Na^+ (or a fall in cell K^+). Both of these predictions require that studies of metabolic inhibition of integumental transport include measurement of intracellular ion concentrations, as well as measurements of tissue ATP levels. Such measurements are seldom carried out, and consequently, mechanistic interpretation of resulting data is difficult. For example, Stewart and Bamford (1975) found that Na-dependent alanine uptake into gills from the clam *M. arenaria* was reduced by 48% by preincubation for 60 min in 2 mM ouabain (an inhibitor of Na,K-ATPase). Riley (1981) reported that the Na-dependent uptake of α-methylglucoside (an analog of D-glucose) into gills from the oyster, *O. edulis*, was not influenced by a 60-min exposure to ouabain. Is one tissue less sensitive to ouabain, or is one transporter less sensitive to cytoplasmic Na^+? Because neither study included measurements of tissue levels of Na^+ or K^+, such questions cannot be answered. The Na,K-exchange pump (Na,K-ATPase) is located in the basal-lateral portion of most epithelial cells, including the basal-lateral membrane of the bivalve gill (Pajor and Wright 1987). This may explain, in some part, the relative insensitivity of some integumental tissues to inhibitors of this process; it may be difficult for compounds such as ouabain to penetrate the integument to reach the basal-lateral aspect of these epithelial cells.

There are several reports of Na-independent integumental transport in marine invertebrates. Siebers (1976) reported that arginine uptake into the oligochaete,

E. albidus was Na-independent, and suggested that this process was one of "facilitated diffusion," i.e., one in which net fluxes occur only in response to the substrate's electrochemical gradient. Such a process would appear to be a liability in an integumental cell, as it would represent, under ordinary conditions, an avenue for the net loss of organic material from the animal to the environment. The transport capacity of this transporter in *E. albidus* is low (5 nmol g^{-1} h^{-1}), suggesting it may not represent a significant efflux pathway. Nevertheless, its role in the physiology of the animal is not clear. Also, prior to and following fertilization of *S. purpuratus* eggs, there is a Na-independent pathway for neutral amino acids (Allemand et al. 1984). By the same reasoning as above, this would appear to represent a facilitated diffusion pathway, and would result in a leak of material from the cell. This prediction has recently been confirmed; Manahan (1986) reported that unfertilized sea urchin eggs rapidly lose glycine (the principal constituent of the cytoplasmic free amino pool). However, concomitant to the activation (upon fertilization) of Na-dependent neutral amino acid uptake, the net loss of glycine gradually declines to be replaced after approximately 20 min with a net uptake of material from sea water. Again, the possible role of the Na-independent pathway is not clear.

Considering these observed effects of Na^+ on net uptake of DOM, it is likely that the general absence of integumental transport processes in freshwater organisms is related to the need for relatively high external concentrations of monovalent cations and inwardly directed ion gradients to support active solute transport. Still, the observations of net carboxylic acid uptake by the freshwater snail, *B. glabrata* (Thomas et al. 1984), warrant further study as an exception to this apparent rule.

6.1.2 Effect of Salinity on Integumental Uptake

Stephens (1964) noted that amino acid uptake into euryhaline annelids decreases sharply as external salinity is reduced, and suggested that integumental transport and the processes associated with osmoregulation may be mutually incompatible. Several subsequent studies have confirmed a correlation between external salinity and uptake of DOM in annelids (Gomme 1981 b; Siebers and Bulnheim 1977) and bivalves (Anderson 1975; Anderson and Bedford 1972; Bamford and Campbell 1976; Stewart and Bamford 1976; Wright et al. 1987). Given the importance of external Na^+ on the activation of integumental uptake noted above, it is tempting to suggest that the basis for this correlation is the fall in Na^+ concentration that occurs as external salinity is reduced. In fact, uptake processes in some systems are more sensitive to acute exposure to reduced salinity than they are to equivalent drops in external $[Na^+]$. For example, acute exposure of gill tissue from *M. californianus* to 60% sea water results in an 80% inhibition in the uptake of taurine and alanine (Wright et al., 1987), while, as is evident from Fig. 5, exposure to sea water in which 40% of the Na^+ has been replaced with Li^+ causes an inhibition of uptake of less than 70% (Wright, 1987). In *N. diversicolor*, however, uptake of D-glucose and L-alanine is acutely depressed by elimination of Na^+, whereas the response to reduced salinity takes much longer to develop (Gomme 1982b). Gomme (1984) has suggested that "the reduced uptake of organic substances in low salinities is probably due to a transition of the epidermis to a different physiological state, somehow related to the transport of inorganic ions, or possibly to cell volume regulation".

6.2 Effect of Divalent Cations on Integumental Transport

A reduction in external salinity results in exposure of the integument to a decrease in elements other than Na^+. Divalent cations, particularly Ca^{2+} and Mg^{2+}, have been linked to a wide variety of regulatory processes. Not surprisingly, these ions have been found to influence integumental transport and permeability. Pierce and Greenberg (1972) reported that exposure of isolated ventricles from the mussel, *M. demissus,* to hypotonic sea water causes an immediate loss of free amino acids from the tissue that quantitatively accounts for the decrease in cellular concentration of these compounds associated with volume regulation. They subsequently (Pierce and Greenberg 1973) reported that the permeability of the ventricle to amino acids is affected by the concentration of Ca^{2+} and Mg^{2+} in the ambient medium. In fact, by maintaining normal concentrations of these divalent cations, the loss of amino acids from this tissue upon exposure to hypotonic sea water is significantly reduced.

Integumental permeability has also been found to be sensitive to divalent cation concentrations. Swinehart et al. (1980) reported that exposure of isolated gill tissue from *M. californianus* to sea water containing reduced concentrations of Ca^{2+} and Mg^{2+} resulted in a loss of primary amines (i.e., fluorescamine positive material), and in an inhibition in the uptake of glycine. The mechanism by which divalent cations influence either the carrier-mediated or passive permeability characteristics of integumental membranes is not clear. Though it is not known if such responses to Ca^{2+} and Mg^{2+} are limited to mollusks or are a common property of the integument of soft-bodied marine invertebrates, the study of the regulatory influence of divalent cations on integumental processes promises to be an important area of future research.

6.3 A General Model of Integumental Transport

The characteristics of the integument that permit the net accumulation of DOM from dilute external solution as well as the maintenance of large intracellular concentrations of small organic solutes (e.g., amino acids and sugars) are summarized in the schema presented in Fig. 6. The external face of the integument (i.e., that exposed to sea water) has a very low passive permeability to small organic molecules. As a consequence, these compounds can be accumulated to high concentrations, through both metabolic and transport pathways, with a comparatively low probability of loss to surrounding sea water via passive diffusion. Carrier-mediated permeability of this membrane is high, but the kinetics of these transporters are such that a significant fraction of the materials lost by passive processes is recovered. Divalent cations (Ca^{2+} and Mg^{2+}) influence both

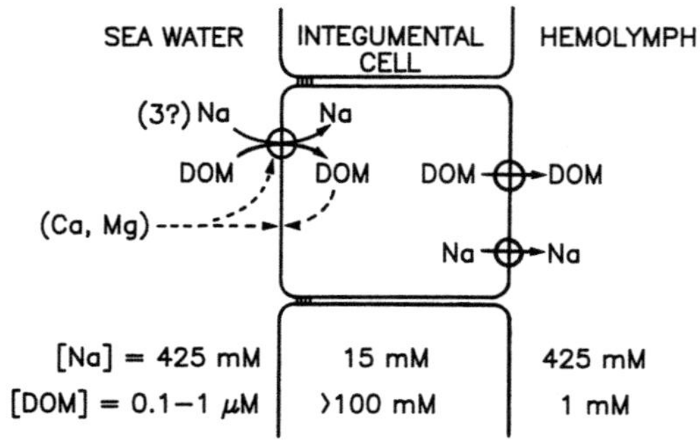

Fig. 6. Proposed model for integumental DOM transport in soft-bodied marine invertebrates. Key points include: low passive permeability to DOM of the external face of integumental cells; this passive component of integumental permeability may be regulated by external Ca^{2+} and Mg^{2+}. There is a high carrier-mediated permeability to DOM via Na-dependent transporters whose kinetic characteristics result in a rapid influx of DOM with little carrier-mediated efflux; carrier-mediated permeability may be regulated by divalent cations as well. Concentrative uptake of DOM is energized by the coupling of several Na ions to the flux of each DOM molecule. A low cytoplasmic Na^+ activity, and hence an inwardly directed electrochemical gradient for Na^+, is maintained through the action of an energy-requiring Na pump (probably a Na, K-ATPase) located at the basal-lateral aspect of integumental cells. Translocation of absorbed DOM to nonintegumental tissues must involve a transport from these cells across the basal-lateral membrane into the hemolymph; this could be a passive process, but there is currently no information concerning the characteristics of this transport

the passive and carrier-mediated permeabilities of the integument, and may play a role in the regulation of exchange of DOM between sea water and surface cells. Activation of transporters requires the relatively high concentrations of Na^+ found in sea water, and the asymmetric distribution of this ion between sea water and the cytoplasm of integumental cells leads to rates of DOM influx that far exceed carrier-mediated efflux. The energy to support the net accumulation of DOM into integumental cells is derived from the inwardly directed electrochemical gradient for Na^+, though this necessitates that multiple Na ions (perhaps three) be coupled to the uptake of each DOM molecule. The Na^+ gradient is maintained under steady-state conditions by the active extrusion of Na^+ from these cells, probably via an active Na^+-pump located in the basal-lateral aspect of integumental cells. The transfer of accumulated DOM to nonintegumental tissues must also occur by efflux from the integument to the hemolymph, across the basal-lateral aspect of this epithelial cell layer. This latter pathway is the least well-understood aspect of the integumental accumulation of DOM.

If the integument of soft-bodied marine invertebrates were merely an inert barrier to the diffusive loss of organic solutes, it would permit the maintenance of the large intracellular concentrations of amino acid and other solutes that are characteristic of these tissues. However, the introduction of a transport process in these cells permits the integument to make an active contribution to the nutritional status of this tissue, and perhaps to the whole organism. The kinetic characteristics of integumental transporters appear to be such that their presence does not significantly increase the risk of loss of endogenous molecules, but instead, actually serves to reduce loss of nutrients from the animal.

7 Conclusions

Transport processes for small, dissolved, organic molecules are a common characteristic of the integument of soft-bodied marine invertebrates. The ability of such processes to accumulate DOM from the very low concentrations found in sea water has been clearly documented, and the rates at which this mode of nutrient acquisition can occur indicate that they are capable of providing a significant supplement to other, "traditional" feeding mechanisms. The quantitative role of integumental uptake in the nutrition of individual organisms is directly related to the concentration of substrate found in the animal's environment. Study of the distribution of DOM in specific habitats, and the concentrations of DOM to which the integument is actually exposed, remains at the heart of our understanding of this issue. Other areas which are not well understood at this time include: (1) the metabolic fate of accumulated materials; (2) the extent to which materials accumulated by the integument are "exported" to other tissues, and how this distribution may be influenced by environmental or physiological factors; and (3) the means by which integumental transport processes are regulated, and the role they play in actively influencing the concentration of small metabolites in the cytoplasm of this tissue.

That the integument of marine invertebrates plays a role in mediating the exchange of nutrient molecules between animal and environment is no longer in question. The exciting prospects for future research lie in assessing the quantitative role of integumental transport in the physiology of these animals.

References

Ahearn GA, Gomme J (1975) Transport of exogenous D-glucose by the integument of a polychaete worm (*Nereis diversicolor* Muller). J Exp Biol 62:243 – 264

Ahearn GA, Townsley SJ (1975) Integumentary amino acid transport and metabolism in the apodus sea cucumber, *Chirodota rigida*. J Exp Biol 62:733 – 752

Aiello EL (1970) Nervous and chemical stimulation of the gill cilia in bivalve molluscs. Physiol Zool 43:60 – 70

Albrechtsen S, Gomme J (1984) Specificity of D-glucose transport by the apical membrane of *Nereis diversicolor* epidermis. Biochim Biophys Acta 770:47 – 54

Allemand D, Renzis G de, Ciapa B, Girard J-P, Payan P (1984) Characterization of valine transport in sea urchin eggs. Biochim Biophys Acta 722:337 – 346

Allemand D, Renzis G de, Maistre C, Girard J-P, Payan P (1985) Uptake of valine and alanine by a neutral amino-acid carrier in sea urchin eggs: cyclic variations in the early clevage stage. J Membr Biol 87:217 – 224

Allemand D, Renzis G de, Payan P, Girard J-P (1986) Regulatory and energetic role of Na^+ in amino acid uptake by fertilized sea urchin eggs. Dev Biol 118:19 – 27

Anderson JW (1975) The uptake and incorporation of glycine by the gills of *Rangia cuneata* (Mollusca: Bivalvia) in response to variations in saliniy and sodium. In: Vernberg FJ (ed) Physiological ecology of estuarine organisms. Univ S Carolina Press, Columbia, pp 239 – 258

Anderson JW, Bedford B (1973) The physiological response of the estuarine clam, *Rangia cuneata* (gray), to salinity. II. Uptake of glycine. Biol Bull 144:229 – 247

Anderson JW, Stephens GC (1969) Uptake of organic material by aquatic invertebrates. VI. Role of epiflora in apparent uptake of glycine by marine crustaceans. Mar Biol 4:243 – 249

Aronson PS (1981) Identifying secondary active solute transport in epithelia. Am J Physiol 240:F1 – F11

Bamford DR (1984) Echinodermata. Permeability and epidermal transport. In: Bereiter-Hahn J, Matoltsy AG, Richards KS (eds) Biology of the integument, vol 1. Invertebrates. Springer, Berlin Heidelberg New York Tokyo, pp 775 – 789

Bamford DR, Campbell E (1976) The effect of environmental factors on the absorption of L-phenylalanine by the gill of *Mytilus edulis*. Comp Biochem Physiol 53A:295 – 299

Bamford DR, Gingles R (1974) Absorption of sugars in the gill of the Japanese oyster, *Crassostrea gigas*. Comp Biochem Physiol 49A:637 – 646

Bamford DR, McCrea R (1975) Active absorption of neutral and basic amino acids by the gill of the common cockle, *Cerastoderma edule*. Comp Biochem Physiol 50A:811 – 817

Bayne BL, Newell RC (1983) Physiological energetics of marine molluscs. In: Saleuddin ASM, Wilbur KM (eds) The mollusca, vol 4. Physiology, part 1. Academic Press, London New York, pp 407 – 515

Bayne BL, Scullard C (1977) Rates of nitrogen excretion by species of *Mytilus* (Bivalvia: Mollusca). J Mar Biol Assoc UK 57:355 – 369

Bayne BL, Thompson RJ, Widdows J (1976) Physiology, vol 1. In: Bayne BL (ed) Marine mussels: their ecology and physiology. Cambridge Univ Press, Cambridge, pp 121 – 206

Bereiter-Hahn J, Matoltsy AG, Richards KS (eds) (1984) Biology of the integument, vol 1. Invertebrates. Springer, Berlin Heidelberg New York Tokyo

Berg WE (1968) Kinetics of uptake and incorporation of valine in the sea urchin embryo. Exp Cell Res 49:379 – 395

Bishop SH (1976) Nitrogen metabolism and excretion:regulation of intracellular amino acid concentration. Estuarine Processes 1:414 – 431

Bishop SH, Ellis LL, Burcham JM (1983) Amino acid metabolism in molluscs. In: Hochachka P (ed) The Mollusca, vol 1. Metabolic chemistry and molecular biomechanics. Academic Press, London New York, pp 243 – 327

Bito LZ (1972) Comparative study of concentrative prostaglandin accumulation by various tissues of mammals and marine vertebrates and invertebrates. Comp Biochem Physiol 43A:65 – 82

Braven J, Evans R, Butler EI (1984) Amino acids in sea water. Chem Ecol 2:11 – 21

Bunde TA, Fried M (1978) The uptake of dissolved free fatty acids from seawater by a marine filter feeder, *Crassostrea virginica*. Comp Biochem Physiol 60A:139 – 144

Burgh ME de (1978) Specificity of L-alanine transport in the spine epithelium of *Paracentrotus lividus* (Echinoidea). J Mar Biol Assoc UK 58:425 – 440

Burgh ME de, West AB, Jeal F (1977) Absorption of L-alanine and other dissolved nutrients by the spines of *Paracentrotus lividus* (Echinoidea). J Mar Biol Assoc UK 57:1031 – 1045

Carlucci AF, Craven DB, Henrichs SM (1985) Surface film microheterotrophs: amino acid metabolism and solar radiation effects on their activities. Mar Biol 85:13 – 22

Castille FL Jr, Lawrence AL (1979) The role of bacteria in the uptake of hexoses from seawater by postlarval penaeid shrimp. Comp Biochem Physiol 64A:41 – 44

Chien PK, Rice MA (1985) Autoradiographic localization of exogenously supplied amino acids after uptake by the polychaete *Glycera dibranciata* Ehlers. Wasmann J Biol 43:60 – 71

Chien PK, Stephens GC, Healey PI (1972) The role of ultrastructure and physiological differentiation of epithelia in amino acid uptake by the bloodworm: *Glycera dibranchiata*. Biol Bull 142:219 – 235

Clements LAJ (1986) Amino acid uptake by a regenerating brittle star. Am Zool 26:42A

Costopulos JJ, Stephens GC, Wright SH (1979) Uptake of amino acids by marine polychaetes under anoxic conditions. Biol Bull 157:434 – 444

Crane RK (1977) The gradient hypothesis and other models of carrier mediated active transport. Rev Physiol Biochem Pharm 78:101 – 163

Crowe JH, Dickson KA, Otto JL, Colon R, Farley KK (1977) Uptake of amino acids by the mussel *Modiolus demissus*. J Exp Zool 202:323 – 332

Davis JP, Stephens GC (1984a) Uptake of free amino acids by bacteria-free larvae of the sand dollar *Dendraster excentricus*. Am J Physiol 247:R733 – R739

Davis JP, Stephens GC (1984b) Regulation of net amino acid exchange in sea urchin larvae. Am J Physiol 247:R1029 – R1037

Davis JP, Keenan CL, Stephens GC (1985a) Na^+-dependent amino acid transport in bacteria-free sea urchin larvae. J Comp Physiol 156:121 – 127

Davis JP, Stephens GC, Rice MA (1985b) Net entry of amino acids into the brittle star *Ophionereis annulata*. Comp Biochem Physiol 81A:899–903
DiDomenico DA, Iverson RL (1977) Uptake of glycolic acid by a marine bivalve. J Exp Mar Biol Ecol 28:243–254
DuPaul WD, Webb KL (1971) Free amino acid accumulation insolated gill tissue of *Mya arenaria*. Arch Int Physiol Biochim 79:327–336
Efford IE, Tsumura K (1973) Uptake of dissolved glucose and glycine by *Pisidium*, a freshwater bivalve. Can J Zool 51:825–832
Epel D (1972) Activation of an Na^+-dependent amino acid transport system upon fertilization of sea urchin eggs. Exp Cell Res 72:74–89
Ferguson JC (1967a) An autoradiographic study of the utilization of free amino acids by starfishes. Biol Bull 133:317–329
Ferguson JC (1967b) Utilization of dissolved exogenous nutrients by the starfishes, *Asterias forbesi* and *Henricia sanguinolenta*. Biol Bull 132:161–173
Ferguson JC (1970) An autoradiographic study of the translocation and utilization of amino acids by starfish. Biol Bull 138:14–25
Ferguson JC (1971) Uptake and release of free amino acids by starfishes. Biol Bull 141:122–129
Ferguson JC (1980a) Fluxes of dissolved amino acids between sea water and *Echinaster*. Comp Biochem Physiol 65A:291–295
Ferguson JC (1980b) The non-dependency of a starfish on epidermal uptake of dissolved organic matter. Comp Biochem Physiol 66A:461–465
Ferguson JC (1982) A comparative study of the net metabolic benefits derived from the uptake and release of free amino acids by marine invertebrates. Biol Bull 162:1–17
Fisher FM Jr, Oaks JA (1978) Evidence for a nonintestinal nutritional mechanism in the Rhynchocoelan, *Lineus ruber*. Biol Bull 154:213–225
Fontaine AR, Chia F-S (1968) Echinoderms: an autoradiographic study of assimilation of dissolved organic molecules. Science 161:1153–1155
Fuhrman JA, Bell TM (1985) Biological considerations in the measurement of dissolved free amino acids in seawater and implications for chemical and microbiological studies. Mar Ecol Prog Ser 25:13–21
Gilles R (1979) Intracellular organic osmotic effectors. In: Gilles R (ed) Mechanisms of osmoregulation in animals. Wiley, New York, pp 111–154
Gomme J (1981a) Recycling of D-glucose in collagenous cuticle: a means of nutrient conservation? J Membr Biol 62:47–52
Gomme J (1981b) D-glucose transport across the apical membrane of the surface epithelium in *Nereis diversicolor*. J Membr Biol 62:29–46
Gomme J (1982a) Laminar water flow, amino acid absorption, and amino acid recycling in the mussel gill. Am Zool 22:898
Gomme J (1982b) Epidermal nutrient absorption in marine invertebrates: a comparative analysis. Am Zool 22:691–708
Gomme J (1984) Annelida: permeability and epidermal transport. In: Bereiter-Hahn J, Matoltsy AG, Richards KS (eds) Biology of the integument, vol 1. Invertebrates. Springer, Berlin Heidelberg New York Tokyo, pp 323–367
Gordon MS (1982) Animal physiology: principles and adaptations, 4th edn. Macmillan, New York
Hammen CS (1979) Metabolic rates of marine bivalves determined by calorimetry. Comp Biochem Physiol 62A:955–959
Hammen CS (1980) Total energy metabolism of marine bivalve mollusks in anaerobic and aerobic states. Comp Biochem Physiol 67A:617–621
Hammen CS (1983) Direct calorimetry of marine invertebrates entering anoxic states. J Exp Zool 228:397–403
Henrichs SM, Farrington JW (1979) Amino acids in interstitial waters of marine sediments. Nature (London) 279:319–322
Henrichs SM, Williams PM (1985) Dissolved and particulate amino acids and carbohydrates in the sea surface microlayer. Mar Chem 17:141–163
Henry RP, Mangum CP (1980) Salt and water balance in the oligohaline clam, *Rangia cuneata*. III. Reduction of the free amino acid pool during low salinity adaptation. J Exp Zool 211: 25–32

Holland ND, Nealson KH (1978) The fine structure of the echinoderm cuticle and the subcuticular bacteria of echinoderms. Acta Zool 59:169 – 185
Holst H, Zebe E (1984) Absorption of volatile fatty acids from ambient water by the lugworm *Arenicola marina*. Mar Biol 80:125 – 130
Johannes RE, Coward SJ, Webb KL (1969) Are dissolved amino acids an energy source for marine invertebrates? Comp Biochem Physiol 29:283 – 288
Jørgensen CB (1975) On gill function in the mussle *Mytilus edulis* L. Ophelia 13:187 – 232
Jørgensen CB (1976) August Putter, August Krogh, and modern ideas on the use of dissolved organic matter in aquatic environments. Biol Rev 51:291 – 328
Jørgensen CB (1982) Uptake of dissolved amino acids from natural sea water in the mussel *Mytilus edulis*. Ophelia 21:215 – 221
Jørgensen CB (1983) Patterns of uptake of dissolved amino acids in mussels (*Mytilus edulis*). Mar Biol 73:177 – 182
Jørgensen NOG (1979) Uptake of L-valine and other amino acids by the polychaete *Nereis virens*. Mar Biol 52:45 – 52
Jørgenson NOG (1980) Uptake of glycine and release of primary amines by the polychaete *Nereis virens* (Sars) and the mud snail *Hydrobia neglecta* Muus. J Exp Mar Biol Ecol 47:281 – 297
Jørgensen NOG (1982) Heterotrophic assimilation and occurrence of dissolved free amino acids in a shallow estuary. Mar Ecol Prog Ser 8:145 – 159
Jørgensen NOG, Kristensen E (1980) Uptake of amino acids by three species of *Nereis* (Annelida: Polychaeta). I. Transport kinetics and net uptake from natural concentrations. Mar Ecol Prog Ser 3:329 – 340
Jørgensen NOG, Lindroth P, Mopper K (1981) Extraction and distribution of free amino acids and ammonium in sediment interstitial water from the Limfijord, Denmark. Oceanogr Acta 4:465 – 474
Krogh A (1934) Dissolved substances as food of aquatic organisms. Biol Rev 6:412 – 442
Langdon CJ (1983) Growth studies with bacteria-free oyster (*Crassostrea gigas*) larvae fed on semi-defined artificial diets. Biol Bull 164:227 – 235
Lesser MP (1986) Bacterial endosymbionts of *Amphipholis squamata:* potential contribution to developmental and embryonic feeding biology. Am Zool 26 – 22A
Lindroth P, Mopper K (1979) High performance liquid chromatographic determination of subpicomole amounts of amino acids by precolumn derivatization with o-phthaldialdehyde. Anal Chem 51:1667 – 1674
Little C, Gupta BL (1968) Pogonophora: uptake of dissolved nutrients. Nature (London) 218:873 – 874
Livingston DR, Widdows J, Fieth P (1979) Aspects of nitrogen metabolism of the common mussel *Mytilus edulis*: adaptation to abrupt and fluctuating changes in salinity. Mar Biol 53:41 – 55
Manahan DT (1983a) The uptake of dissolved glycine following fertilization of oyster eggs, *Crassostrea gigas* (Thunberg). J Exp Mar Biol Ecol 68:53 – 58
Manahan DT (1983b) The uptake and metabolism of dissolved amino acids by bivalve larvae. Biol Bull 164:236 – 250
Manahan DT (1986) Net fluxes of amino acids to and from sea water following fertilization of sea urchin eggs. Am Zool 26:5A
Manahan DT, Arnold KE (1983) Microspatial variability of free amino acids and ammonia in sea water. Trans Am Geophys Un (EOS) 64:1020
Manahan DT, Crisp DJ (1982) The role of dissolved organic material in the nutrition of pelagic larvae: amino acid uptake by bivalve veligers. Am Zool 22:636 – 646
Manahan DT, Crisp DJ (1983) Autoradiographic studies on the uptake of dissolved amino acids from sea water by bivalve larvae. J Mar Biol Assoc UK 63:673 – 682
Manahan DT, Jaeckle WB (1986) Balancing the energy budgets of pelagic larvae: the importance of dissolved organic material. Trans Am Geophys Un (EOS) 67:992
Manahan DT, Richardson K (1983) Competition studies on the uptake of dissolved organic nutrients by bivalve larvae (*Mytilus edulis*) and marine bacteria. Mar Biol 75:241 – 247
Manahan DT, Wright SH, Stephens GC, Rice MA (1982) Transport of dissolved amino acids by the mussel, *Mytilus edulis:* demonstration of net uptake from natural sea water. Science 215: 1253 – 1255
Manahan DT, Wright SH, Stephen GC (1983a) Simultaneous determination of net uptake of 16 amino acids by a marine bivalve. Am J Physiol 244:R832 – R838

Manahan DT, Davis JP, Stephens GC (1983b) Bacteria-free sea urchin larvae: selective uptake of neutral amino acids from sea water. Science 220:204–206
McCammon HM, Reynolds WA (1976) Experimental evidence for direct nutrient assimilation by the lophophore of articulate brachiopods. Mar Biol 34:41–51
McWhinnie MA, Johanneck R (1966) Utilization of inorganic and organic carbon compounds by antarctic zooplankton. Antarct JUS 1:210
McWhinnie MA, Urbanski RJ (1971) Absorption of soluble organic compounds by polar marine zooplankton. Antarct JUS 6:156–157
Mopper K, Lindroth P (1982) Diel and depth variations in dissolved free amino acids and ammonium in the Baltic Sea determined by shipboard HPLC analysis. Limnol Oceanogr 27:336–347
Nell JA, Skeel ME, Dunkley P (1983) Uptake of some dissolved organic nutrients by the Sydney rock oyster, *Saccostrea commercialis*. Mar Biol 74:313–318
Nishioka D, Killian CE, McGwin-Scully NF (1985) Increased uptake of nucleosides in the activation of sea urchin eggs. In: Gilles R, Gilles-Baillien M (eds) Transport processes, iono- and osmoregulation. Current comparative approaches. Springer, Berlin Heidelberg New York Tokyo, pp 303–325
North BB (1975) Primary amines in California coastal waters: utilization by phytoplankton. Limnol Oceanogr 20:20–27
Pajor AM, Wright SH (1987) L-alanine uptake in brush border membrane vesicles from the gill of a marine bivalve. J Membr Biol 96:209–223
Parsons TR (1975) Particulate organic carbon in the sea. In: Riley JP, Skirrow G (eds) Chemical oceanography, vol 2. Academic Press, London New York, pp 365–383
Pearse JS, Pearse VB (1973) Removal of glycine from solution by the sea urchin *Strongylocentrotus purpuratus*. Mar Biol 19:281–284
Péquignat E (1966) 'Skin digestion' and epidermal absorption in irregular and regular urchins and their probable relation to the outflow of spherule-coelomocytes. Nature (London) 210: 397–399
Péquignat E (1972) Some new data on skin-digestion and absorption in urchins and sea stars (*Asterias* and *Henricia*). Mar Biol 12:28–41
Péquignat E (1973) A kinetic and autoradiographic study of the direct assimilation of amino acids and glucose by organs of the mussel *Mytilus edulis*. Mar Biol 19:227–244
Pierce SK, Greenberg MJ (1972) The nature of cellular volume regulation in marine bivalves. J Exp Biol 57:681–692
Pierce SK, Greenberg MJ (1973) The initiation and control of free amino acid regulation of cell volume in salinity-stressed marine bivalves. J Exp Biol 59:435–446
Pilsum JF van, Taylor D, Bans L (1975) Studies on the uptake of creatine from sea water by the marine annelid, *Glycera dibranchiata*. Comp Biochem Physiol 51A:611–617
Poulet SA, Martin-Jezequel V, Delmas D (1985) Gradient of dissolved free amino acids and phytoplankton in a shallow bay. Hydrobiologia 121:11–17
Preston RL (1986) Uptake and metabolism of D-alanine by marine invertebrates. Am Zool 26:74A
Preston RL, Stevens BR (1982) Kinetic and thermodynamic aspects of sodium-coupled amino acid transport by marine invertebrates. Am Zool 22:709–721
Putter A (1909) Die Ernährung der Wassertiere und der Stoffhaushalt der Gewässer. Fisher, Jena
Rice MA, Stephens GC (1987) Influx and transepithelial flux of amino acids in the mussel, *Mytilus edulis* L. J Exp Biol (in press)
Rice MA, Wallis K, Stephens GC (1980) Influx and net flux of amino acids into larval and juvenile European flat oysters, *Ostrea edulis* (L.). J Exp Mar Biol Ecol 48:51–59
Richards KS, Arme C (1979) Transintegumentary uptake of amino acids by the lumbricid earthworm *Eisenia foetida*. Comp Biochem Physiol 64A:351–356
Richards KS, Arme C (1980a) Transintegumentary uptake of D-galactose, D-fructose and 2-deoxy-D-glucose by the lumbricid eathworm *Eisenia foetida*. Comp Biochem Physiol 66A:209–214
Richards KS, Arme C (1980b) Transintegumentary uptake of sodium acetate by the lumbricid earthworm, *Lumbricus rubellus*. Comp Biochem Physiol 67A:403–407
Riley RT (1981) α-Methylglucoside transport by the gill of the oyster *Ostrea edulis*. Biol Bull 161:416–425
Roe P, Crowe JH, Crowe LM, Wickham DE (1981) Uptake of amino acids by juveniles of *Carcinonemertes errans* (Nemertea). Comp Biochem Physiol 69A:423–427

Schlichter D (1975) The importance of dissolved organic compounds in sea water for the nutrition of *Anemonia sulcata* Pennant (Coelenterata). In: Barnes H (ed) Proc 9th Eur Mar Biol Symp. Aberdeen Univ Press, Aberdeen, pp 395–405

Schlichter D (1978) On the ability of *Anemonia sulcata* (Coelenterata: Anthozoa) to absorb charged and neutral amino acids simultaneously. Mar Biol 45:97–104

Schlichter D (1984) Cnidaria: permeability, epidermal transport and related phenomena. In: Bereiter-Hahn J, Matoltsy AG, Richards KS (eds) Biology of the integument, vol 1. Invertebrates. Springer, Berlin Heidelberg New York Tokyo, pp 79–95

Schmidt-Nielsen K (1983) Animal physiology. Adaptation and environment, 3rd edn. Cambridge Univ Press, London

Schneider EG (1985) Activation of Na^+-dependent transport at fertilization in the sea urchin: requirements of both an early event associated with exocytosis and a later event involving increased energy metabolism. Dev Biol 108:152–163

Segel IH (1975) Enzyme kinetics. Wiley Sons, New York

Sepers ABJ (1977) The utilization of dissolved organic compounds in aquatic environments. Hydrobiologia 52:39–54

Siebers D (1976) Absorption of neutral and basic amino acids across the body surface of two annelid species. Helgol Wiss Meeresunters 28:456–466

Siebers D (1982) Bacterial-invertebrate interactions in uptake of dissolved organic matter. Am Zool 22:723–733

Siebers D, Bulnheim HP (1977) Salinity dependence, uptake kinetics, and specifity of amino acid absorption across the body surface of the oligochaete annelid, *Enchytraeus albidus*. Helgol Wiss Meeresunters 29:473–492

Siebers D, Winkler A (1984) Amino-acid uptake by mussels, *Mytilus edulis*, from natural sea water in a flow-through system. Helgol Wiss Meeresunters 38:189–199

Smith DF (1986) Small-scale spatial heterogeneity in dissolved nutrient concentrations. Limnol Oceanogr 31:167–171

Smith DF, Meyer DL, Horner SMJ (1981) Amino acid uptake by the comatulid crinoid, *Cenometra bella* (Echinodermata) following evisceration. Mar Biol 61:207–213

Southward AJ, Southward EC (1968) Uptake and incorporation of labelled glycine by pogonophores. Nature (London) 218:875–876

Southward EC (1984) Pogonophora. In: Bereiter-Hahn J, Matoltsy AG, Richards KS (eds) Biology of the integument, vol 1. Invertebrates. Springer, Berlin Heidelberg New York Tokyo, pp 376–388

Stephens GC (1960) Uptake of glucose from solution by the solitary coral, *Fungia*. Science 131: 1532

Stephens GC (1962) Uptake of organic material by aquatic invertebrates. I. Uptake of glucose by the solitary coral, *Fungia scutaria*. Biol Bull 123:648–659

Stephens GC (1963) Uptake of organic material by aquatic invertebrates. II. Accumulation of amino acids by the bamboo worm, *Clymenella torquata*. Comp Biochem Physiol 10:191–202

Stephens GC (1964) Uptake of organic material by aquatic invertebrates. III. Uptake of glycine by brackish-water annelids. Biol Bull 126:150–162

Stephens GC (1968) Dissolved organic matter as a potential source of nutrition for marine organisms. Am Zool 8:95–106

Stephens GC (1972) Amino acid accumulation and assimilation in marine organisms. In: Campbell JW, Goldstein L (eds) Nitrogen metabolism and the environment. Academic Press, London New York, pp 155–184

Stephens GC (1975) Uptake of naturally occurring primary amines by marine annelids. Biol Bull 149:397–407

Stephens GC (1982) Recent progress in the study of "Die Ernährung der Wassertiere und der Stoffhaushalt der Gewässer". Am Zool 22:611–619

Stephens GC, Schinske RA (1961) Uptake of amino acids by marine invertebrates. Limnol Oceanogr 6:175–181

Stephens GC, Volk MJ, Wright SH, Backlund PS (1978) Transepidermal transport of naturally occurring amino acids in the sand dollar, *Dendraster excentricus*. Biol Bull 154:335–347

Stevens BR, Preston RL (1980a) Sodium-dependent steady-state L-alanine accumulation by the body wall of *Glycera dibranciata*. J Exp Zool 212:139–146

Stevens BR, Preston RL (1980b) The transport of L-alanine by the integument of the marine polychaete, *Glycera dibranchiata*. J Exp Zool 212:119–127
Stevens BR, Preston RL (1980c) The effect of sodium on the kinetics of L-alanine influx by the integument of the marine polychaete *Glycera dibranchiata*. J Exp Zool 211:129–138
Stewart MG (1977) The uptake and utilization of dissolved amino acids by the bivalve *Mya arenaria* (L.). In: McKlusky DS, Berry AJ (eds) Physiology and behavior of marine organisms. Pergamon Press, New York, pp 165–176
Stewart MG (1978) Kinetics of neutral amino-acid transport by isolated gill tissue of the bivalve *Mya arenaria* (L.). J Exp Mar Biol Ecol 32:39–52
Stewart MG (1979) Absorption of dissolved organic nutrients by marine invertebrates. Oceanogr Mar Biol Annu Rev 17:163–192
Stewart MG (1981) Kinetics of dipeptide uptake by the mussel *Mytilus edulis*. Comp Biochem Physiol 69A:311–315
Stewart MG (1984) Mollusca. Permeability and epidermal transport. In: Bereiter-Hahn J, Matoltsy AG, Richards KS (eds) Biology of the integument, vol 1. Invertebrates. Springer, Berlin Heidelberg New York Tokyo, pp 486–501
Stewart MG, Bamford DR (1975) Kinetics of alanine uptake by the gills of the soft shelled clam *Mya arenaria*. Comp Biochem Physiol 52A:67–74
Stewart MG, Bamford DR (1976) The effect of environmental factors on the absorption of amino acids by isolated gill tissue of the bivalve, *Mya arenaria*. J Exp Mar Biol Ecol 24:205–212
Swinehart JH, Crowe JH, Giannini AP, Rosenbaum DA (1980) Effects of divalent cations on amino acid and divalent cation fluxes in gills of the bivalve mollusc, *Mytilus californianus*. J Exp Zool 212:389–396
Taylor AG (1969) The direct uptake of amino acids and other small molecules from sea water by *Nereis virens* Sars. Comp Biochem Physiol 29:243–250
Testerman JK (1972) Accumulation of free fatty acids from sea water by marine invertebrates. Biol Bull 142:160–177
Thomas JD, Sterry PR, Patience RL (1984) Uptake and assimilation of short chain carboxylic acids by *Biomphalaria glabrata* (Say), the freshwater pulmonate host of *Schistosoma mansoni* (Sambon). Proc R Soc London Ser B 222:447–476
Turner RJ (1983) Quantitative studies of cotransport systems: models and vesicles. J Membr Biol 76:1–15
Wheatley DN, Inglis MS (1984) Interactions of structurally similar amino acid molecules during uptake into the intracellular acid-extractable pool. Mol Physiol 5:85–98
Williams PJ le B (1975) Biological and chemical aspects of dissolved organic material in sea water. In: Riley JP, Skirrow G (eds) Chemical oceanography, vol 2. Academic Press, London New York, pp 301–363
Williams PM, Oeschager H, Kinney P (1969) Natural radiocarbon activity of the dissolved organic carbon in the northeast Pacific Ocean. Nature (London) 224:256–258
Winne D (1973) Unstirred layer, source of biased Michaelis constant in membrane transport. Biochim Biophys Acta 298:27–31
Woodwell GM, Whittaker RH, Likens WA, Delwiche CC, Botkin DB (1978) The biota and world carbon budget. Science 199:141–146
Wright SH (1979) Effect of activity of lateral cilia on transport of amino acids in gills of *Mytilus californianus*. J Exp Zool 209:209–220
Wright SH (1982) A nutritional role for amino acid transport in filter-feeding marine invertebrates. Am Zool 22:621–634
Wright SH (1985) Multiple pathways for amino acid transport in *Mytilus* gill. J Comp Physiol B 156:259–267
Wright SH (1987) Alanine and taurine transport by the gill epithelium of a marine bivalve: effect of Na^+ on influx. J Membr Biol 95:37–45
Wright SH, Secomb TW (1984) Epidermal taurine transport in marine mussels. Am J Physiol 247:R346–R355
Wright SH, Secomb TW (1986) Epithelial amino acid transport in marine mussels: role in net exchange of taurine between gills and sea water. J Exp Biol 121:251–270
Wright SH, Stephens GC (1977) Characteristics of influx and net flux of amino acids in *Mytilus californianus*. Biol Bull 152:295–310

Wright SH, Stephens GC (1978) Removal of amino acid during a single passage of water across the gill of marine mussels. J Exp Zool 205:337 – 352
Wright SH, Johnson TL, Crowe JH (1975) Transport of amino acids by isolated gills of the mussel *Mytilus californianus* Conrad. J Exp Biol 62:313 – 325
Wright SH, Becker SA, Stephens GC (1980) Influence of temperature and unstirred layers on the kinetics of glycine transport in isolated gills of *Mytilus californianus*. J Exp Zool 214:27 – 35
Wright SH, Kippen I, Wright EM (1982) Stoichiometry of Na^+-succinate cotransport in renal brush-border membranes. J Biol Chem 257:1773 – 1778
Wright SH, Hirayama B, Kaunitz JD, Kippen I, Wright EM (1983) Kinetics of sodium-succinate cotransport across renal brush border membranes. J Biol Chem 258:5456 – 5462
Wright SH, Southwell UM, Stephens GC (1984) Autoradiographic analysis of amino acid uptake by the gill of *Mytilus*. J Comp Physiol 154:249 – 256
Wright SH, Secomb TW, Bradley TJ (1987) Apical membrane permeability of *Mytilus* gill: influence of ultrastructure, salinity, and competitive inhibitors on amino acid fluxes. J Exp Biol 129:205 – 230
Youg JD, Ellory JC (1977) Red cell amino acid transport. In: Ellory JC, Lew VL (eds) Membrane transport in red cells. Academic Press, London New York, pp 301 – 325
Zurburg W, Zwaan A de (1981) The role of amino acids in anaerobiosis and osmoregulation in bivalves. J Exp Zool 215:315 – 325

Chapter 6

The Fate of Stable Pollutants – Heavy Metals and Organochlorines – in Marine Organisms

J.M. Bouquegneau[1] and *C. Joiris*[2]

Contents

1 Introduction

Ecotoxicology can be defined as the study of the contamination of ecosystems (i.e., input, concentration, distribution, and speciation of the pollutant in the ecosystem) by natural and artificial pollutants, their mechanisms of accumulation (i.e., uptake, elimination, and detoxification) and their effects on living organisms (see Ramade 1977; Ravera 1984; Bouquegneau et al. 1985a).

The first observations on the contamination of ecosystems suggested that the transfer of stable pollutants mainly occurs through alimentation, with bioconcentration from a trophic level to the next one along the pyramid of biomasses (see Bouquegneau 1980). Although such a picture is schematically acceptable for some toxicants in terrestrial food webs, the situation is, however, much more complicated for a wider approach, as illustrated by the case of silver concentrations in mollusks from different origins.

Figure 1 shows a synthesis of the results obtained by Martoja et al. (1985) representing a survey of silver levels in some marine, freshwater, and land mollusks, all collected from presumably unpolluted sites. High loads of silver were found in marine species: the metal was detectable but in low concentration in terrestrial ones and undetectable in freshwater species, independent of either the phyletic relations or the position in the food web of the animals. It is worth

1 Laboratory for Oceanology, University of Liège, B-4000 Liège, Belgium
2 Laboratory for Ecotoxicology, Free University of Brussels (V.U.B.), B-1050 Brussels, Belgium

Advances in Comparative and Environmental Physiology. Vol. 2

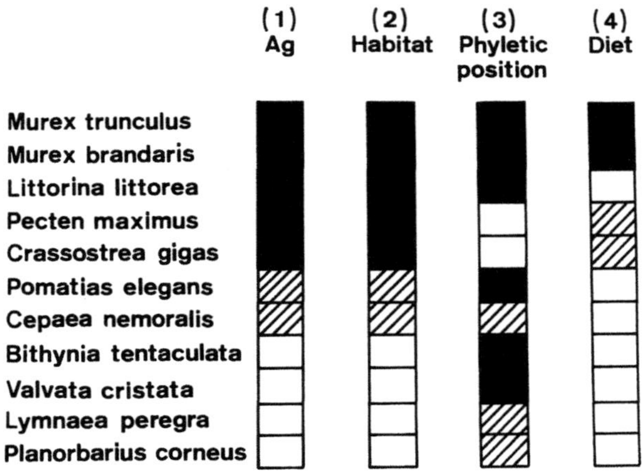

Fig. 1. Comparison between the silver load of 11 mollusk species, their phyletic relations, their habitat and their position in the food web. (Compiled from Martoja et al. 1985).
■ *1* High Ag content (more than 1 µg g^{-1}) or Ag detected by X-ray microanalysis; *2* marine species; *3* Prosobranchia gastropod; *4* carnivorous.
▨ *1* Ag detectable, but not by X-ray microanalysis; *2* terrestrial species; *3* pulmonate gastropod; *4* filter feeder.
□ *1* Undetectable silver content; *2* freshwater species; *3* bivalve; *4* herbivorous

noting that in unpolluted environments, the silver content of fresh waters is about three times higher than in marine ones, and it can be concluded that in the studied species the silver load results from a physiological convergence phenomenon: marine mollusks (except *Crassostrea gigas*) store the silver bound to proteins as Ag_2S in the lysosomes of the connective tissues, which leads to an important accumulation of the metal. The excretion of the silver accumulated by terrestrial mollusks occurs via another route and freshwater mollusks do not accumulate silver in detectable amounts.

It therefore appears that, in addition to an indirect uptake from food, the contamination of aquatic organisms further depends on their own physiological characteristics and also on the importance of a direct uptake from water, as demonstrated later in this chapter.

When considering heavy metals in their inorganic forms, no increase in the metal loads along aquatic food chains can be detected (Bouquegneau et al. 1976). On the contrary, the following increasing sequences of heavy metals concentration are often shown:

water < fish < sediments < aquatic invertebrates (Prosi 1979).

water < fish < plankton < sediments and macroinvertebrates (Bouquegneau et al. 1982a).

When considering the effects of pollutants on living organisms, effects can be detected in situ, at individual, population, and community levels, when the detection techniques are sensitive enough. However, it is regrettable that the assessment of the impact of the pollution often results from the study of the effects of pollutants in terms of mortality tests (e.g., LC50), from which it generally appears that the lethal concentrations during short-term experiments are much higher than the concentration ranges encountered in the environment.

This chapter deals both with the mechanisms of tranfer and sequestration of stable pollutants in marine organisms and emphasizes the difficulty of assessing their effects at the ecosystem level.

2 Spatial and Temporal Variations of the Contamination Level of Marine Organisms

The contamination level of marine organisms obviously presents spatial and temporal variations related not only to the concentration of pollutant in their environment, but also to the time of exposure, to seasonal variations, and to biological characteristics such as age, sex, feeding habits.

2.1 Contamination Level of the Environment

Among many examples, Table 1 shows that the level of heavy metals in the environment affects the metal load of the organisms which are living in it. However, a clear difference appears between essential and nonessential metals: zinc and copper are generally much better controlled than cadmium, for example. Moreover, high loads of essential heavy metals can be found in species due to their peculiar physiology, e.g., copper loads of species using hemocyanin as oxygen carrier and iron loads of species using hemoglobin.

Table 1. Cadmium, zinc, and copper content ($\mu g\ g^{-1}$ wet wt.) of three mollusk species collected in different environments (Noel-Lambot et al. 1978)

Species	Lightly polluted medium	Highly polluted medium
Patella vulgata	Polperro (England)	Portishead (England)
Cadmium	0.8	116.9
Zinc	28.1	51.7
Copper	3.6	4.0
Purpura lapillus	Cap Gris-Nez (France)	Weston-super-Mare (England)
Cadmium	11.4	84.6
Zinc	114.8	315.2
Copper	8.6	125.0
Littorina littorea	Polperro (England)	Weston-super-Mare (England)
Cadmium	0.7	10.7
Zinc	27.3	52.1
Copper	23.0	44.7

2.2 Time of Exposure and Biological Characteristics

Numerous examples showing the existence of a close correlation between the pollutant content and the age of aquatic organisms are available in the literature. Once again, when considering heavy metals, essential and nonessential ones can sometimes be distinguished, as shown in Fig 2: in limpets collected from Weston-super-Mare, the cadmium content of soft tissues increases with the age of the animal, while the zinc and copper contents decrease with age (Noel-Lambot et al. 1978). In limpets, copper from hemocyanin catabolism is either used again or excreted out of the tissues, which explains why there is no increase with age. However, in species, where copper from hemocyanin catabolism is stored as copper sulfide in conjunctive cells of the visceral mass, there is a positive correlation between age and copper content of the animals (Fig.3; Bouquegneau and Martoja 1982).

Here again, it appears that adequate knowledge of the natural physiology of organisms is necessary in order to interpret observations on pollutant contamination found in the environment.

The influence of both age and sex on the contamination by organochlorines (PCBs and DDT) clearly appears, for instance, in the study of antarctic dolphins (Fig. 4). The important decrease in the levels in females as a function of age could

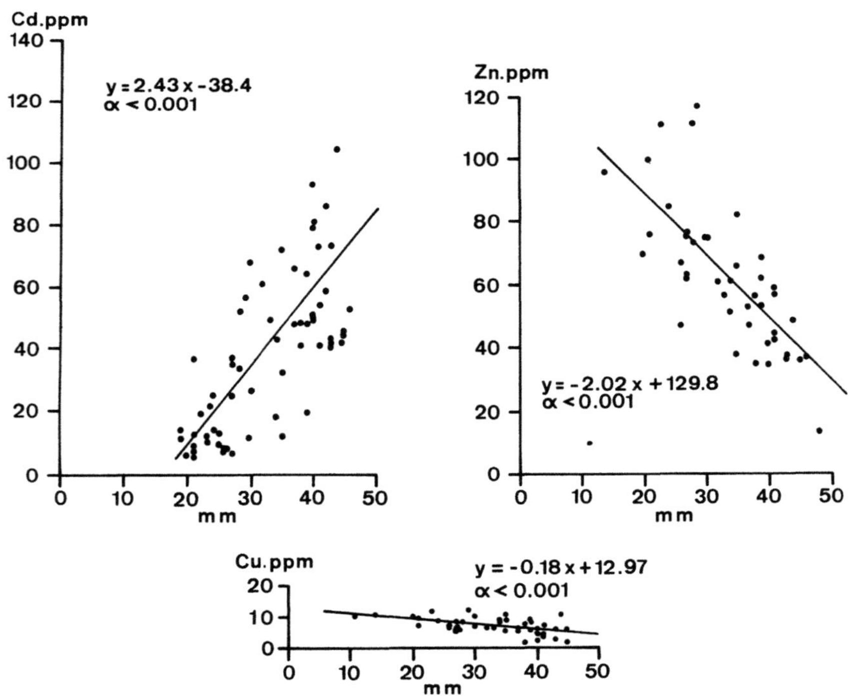

Fig. 2. Limpet, *Patella vulgata*. Relation between metal concentration in soft tissue (μg g^{-1} wet wt.) and shell length for Cd, Zn, and Cu in limpets collected from Weston-super-Mare (Noel-Lambot et al. 1980)

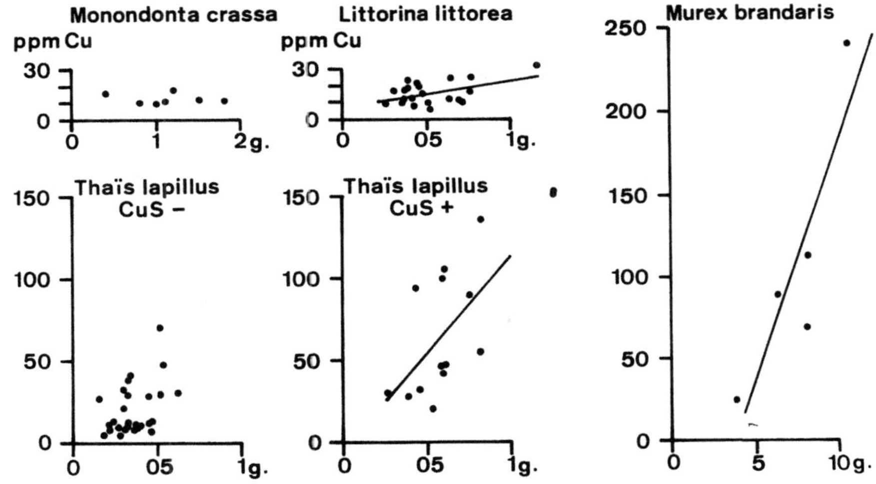

Fig. 3. Copper concentration (μg g^{-1} wet wt.) versus weight of soft tissues in four species of gastropods. In the case of *Thaïs lapillus*, a distinction has been established between animals either containing or not containing copper sulfide in the visceral mass (indicated by CuS^+ and CuS^-) (Bouquegneau and Martoja 1982)

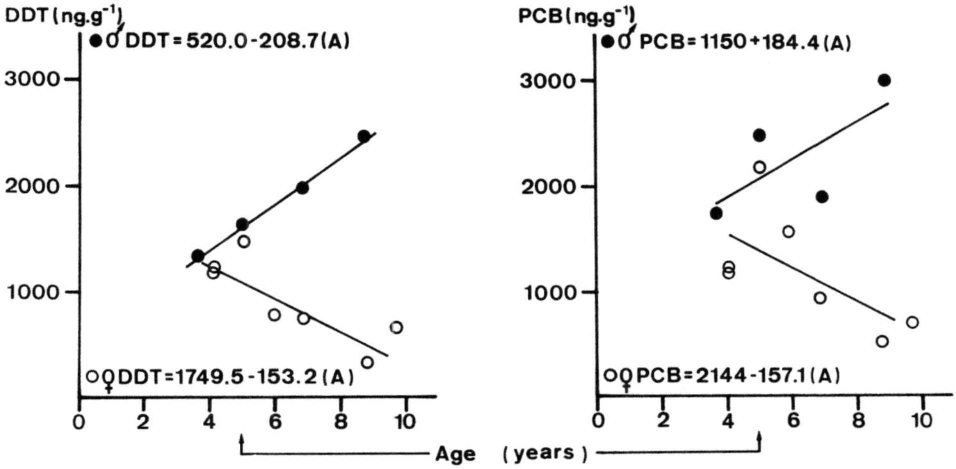

Fig. 4. Commerson's dolphin. Variation with age of the contaminant levels of DDT and PCB (Abarnou et al. 1986)

be due to a higher mobilization of fat during reproduction and to an important transfer of lipids, together with their loads of lipophilic residues, to the offspring (Abarnou et al. 1986).

The importance of feeding habits can be illustrated by comparing the load of various seabirds from the North Sea by PCBs (Delbeke and Joiris 1985). In the muscles, for instance, the contamination varies from 37 μg g^{-1} dry wt. for species eating mainly fish and zooplankton, like the fulmar (*Fulmarus glacialis*), to 8 μg g^{-1} dry wt. for fish-eating birds like the common guillemot (*Uria aalge*). These differences cannot be explained by a simple difference in the contamination of the food, since zooplankton and fish show comparable levels of PCBs. The observed differences thus imply physiological causes as well (see also Fig. 13).

2.3 Seasonal Variations

Within 1 year, many abiotic and biotic events are able to modify the pollutant load of marine organisms. The first event is temperature whose effects will be described below. Spring blooms of phytoplankton can reduce the availability of dissolved pollutants. Starvation and variations in the weight of gonads during the reproductive cycle can also be responsible for annual variations of the pollutant content of the animals. Particularly in the case of organometal compounds and organochlorine residues, the seasonal difference in fat content influences their storage capacity.

Mussels have been the most studied with regards to the seasonal variations of their pollutant content, since it is well known that they are good accumulators of many toxicants from the aquatic environment. This led many authors to use mussels as an indicator of pollution by trace metals (see e.g., Phillips 1976a, b; Goldberg et al. 1978). Many works (e.g., Luten et al. 1986) effectively show that it is quite possible to detect pollution gradients in a given area. However, the important interactions between metal burdens, seasonal variations in soft tissue weights, sexual maturation, age, sex, and intracellular speciation of the metals (Cossa et al. 1979; Orren et al. 1980; Latouche and Mix 1981; Ray 1984; Luten et al. 1986b) make valid comparisons very difficult when dealing with large geographical areas and large time spans.

Among other species, e.g., works by Luoma and Cain (1979), Strong and Luoma (1981), and Cain and Luoma (1986) also clearly show that the whole body concentrations of copper, zinc, and silver in clams (*Macoma balthica*) in the South San Francisco Bay fluctuate seasonally and from year to year. Here again, an interaction between the seasonally changing tissue weight and the trace metal concentrations has been pointed out.

Bouquegneau and Martoja (1987) have described the seasonal variation of the cadmium content of *Murex trunculus* in the cadmium unpolluted bay of Calvi (Corsica) and suggested that this metal could be implicated in shell formation, which could explain the observed important cadmium fluctuations of the visceral mass.

Finally, specific characteristics of some species like the change in diet during the development of fish or the molt of crustaceans can affect their pollutant load. For example, *Carcinus maenas* has been shown to lose 60% of the total 65Zn with every discarded exoskeleton (Renfro et al. 1975).

3 Fate of Pollutants in Marine Organisms

The modeling of the distribution of toxic substances in marine ecosystems requires knowledge, not only of the ecology of the system, but also of the physio-

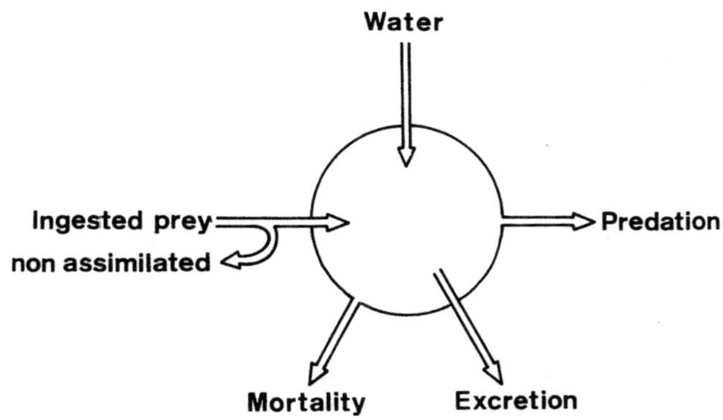

Fig. 5. Model of contamination at a trophic level in an aquatic ecosystem

logical mechanisms implied in the bioaccumulation of pollutants. Such a model is illustrated in Fig. 5 and the corresponding differential equations describing the evolution of the concentration of a toxic substance in the biomass at a trophic level have been described by Jørgensen (1983). Its use requires ecological data (biomasses and fluxes), pollutant concentrations of the different compartments in the field, and ecotoxicological data which are to be determined in the laboratory (i.e., the rates of direct uptake from water, of indirect uptake from food, and of excretion).

3.1 Mechanisms of Entry

At the first trophic level, only direct accumulation from the water occurs. When considering phytoplankton, it appears that the detree of pollutant association with the cells is generally directly proportional to the external concentration. Dead cells accumulate metals comparable to living ones, indicating that the initial association of metal with the cell is governed by adsorption. A very rapid initial uptake rate of the metal therefore occurs, but it is followed by a period of slow and sustained uptake, indicating entry of the pollutants inside the cells. In culture, the uptake is dependent on the growth phase and light conditions (Cossa 1976; McLean and Williamson 1977; Fisher et al. 1984). An internal sequestering of heavy metals in a harmless form (probably sulfide precipitates) has been described in *Dunaliella tertiolecta* (Davies 1976).

Due to their large surface/volume ratio and to their relatively short life span, phytoplankton cells can significantly affect the concentration of dissolved heavy metals in aquatic ecosystems (Morris 1971; Knauer and Martin 1973; Schultz-Baldes and Lewin 1976; see Prosi 1979).

The accumulation of organochlorines by phytoplankton has also been explained by adsorption, especially on small particles (see Duinker 1986). However, when the results are expressed not only on a fresh weight but also on a lipid basis, the existence of absorption and partition mechanisms within some cellular components such as lipids clearly appears: this results in a good correlation between the loading of particles by PCBs and their lipid content (Delbeke and Joiris 1985, 1986).

Marine macrophytes (macroalgae and phanerogames) are also able to concentrate heavy metals and their metal content also depends on adsorption phenomena and is also directly related to the metal concentrations in the water. They have been therefore recommended for use as indicators of metal pollution (Bryan 1971; Prosi 1979; Cristiani et al. 1980; Augier et al. 1984).

At higher trophic levels, it is often quite difficult to determine the relative importance of pollutant uptake from food, from other particulate material, and from solution. The relative importance of each route varies from pollutant to pollutant and from animal to animal so that each chemical and organism must be considered separately, as shown by the abundant literature reviewed by Bryan (1979) and Coombs (1980).

The contamination of zooplankton by PCBs is sometimes explained by both a direct and indirect uptake (see review by Harding 1986). From our own data, however, by comparing the results expressed in different units, we came to the conclusion that all the results could be explained by indirect contamination (Delbeke and Joiris 1985, 1987).

The percentage of ingested pollutants assimilated from food by aquatic organisms can be accurately determined by means of a laboratory setup which allows one to simultaneously follow the pollutant concentration in animals fed with contaminated prey and that of animals living in the same water but without food, thus taking into account the direct contamination from the water (Bouquegneau et al. 1979). Table 2 shows some percentages of ingested heavy metals assimilated by some aquatic organisms.

In the experiments carried out by Bouquegneau et al. (1979), the fact that only part of the ingested pollutant was assimilated clearly led to an important secondary direct contamination of the animals. Several physiological mechanisms can be implicated in such a direct uptake of pollutants by marine

Table 2. Percentage of ingested heavy metals assimilated by some aquatic organisms

Consumers	Cd	Cu[a]	Hg	Zn[a]	Reference
Crustacea					
Artemia salina	3.5	ND	5.7 – 28.8	ND	Bouquegneau et al. (1979)
Daphnia magna	–	–	6 – 58	–	Boudou and Ribeyre (1981)
Leptomysis sp.	8.0	–	19 – 26	–	Bouquegneau et al. (unpubl.)
Praunus sp.	7.0	–	20 – 43	–	Bouquegneau et al. (unpubl.)
Tunicata					
Ascidiella aspersa	ND	–	–	–	Kayser (1982)
Ciona intestinalis	2.0 – 17.4	–	–	–	Kayser (1982)
Fish					
Fish	–	–	10 – 15	–	Jernelov and Lann (1971)
Lebistes reticulata	0.1	–	0.1 – 53.2	10.6	Bouquegneau et al. (1979)
Serranus cabrilla	0.8	–	–	–	Bouquegneau et al. (1979)

[a] ND = not detectable; – not measured.

organisms: diffusion (facilitated or not) and active transport of dissolved pollutants, and endocytosis for particulate and colloïdal species.

Such processes as endocytosis in shellfish have been described by George and co-workers utilizing electron microscopy combined with X-ray microprobe analysis. An engulfment of the particulate material by the epithelial cell membrane occurs, which then pinches off to form membrane-limited vesicles within the cell (in Coombs 1980).

In gill-bearing marine animals, the gills are clearly the main site of entry of dissolved pollutants. Many studies have been devoted to the physiological mechanisms involved in such an uptake and have been reviewed by Coombs (1980). Among more recent papers, let us note the one by Carpene and George (1981) which shows that the cadmium uptake by gills of *Mytilus edulis* is directly proportional to the external Cd concentration, which shows for the authors the importance of diffusion. However, accumulation was facilitated by intracellular binding and sequestration. On the other hand, Bouquegneau et al. (1982b) have shown the existence of an inorganic mercury carrier in the gill epithelium of the teleost, *Serranus cabrilla*, since there is a Michaelis-Menten relation between the rate of entry of the pollutant and its concentration in the surrounding medium.

The direct accumulation of metals from water as a function of time can be described by the following equation (Pentreath 1975):

$$Ct = Css\,(1 - e^{-Kt})\,,$$

where Ct = concentration of the pollutant at time t, Css = steady-state concentration, and $K = 0.693/t_{b1/2}$, $t_{b1/2}$ is the elimination halftime.

The shape of the curve depends on both the rate of elimination (biological halftime) and the initial rate of uptake of the pollutant. The rate of elimination can be rather easily determined in the laboratory by placing contaminated animals in unpolluted water and following the decrease of the contamination as a function of time, and many data are available. In contrast, when considering the uptake rate of toxicants, only few data can be found in the literature, and Jørgensen (1983) has proposed the use of the concentration factor (CF). Indeed, at steady state, CF is equal to the rate of uptake divided by the rate of excretion of the pollutant by the animal. This parameter (CF) is easier to determine in the laboratory and abundant data are also available in the literature.

The use of the laboratory determined CF for the modeling of the fate of toxic substances in marine ecosystems, however, requires some comments:

1. The concentration factor is not constant, even when considering one given pollutant for a single species. As a rule, it seems to decrease when the concentration of the pollutant increases in the medium (see, for example, Bouquegneau et al. 1985b).
2. Such a decrease is only of value as far as no storage mechanism has been initiated from a threshold concentration in the organism (as will be shown below, Fig. 11), so that in the laboratory, it is necessary to determine the CF in a concentration range close to that of the studied environment.

Considering the concentration factor, Moriarty (1986) made some remarks which have to be kept in mind: more than one source of pollutant occurs in the

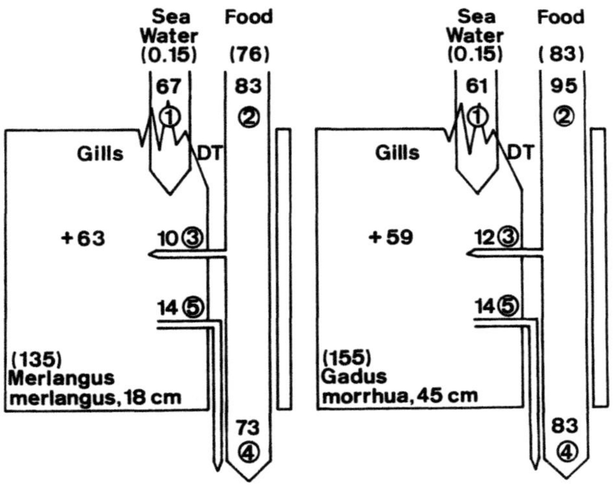

Fig. 6. Mercury accumulation in *Merlangus merlangus* and *Gadus morrhua* caught off the Belgian coastal zone of the North Sea (Bouquegneau and Noel-Lambot 1977). Concentrations: ngHg g^{-1} wet wt. Fluxes: 10^{-8} ngHg g^{-1} wet wt. day^{-1}. *1* Mercury absorbed from water; *2* mercury ingested from the food; *3* mercury assimilated from the food; *4* mercury unassimilated; *5* mercury excreted

environment; the use of inconsistent units of concentration (dry wt., wet wt.) may lead to misleading conclusions (see Sect. 4); steady-state concentrations may take a long time before becoming established; etc.

However, when sufficient ecological and physiological data are available, models of pollutant transfer in marine systems can be described, for example, in the case of mercury in cods (*Gadus morrhua*) and whitings (*Merlangus merlangus*) from the Belgian coastal zone of the North Sea (Fig. 6). As pointed out by Elskens and Goeyens (1985), such a model illustrates that the contribution of direct accumulation is, in this case, clearly greater than that of indirect accumulation, and that a large quantity of pollutant passes through the digestive system and emerges in the form of rejections whose mercury content is highly concentrated in comparison to those of prey. Many similar cases have been discussed by Bouquegneau and Noel-Lambot (1977).

Data on the PCB levels in fish concern various species: commercial and by catch, pelagic and demersal, herbivorous and carnivorous fish of different sizes. Their PCB levels are related to their lipid content, which suggests an important role of adsorption, absorption, and partitioning of PCBs in fish (Delbeke and Joiris 1987), in agreement with other studies in the field (Schneider 1981; Boon and Duinker 1985; Doursma et al. 1986) and in the laboratory (Verschueren 1983). Other data, however, indicate the importance of indirect contamination as well, like the higher concentration in fish than in suspended matter, or an increasing contamination with length (with age): such aspects cannot be understood in the frame work of a simple partition mechanism.

Many factors, other than the simple concentration of pollutant in the surrounding water and in the prey influence the accumulation of pollutants in marine organisms: the nature of the pollutant, its chemincal form (speciation),

Table 3. Concentrations of mercury (ng g^{-1} wet. wt.) and quantity of metal accumulated in eels contaminated for 1 month in sea water containing 1 ng g^{-1} mercury as $HgCl_2$ or CH_3HgCl. (After Bouquegneau 1975)

	Controls	Contamination by	
		$HgCl_2$	CH_3HgCl
Muscles	220	345	750
Skin	50	495	250
Digestive tract	100	290	660
Gills	80	1450	1130
Liver	150	500	1540
Kidney	90	790	1080
Whole body	200	390	720
Hg accumulated in fish (ng g^{-1} wet wt):		190	520

physicochemical factors (pH, Eh, temperature, and salinity), and ecological (biomass) characteristics of the environment and some physiological adaptations which organisms are able to develop (extracellular and intracellular traps).

The most striking example of the importance of the speciation of a pollutant with regard to its rate of accumulation, its fate, and its toxicity, is the effect of mercury, whether it is present in the environment in an inorganic (e.g., $HgCl_2$) or an organic form (e.g., CH_3HgCl). Table 3 shows, for example, the distribution of mercury in the different organs of the sea water-adapted eel *Anguilla anguilla* contaminated for 1 month by water containing 1 μg l^{-1} mercury as either $HgCl_2$ or CH_3HgCl.

It first appears that the lipophilic methylmercuric compound is accumulated 2.5 times more repidly than the inorganic one. However, the main route of entry of both toxicants is through the gills (Bouquegneau 1975) which contain more mercury in the case of $HgCl_2$ contamination. This can be explained by the fact that inorganic mercury induces the synthesis of metallothionein in the gills (Bouquegneau et al. 1975). Methylmercury, not retained in the gill tissue, readily passes into the other organs via the blood.

The mercury content of the liver is three times higher in CH_3Hg^+ contaminated animals but only 1.3 times higher when considering the kidney: methylmercury is mainly accumulated in the liver where it is excreted via the bile, while inorganic mercury is mainly accumulated in the kidney where it is excreted via the urine.

When considering the indirect contamination by mercury (Table 4), the difference between the uptake of both compounds is still more evident: the lipophilic methylmercury passes much more efficiently through the intestinal wall than inorganic mercury. From Boudou and Ribeyre's results (1981), it appears that these accumulation rates are stable as a function of time. However, for methylmercury, a stimulating effect of a temperature increase has been detected. The fact that the lipophiliaty of a toxicant leads to a high rate of absorption through the digestive route is confirmed by the study of the indirect

Table 4. Percentages of ingested mercury assimilated by some aquatic organisms depending on whether the prey have been directly contaminated by $HgCl_2$ or CH_3HgCl

Food chain	$HgCl_2$	CH_3HgCl	Reference
Dunaliella bioculata-Artemia salina	5.7	28.8	Bouquegneau et al. (1979)
Artifical food, *Praunus* sp.	20.0	43.0	Bouquegneau et al. (1985a)
Artificial food, *Leptomysis* spp.	19.0	26.0	Bouquegneau et al. (1985a)
Chlorella vulgaris-Daphnia magna (10 °C)	6.0	21.0	Boudou and Ribeyre (1981)
Chlorella vulgaris-Daphnia magna (18 °C)	6.0	58.0	Boudou and Ribeyre (1981)
Tubifex tubifex-Lebistes reticulata	0.7	45.2	Bouquegneau et al. (1979)

uptake of p,p′-DDT by the marine planktonic copepod *Calanus finmarchicus* which shows a 60–70% retention of the ingested DDT. However, under bloom conditions (high phytoplankton concentration), the absorption rate decreased to 10% (Harding et al. 1981).

Recently, Nelson and Donkin (1985) have pointed out the importance of chemical speciation in the processes of bioaccumulation, taking into account the numerous papers devoted to this problem.

The PCB patterns found in marine organisms are no longer similar to the original mixtures (such as Arochlor 1254) (Abarnou et al. 1986; Delbeke and Joiris 1985). This quantitative difference may be related to slight alterations in liposolubility and/or degradability of the individual components, showing their effects after a long stay in the sea. Thus, different authors no longer expressed results in relation to the whole standard mixture, but used individual congeners (i.e., some of the peaks from the chromatogram), generally four to six (see Duinker 1980). It must, however, be stressed that if this approach is much more precise, the results are only expressed for these congeners, and it is more and more difficult to obtain an estimation for the concentration of the complete PCB group of residues.

The influence of pollutant speciation thus clearly depends on the subsequent changes which occur in their physicochemical properties. This is also true when comparing different pollutants.

When considering the organic ones, it seems clear that their liposolubility and their level of content in the prey determine the pollutant loads. For example, Veith et al. (1980) have shown a close correlation between bioconcentration factors of organic chemicals and the n-octanol-water partition coefficients.

When considering heavy metals, Simkiss (1984) has suggested that, at variance with "class a" hard acid metals (Na, K, Ca, Mg) which dissociate in water to form charged and hydrated particles, the "class b" metals (e.g., Cu, Zn, Cd, Hg) present the phenomenon of ion pair formation which confers to heavy metals a property of lipid solubility which could explain the high rates of penetration of these pollutants across cell membranes.

Spaargaren and Ceccaldi (1984) have described relations between the elementary chemical composition of marine organisms and that of sea water. They showed, e.g., that there is a correlation between the concentration factor of an element and its residence time in the oceans.

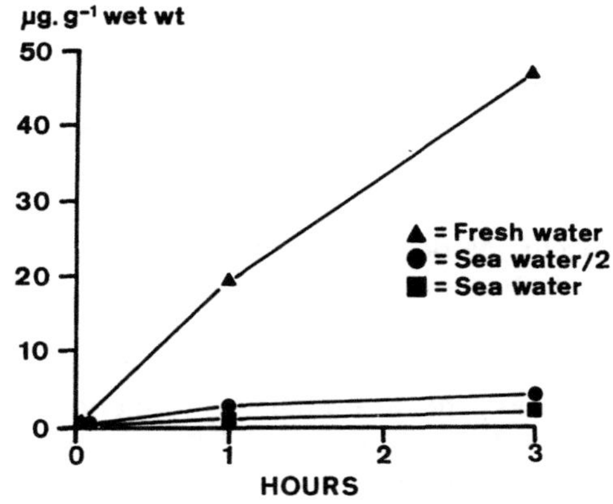

Fig. 7. Direct accumulation of mercury (as $HgCl_2$, 10 µg g^{-1} Hg in the water) by the whole body of *Eriocheir sinensis* (Bouquegneau, Hibaude and Gilles, unpubl.)

Much further research is needed in these promising directions in order to develop general laws that would allows one to integrate the actual data within a general explanatory framework to predict the fate of pollutants in marine environments.

Physicochemical characteristics of the environment influence the pollutant loads of aquatic organisms, because they affect their speciation, the physiology of the organisms, or both. The example of the effect of temperature on the assimilation rate of methylmercury has been noted previously (Table 4). An increase of the water temperature generally increases the direct accumulation of heavy metals as well (Vernberg et al. 1974; Fischer 1986).

Salinity can have a drastic effect on the accumulation of heavy metals by aquatic organisms, as shown in Fig. 7 in the case of the direct uptake of inorganic mercury by the euryhaline crab *Eriocheir sinensis:* the lower the salinity, the higher the rate of uptake of the pollutant. In teleosts and in invertebrates, a much more important accumulation of cadmium at lower salinities has been described (e.g., Phillips 1976a; Vernberg et al. 1977; George et al. 1978; Wright and Brewer 1979). This phenomenon most probably results from an increase of the water influx at low salinities.

Finally, the presence of many toxicants may also alter the accumulation of one of them. For example, Eisler and Gardner (1973) have shown that a high zinc content of the water decreased the accumulation of cadmium by teleosts.

3.2 Elimination and Storage

The presence of heavy metals in abnormal concentrations in the environment can sometimes be considered as a stress to which the animal may be able to respond by developing one or several physiological mechanisms which may lead to either a decrease of the rate of entry of the pollutant or an increase which is generally accompanied by an enhanced tolerance to the pollutant.

3.2.1 Mechanisms Limiting Bioaccumulation

Pollutants may be lost by diffusion or excretion through the general body surface. Excretion into feces via the intestine or the liver, into the urine via the kidney, and directly into the external medium via the gills occurs as well (Bryan 1979). Increased excretion may be a response to increased uptake as demonstrated by George and Coombs (1977), but a decreased absorption may also occur, as demonstrated by Bryan and Hummerstone (1973) in the case of the polychaete *Nereis diversicolor* when exposed to copper and zinc.

Another physiological mechanism which decreases the bioaccumulation of pollutants is a high production of mucus. Several authors have shown an effect of heavy metals on mucus production in teleost fishes (e.g., Lock and van Overbeeke 1981). Radoux and Bouquegneau (1979) and Bouquegneau et al. (1979) have shown that increase mucus production by gills of *Serranus cabrilla* in response to $HgCl_2$ contamination, and its subsequent delamination, highly decreased the rate of entry of the pollutant into the animal.

In several species of sea water fish, another kind of extracellular trap has been described by Noel-Lambot (1981). White mucous corpuscles exist in the intestinal lumen of unfed teleosts and are regularly evacuated by the anus. In fish directly contaminated with cadmium, zinc, or copper, these corpuscles are found to contain high concentrations of those metals and, although their weight is small, they carry a large part of the total metals found in the animals. The presence of such corpuscles, which occurs in the natural environment, directly accumulating heavy metals, greatly limits their entry through the intestine wall by decreasing the pollutant concentration of the intestinal liquid.

A similar phenomenon has been described at the level of the mucous secretions of the branchial pharynx and of the intestine of tunicates which accumulate large quantities of cadmium and are regularly ejected by the exhalant siphon (Noel-Lambot et al. 1980a).

3.2.2 Mechanisms Promoting Bioaccumulation

In contrast to the mechanisms described above, two further mechanisms lead to important concentrations of pollutants in organisms because their development generally results in decreased elimination rates. Heavy metals, for example, can be stored and detoxified by marine organisms either by binding to specific proteins or by a compartmentation process within membrane-limited vesicles (see review by Bouquegneau et al. 1984).

3.2.2.1 Storage of Metals by Binding to Specific Proteins. Among metal-binding proteins, metallothioneins have been the most studied in marine organisms. They are soluble and heat-stable, and the molecular weight of the apoprotein is about 6,800 daltons when estimated by amino acid analyses. They are characterized by an unusually high cysteine content and the absence of aromatic amino acids and histidine (see review by Roesijadi 1981).

However, some proteins are similar to metallothioneins but have significant differences and others, which also bind heavy metals, are totally unrelated (see Stone and Overnell 1985).

Metallothioneins, and metallothionein-like proteins, are widely distributed. They have been reported notably in several marine vertebrates and invertebrates, including Mollusca, Crustacea, and polychaetes (see Roesijadi 1981; Bouquegneau et al. 1984).

Metals which bind to metallothioneins are bismuth, cadmium, copper, gold, mercury, silver, and zinc, some of them being able to induce the synthesis of the protein. The order of binding affinity is Zn-Cd-Cu-Hg (Holt et al. 1980; Funk et al. 1987), so that toxic metals such as cadmium and mercury are able to displace essential ones from the protein, leading to an eventual increase of metallothionein synthesis. As suggested by Noel-Lambot et al. (1978), there may exist a threshold concentration for metals in tissues above which metallothioneins are synthesized. As shown in Fig. 8 (Noel-Lambot et al. 1980b), there is a close relationship between the quantity of cadmium bound to metallothioneins and the concentration of the total cadmium in the tissue from 14 $\mu g\ g^{-1}$ wet wt. on, while the cadmium in the high molecular weight protein fractions of the cytosol

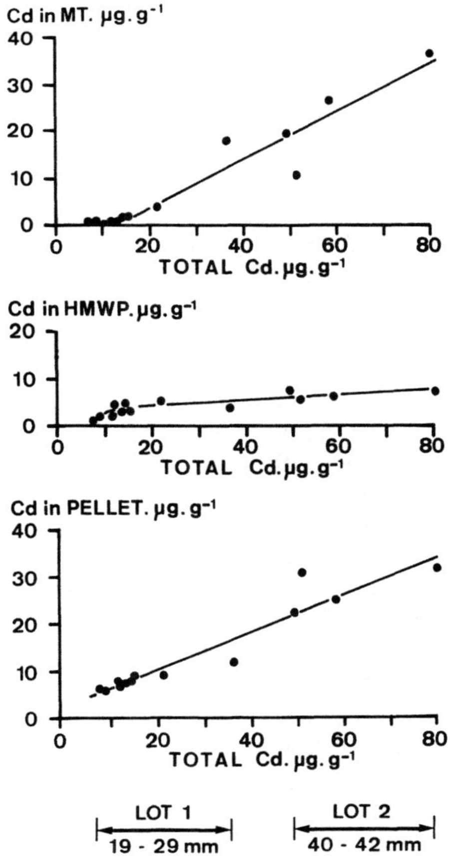

Fig. 8. Relation between cadmium concentration in whole soft parts of limpets collected from Weston-super-Mare and the concentration of cadmium associated with metallothioneins (*MT*), high molecular weight proteins (*HMWP*), and the insoluble fraction of the tissues (*PELLET*) (Noel-Lambot et al. 1980b)

remains constant in limpets collected in the Bristol Channel. More recently, Tohyama et al. (1986) have shown for the harbor seal a relationship between the increased level of metallothionein in the liver and kidney, their metal loads, and age.

Such observations explain the detoxifying effect of metallothioneins which has been demonstrated in aquatic animals by showing a correlation between the appearance of the protein following a pretreatment with sublethal doses of metals and an enhanced tolerance to mercury, cadmium, copper, or zinc (Bouquegneau et al. 1975; Bouquegneau 1979; Roesijadi et al. 1982; Young and Roesijadi 1983; Roesijadi 1983; Duncan and Klaverkamp 1983). Jayasekara et al. (1986) have recently shown a tolerance to cadmium in brine shrimp related to the induction of metallothionein. In this case, it was noted that cadmium was also found to accumulate in other proteins, thus the authors concluded that the high tolerance of this animal cannot be adequately explained by the induction of the metallothionein alone. It is worth noting that such an effect of a preexposure is not generalizable to every metal and every animal (see results and discussion by Weis and Kim 1986).

Therefore, it seems unlikely that the primary function of metallothioneins is the protection against heavy metals, but rather the extracellular and intracellular control of zinc and copper metabolism. More generally, metallothioneins would be involved in the control of normal cellular metabolism and in cellular adaptation to various types of stress (see Karin 1985).

Whatever the case, the role of metallothioneins in heavy metal accumulation and elimination has been clearly shown by Viarengo et al. (1985), and the increase in the protein concentration in exposed animals leads to important loads of the toxicants in their tissues (see Noel-Lambot et al. 1980a).

3.2.2.2 Storage of Metals in Lysosomes and Spherocrystals. Intralysosomal accumulation of metals is well documented (see Sternlieb and Goldfischer 1976), and in marine organisms, the phenomenon appears to be widespread (see Bouquegneau et al. 1984).

In the case of the mussel, for instance, metal-containing lysosomes of the kidney have been characterized (George et al. 1982): they contain about 10% (on a dry wt. basis) of the inorganic components, principally iron, zinc, and calcium, associated with sulfur and phosphorus together with lipofuscin. As illustrated in Fig. 9 (George 1983), the uptake of metals in controlled by the concentration of free ligands (aminoacids, proteins) and for example, induces the synthesis of metallothionein (MT). Primary lysosomes take up metal complexes to form secondary lysosomes, proteins are hydrolyzed, and metal, after being released under acid conditions, binds to ligand X in the lysosome. In tertiary lysosomes, some metal can be exchanged between sterically available ligands, whereas the remainder is physically trapped by lipid peroxidation in the formation of lipofuscin pigments (George 1983).

In the case of the visceral mass of the winkle *Littorina littorea,* the presence of copper, precipitated in the form of clusters of copper sulfide needles inside the lysosomes of the pore cells of connective tissue, has been described by Martoja et al. (1980).

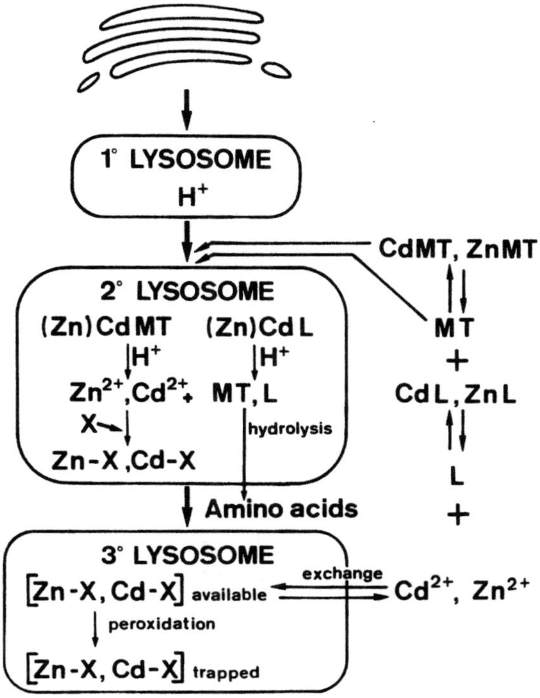

Fig. 9. Pathway of Cd and Zn metabolism in the kidney cell of *Mytilus* (Georges 1983). See comments in text

The presence of copper sulfide in residual bodies of pore cells which are not excreted from the animal is a factor which favors high copper loads in the animal, as shown in *Murex brandaris* by Bouquegneau and Martoja (1982) where there is a close relationship between age and copper content (see Fig 3). In *Murex trunculus*, in contrast, copper from hemocyanin degradation results in the formation of copper sulfide in the lysosomes as well (M. Martoja et al. 1984), but the storage is not definitive and no correlation between copper content and age can be detected (Martoja and Bouquegneau in prep.).

Other metals, such as cadmium (M. Martoja et al. 1984) or silver (Martoja et al. 1985; George et al. 1986), have been shown to be associated with sulfur in lysosomes.

As such phenomena might lead to very high metal contents in the animals, their ecotoxicological impact should be studied.

In contrast to lysosomes, spherocrystals occur in invertebrates only. They are characterized by concentric layers and a radial organization. They also differ from lysosomes by the fact that there is never any supply of hydrolytic enzymes in their structure; whereas in lysosomes there is a degradation of inorganic matter, in spherocrystals. there is a synthesis of an organic network on which the inorganic matter can be precipitated as phosphates, carbonates, or oxalates (see, e.g., Martoja and Truchet 1983; Overnell 1983). The relative roles of spherocrystal-containing cells and of pore cells (containing lysosomes) have been reviewed by Richardot (1979). Once again, these structures appear to be widespread (but only in invertebrates), and they might lead to high metal loads in organisms (see Bouquegneau et al. 1984). In this regard, Martoja and R. Martoja (1984) have emphasized the difficulty to dissociate, at lysosomal and spherocrystal levels, the sequestered metal loads resulting from pollution, from natural physiological processes.

3.2.2.3 Storage of Organochlorine Residues. As mentioned earlier, the results on PCB accumulation in fish do not provide and unambiguous interpretation, some data being in favor of a direct contamination plus partition in the lipids, other observations, like increasing levels with increasing length (i.e., increasing age), do not fit in such a simple scheme. The results on mercury levels are very similar, thus leading to a similar interpretation. It is suggested that organic residues like PCBs and organic mercury could be stored in some special lipid structures (like fat globules) and could so escape the simple partition equilibrium between water and "normal" lipids.

3.3 Ecotoxicological Significance

The mechanisms described above illustrate the large diversity of responses of aquatic organisms to their environment. This is further demonstrated in Figs. 10 and 11.

Figure 10 shows the differences in the kinetics of accumulation of mercury by three teleosts: *Anguilla anguilla, Serranus cabrilla,* and *Crenilabrus ocellatus.* Three quite different curve shapes can be observed: the accumulation of mercury by the eel is quite linear, that ot *Serranus* approaches rapidly an equilibrium, and that of *Crenilabrus* is at first slow, then increases significantly from the fourth to the seventh day, and afterwards, reaches a steady state. The following mechanisms explain the diversity of the responses:

1. The initial rate of mercury uptake by eel gills is fast, the threshold of synthesis of metallothioneins is quickly reached, and a very slow excretion of the pollutant occurs (Bouquegneau et al. 1985a).

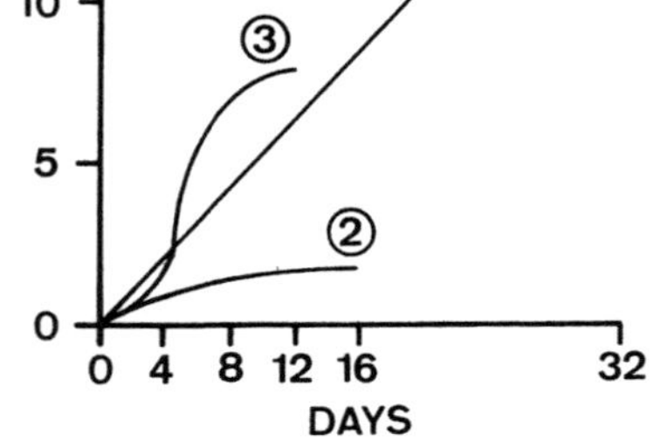

Fig. 10. Accumulation of mercury from sea water containing 100 µgHg l^{-1} as $HgCl_2$. *1 Anguilla anguilla* (Bouquegneau 1975); *2 Serranus cabrilla* (Radoux and Bouquegneau 1979); *3 Crenilabrus ocellatus* (Bouquegneau and Velissarides, unpubl.)

2. The initial rate of uptake by *Serranus* is also fast, but a mechanism of rapid excretion by the gills occurs (i.e., intense mucus production) which limits the entry of the pollutant (Radoux and Bouquegneau 1979). No metallothionein can be detected (Bouquegneau, unpubl.).
3. The initial rate of uptake by *Crenilabrus* is slow, but a synthesis of metallothioneins occurs from the fourth to the seventh day of contamination (Bouquegneau and Velissarides, unpubl.).

Thus, when storage mechanisms such as metallothionein synthesis develop, there is clearly no tendency for equilibrium. This is again clearly illustrated in Fig. 11 which shows theoretically the expected kinetics of accumulation of cadmium by two mollusks living in three different environments. In unpolluted, natural sea water, relatively low levels of cadmium are observed in *Patella vulgata* and in *Littorina littorea,* since the threshold level for metallothionein synthesis is not reached. In the polluted water of the Severn Estuary, the accumulation of cadmium is much more rapid in limpets than in periwinkles: a high rate of synthesis of metallothionein occurs in limpets but is not detectable in periwinkles. In the laboratory, however, when the cadmium concentration of the water is sufficient to induce the synthesis of the protein in periwinkles and to increase it in limpets, the accumulation of the metal is high, and higher in periwinkles than in limpets. These examples illustrate that no steady state can be approached once the threshold of metallothionein synthesis is reached.

For us, it is clear that detoxification mechanisms such as those described above allow organisms to withstand increases in heavy metal concentrations in the ranges generally encountered in polluted environments. As the storage of heavy metals under a detoxified form considerably changes the processes of elimination of these either essential or nonessential metals, we think that this should be taken into account in the mathematical models as illustrated in Fig. 12 (cf. Fig. 5). However, in the specific case described in Fig. 6, only background levels of metallothioneins occurred, so that the Fig. 5 model remains adequate.

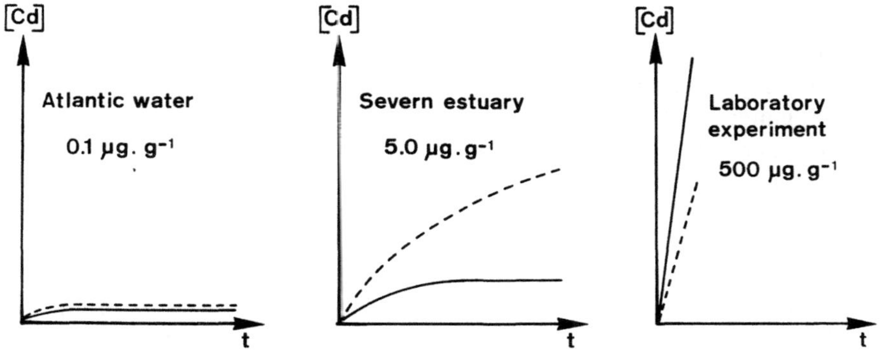

Fig. 11. Theoretically expected kinetics of accumulation of cadmium by the limpet *Patella vulgata* (– – –) and by the periwinkle *Littorina littorea* (———) in three different environments. (From data by Noel-Lambot et al. 1978)

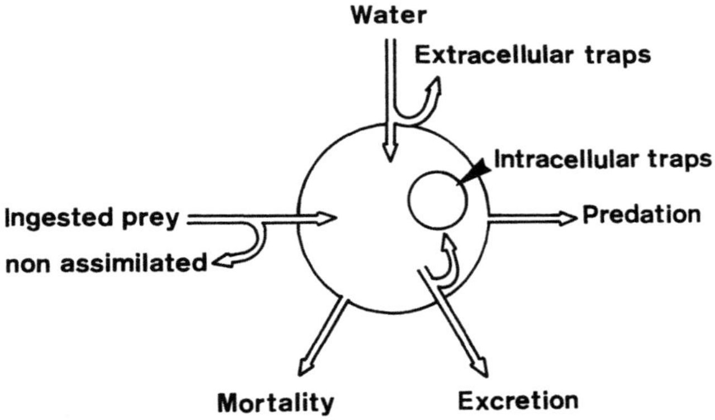

Fig. 12. Model of contamination of a trophic level in an aquatic ecosystem: revised version of Fig. 5 when extra- and intracellular traps are considered

Such concepts led Bascom (1981) to the idea that inorganic metals in the sea, even in regions with high concentrations, are not a hazard to sea animals or to persons who consume them. He based his arguments on the following considerations:

1. The concentrations of metals in the food webs show that high and low values of metal levels in macroinvertebrates are randomly distributed between contaminated and uncontaminated control regions.
2. The existence of protective mechanisms in sea animals, allowing them to withstand heavy metals concentrations increases. Besides, he considers that the generally used toxicity tests are inadequate.

If we agree with his statement of the inadequacy of most toxicity data, we cannot agree with him totally with regards to the other two considerations:

1. The storage mechanisms, although generally considered as a protective mechanism, may be responsible for high metal loads in some organisms. The transfer and the toxicity of such sequestrated metals to higher levels of the food chain are unknown.
2. The high loads of heavy metals encountered in invertebrates can often be explained by their peculiar physiology, and natural processes of storage may mask an accumulation due to pollution.
3. The sediments of coastal and estuarine heavy metal contaminated areas sometimes contain very high levels of trapped pollutant. Who can guarantee that the release of these metals following a change of the local physicochemical characteristics would be harmless?
4. It is well known that some inorganic metals such as mercury are methylated following microorganism activity of marine ecosystems and we think that the harmful effect of methylmercury in the bay of Minamata, for instance, does not need to be demonstrated.

4 Stable Pollutants at the Ecosystem Level, Biological Indicators and Monitoring

The assessment of the pollution level of ecosystems requires knowledge of both the circulation of pollutants (concentrations and fluxes in and between the different compartments, sinks, and sources) and their potential and effective effects. This requires not only purely toxicological information, but also ecological data. As long as this step is not achieved, the approach is only a purely toxicological one and thus does not concern ecotoxicology, even if this term is often misused from this viewpoint. As shown earlier, a certain knowledge of the physiology of organisms is necessary for understanding the meaning of the data on stable pollutant concentration. Thus the physiology of the ecosystems must be known in order to understand the concentrations in the main compartments and the flows between them.

When discussing ecotoxicological data gathered in the southern North Sea, for instance, one clearly has to know that its ecological structure leads only to a limited consumption of primary products by zooplankton grazing (20%), the main part being recycled by planktonic (40%) and benthic (40%) microheterotrophs (Joiris et al. 1982). The interpretation of results on stable pollutant levels in the different compartments requires knowledge of the basic contamination mechanisms: indirect or direct, adsorption and absorption, partition into the lipids, etc. A first step consists in simply expressing the pollutant concentrations in three different ways: on a dry, fresh, or carbon weight basis, on a lipid weight basis, and reference to 1 m^3 sea water, taking into account the respective biomasses (Fig. 13, Bouquegneau et al. 1985a).

By comparing a series of data expressed in such a way, one can easily detect which mechanism is playing an important role. If not, wrong conclusions could be drawn. The concentration of PCBs on suspended matter, for instance, is directly dependent on the amount of particles: if this amount increases during phytoplanktion growth, for example, their contamination per dry weight decreases and such a variation does not have any ecotoxicological significance (Duinker 1986; Delbeke and Joiris 1987, Fig. 14). The only significant estimation, in this case, is the total load per water volume.

Similar cases exist in which the contamination of a system cannot be described by simple concentration data but rather by a complete curve: PCB load of sediments as a functtion of their clay content (Sawhney et al. 1981), PCB load of fish as a function of their fat content (Schneider 1981) or, in terrestrial ecosystems, organochlorine contamination of raptors as a function of the relative importance of birds in their diet (Joiris and Delbeke 1985).

These, and all other considerations developed in the previous sections of this chapter (mainly the interferences between the level of pollution and the physiology of organisms; for a discussion of this subject, see, e.g., Mason and Simkiss 1983; Simkiss and Taylor 1981; Fischer 1986, who has notably pointed out the importance of growth in the mussel watch concept) make us skeptic of the usefulness of the utilization of classical biological indicators for the monitoring of pollution. Many organisms, or parts of organisms, have been proposed as

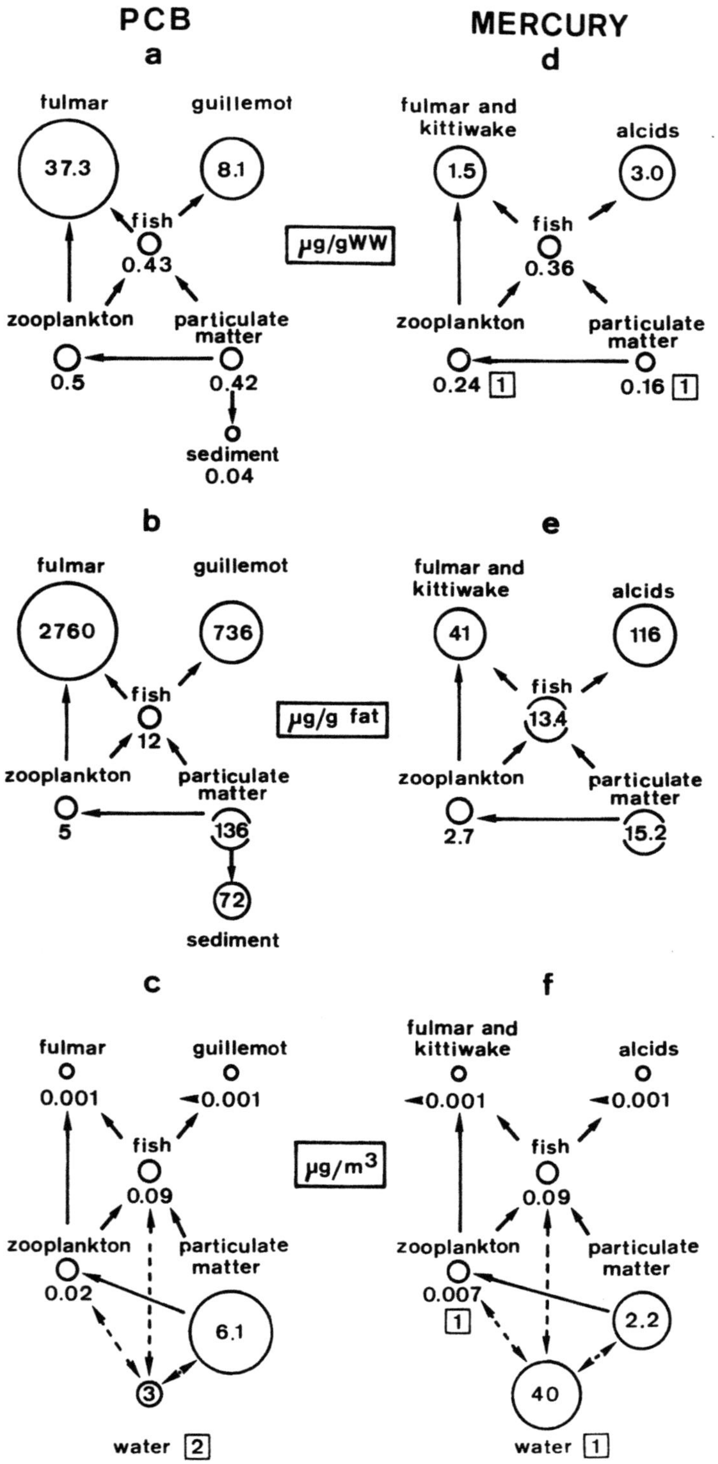
PCB
a
fulmar
guillemot
37.3
8.1
fish
0.43
zooplankton
particulate matter
0.5
0.42
sediment
0.04
µg/gWW
MERCURY
d
fulmar and kittiwake
alcids
1.5
3.0
fish
0.36
zooplankton
particulate matter
0.24 1
0.16 1
b
fulmar
guillemot
2760
736
fish
12
zooplankton
particulate matter
5
136
72
sediment
µg/g fat
e
fulmar and kittiwake
alcids
41
116
fish
13.4
zooplankton
particulate matter
2.7
15.2
c
fulmar
guillemot
0.001
0.001
fish
0.09
zooplankton
particulate matter
0.02
6.1
3
water 2
µg/m3
f
fulmar and kittiwake
alcids
0.001
0.001
fish
0.09
zooplankton
particulate matter
0.007
1
2.2
40
water 1

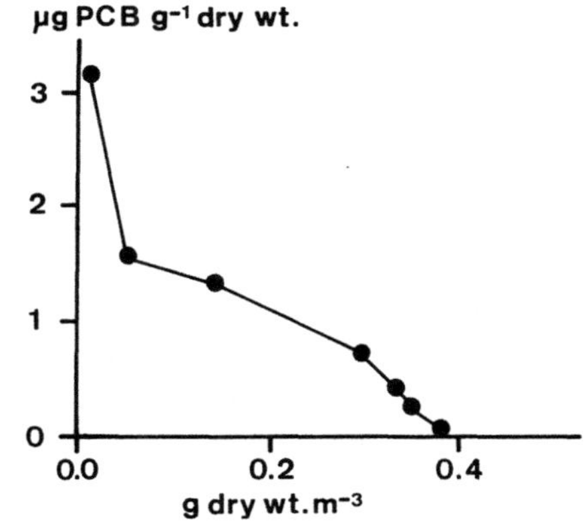

Fig. 14. Relation between PCB levels (µg g^{-1} dry wt.) and particle concentration (g dry wt. m^{-3}) of net plankton sampled at different stations in the North Sea. (After Delbeke and Joiris 1987)

biological indicators of pollution level ("mussel watch", Goldberg et al. 1978; fish, Essink 1985; shell content of mollusks, Fischer 1983; metallothionein content of selected organs, Roch et al. 1982; etc.). However, in our opinion, these investigations could remain of value at a local scale, reflecting, for instance, pollution gradients.

Recent studies increasingly tend to deal with the effects of pollutants at the ecosystem level (for reviews, see Sheehan et al. 1984, 1985). When considering biological indicators, studies dealing with the determination of toxicant concentrations in organisms ("mussel watch") should be replaced by biochemical and physiological studies of the effects of stress and pollution on marine organisms (for a review, see Bayne et al. 1985). In this regard, the evaluation of the scope for growth (i.e., the energy available to an animal for growth and reproduction) of mussels ("mussel stress") has already supplied promising results. For instance, Donkin and Widdows (1986) have shown the close correlation between the contamination of mussels by aromatic hydrocarbons and their scope for growth. Here again, the comparisons concern relatively small areas but, at larger scales, problems such as the providing of standard mussel populations will have to be solved.

The general effects of pollution, resulting notably from an eutrophication of the marine environment, are quite easy to detect. One of them is a change of the food web structure, resulting in reduced fish production in the ecosystem: in unstressed oligotrophic areas, large diatoms are grazed by large zooplankton, which is eaten by fish; in stressed areas, small flagellates are grazed by small zooplankton which is eaten by Medusae and Ctenophores (see Gray 1982). In the eutrophicated coastal area of the Southern Bight of the North Sea (e.g., see Joiris et al. 1982), the spring phytoplanktonic bloon is dominated by *Phaeocystis poucheti,* rarely consumed by zooplankton and most of the production is recycled by bacterial heterotrophic activity.

Fig. 13a–f. Biomagnification of PCBs (*a–c*) and mercury (*d–f*) through the trophic web of the North Sea. (After Bouquegneau et al. 1985a). Contamination levels denoted by *figures in circles: a,d* µg g^{-1}dry weight; *b,e* µg g^{-1} fat weight; *c,f* µg m^{-3}. For seabirds and fish, the contamination levels of the muscles are considered. (*1*) Decadt (1985); (*2*) Geyer et al. (1984)

It is, however, often difficult to point out the effect of one or some stable pollutants in a marine environment. For example, in the case of visceral granulomatous nodules and skin ulcerations of fish inhabiting polluted areas, the presence of heavy metals in these structures suggest their possible effect in fish diseases (Vandorpe and Smigielski 1978). However, and evident relationship between heavy metals and fish diseases cannot be detected after careful examination (Martoja et al. 1984).

Another example is given by Miller and Pondick (1984) who have described the heavy metal levels in the gastropod *Nucella lapillus* from sites with normal and penis-bearing females. The presence of the penis and its pattern of distribution in the populations suggest the involvement of heavy metal pollution associated with boating activity. However, results by Kantor (1984) on *Buccinum undatum* and by Martoja and Bouquegneau (unpubl.) on *Murex trunculus* show that this phenomenon is, in both Prosobranchia, independent of pollution. It may not be an anomaly but a normal evolutive process towards hermaphroditism, perhaps accelerated by an alteration of the environment, as has previously occurred in pulmonates and Opistobranchia.

This last example, once again, points out the necessity of adequate knowledge of the biology of marine organisms prior to an estimation of their responses (accumulation and effects) to pollutants in the field. Further fundamental research is needed in this direction rather than random investigations in the environment.

Acknowledgments. We are grateful to K. Delbeke for useful discussions. One of us (J.M.B) is a "Chercheur Qualifié" of the Belgian National Science Foundation (F.N.R.S.).
This paper is dedicated to Professor A. Distèche at the occasion of the end of his official career.

References

Abarnou A, Robineau D, Michel P (1986) Contamination par les organochlorés des dauphins de Commerson des îles Kerguelen. Oceanol Acta 9:19–29

Augier H, Gilles G, Ramonda G (1984) L'herbier de *Posidonia oceanica* et la pollution par le mercure sur le littoral des Bouches-de-Rhone et du Var (France). Int Workshop *Posidonia Oceanica* beds. In: Boudouresque CF, Jeudy de Grissac A, Oliver J (eds), GIS Posidonie Publ 1. Fr, pp 399–406

Bascom W (1981) The non-toxicity of metals in the sea. MTS J 17 (no 1):59–66

Bayne BL, Brown DA, Burns K, Dixon DR, Ivanovici A, Livingstone DR, Lowe DM, Moore MN, Stebbing ARD, Widdows J (1985) The effects of stress and pollution on marine animals. Praeger, New York, 384 pp

Boon JP, Duinker JC (1985) Kinetics of individual polychlorinated biphenyl (PCB) components in juvenile sole (*Solea solea*) in relation to concentrations in water and to lipid metabolism under conditions of starvation. Aquat Toxicol 7:119–134

Boudou A, Ribeyre F (1981) Comparative study of the trophic transfer of two mercury compounds – $HgCl_2$ and CH_3HgCl – between *Chlorella vulgaris* and *Daphnia magna*. Influence of temperature. Bull Environ Contam Toxicol 27:624–629

Bouquegneau JM (1975) L'accumulation du mercure et ses effets physiologiques chez *Anguilla anguilla* et *Myoxocephalus scorpius*. Doct Thesis, Univ Liège, Belgium

Bouquegneau JM (1979) Evidence for the protective effect of metallothioneins against inorganic mercury injuries to fish. Bull Environ Contam Toxicol 23:218–219

Bouquegneau JM (1980) Chemical pollutants in the terrestrial environment. In: Gilles R (ed) Animals and environmental fitness, vol 1. Invited lectures. Pergamon Press, Oxford New York, pp 303–314

Bouquegneau JM, Martoja M (1982) La teneur en cuivre et son degré de complexation chez quatre gastéropodes marins. Données sur le cadmium et le zinc. Oceanol Acta 5 (2):219–228

Bouquegneau JM, Martoja M (1987) Seasonal variation ot the cadmium content of *Murex trunculus* in a non-cadmium polluted environment. Bull Environm Contam Toxicol 39:69–73

Bouquegneau JM, Noel-Lambot F (1977) L'accumulation du mercure à partir de l'eau et de la nourriture chez les poissons marins. Rev Int Oceanogr Med 48:107–116

Bouquegneau J M, Gerday Ch, Disteche A (1975) Fish mercury binding thionein related to adaptation mechanisms. FEBS Lett 55:173–177

Bouquegneau J M, Noel-Lambot F, Disteche A (1976) Le problème de l'intoxication directe et indirecte par les métaux lourds. In: Nihoul JCJ, Disteche A (eds) Contamination des produits de la mer. Programme Natl Rech Dev. Environnement Eeu-Proj Mer Serv Premier Ministre, Program Politique Sci, Bruxelles pp 266–292

Bouquegneau JM, Noel-Lambot F, Disteche A (1979) Fate of heavy metals in experimental aquatic food chains: uptake and release of mercury and cadmium by some marine organisms; role of metallothioneins. ICES – CM 1979/E : 58. Mar Environ Qual Comm

Bouquegneau JM, Noel-Lambot F, Verthe C, Disteche A (1982a) The accumulation of heavy metals in marine organisms. ICES – CM 1982/E:41. Mar Environ Qual Comm

Bouquegneau JM, Radoux D, Bay D (1982b) Mise en évidence de l'existence d'un transporteur de mercure au niveau de l'épithélium branchial de *Serranus cabrilla*. Bull Soc R Sci Liege 11–12; 384–386

Bouquegneau JM, Martoja M, Truchet M (1984) Heavy metal storage in marine animals under various environmental conditions. In: Bolis L, Zadunaisky J, Gilles R (eds) Toxins, drugs and pollutants in marine animals. Springer, Berlin Heidelberg New York Tokyo, pp 147–160

Bouquegneau JM, Joiris C, Delbeke K (1985a) Marine ecotoxicology: field and laboratory approaches. In: Grieken R van, Wollast R (eds) Progress in Belgian oceanographic research. Univ Antwerp, Antwerpen, Wilrijk, Belgium, pp 368–379

Bouquegneau JM, Verthe C, Moureau Z, Mania B, Baelen J van, Ben D van der, Cogneau M, Vandecasteele CM (1985b) Accumulation of technetium by four marine mollusks and transfer to a predatory fish. In: Grieken R van, Wollast R (eds) Progress in Belgian oceanographic research. Univ Antwerp, Antwerpen, Wilrijk, Belgium, pp 380–392

Bryan GW (1971) The effects of heavy metals (other than mercury) on marine and estuarine organisms. Proc R Soc London Ser B 177:389–410

Bryan GW (1979) Bioaccumulation of marine pollutants. Philos Trans R Soc London Ser B 286:483–505

Bryan GW, Hummerstone LG (1973) Adaptation of the polychaete *Nereis diversicolor* to estuarine sediments containing high concentrations of zinc and cadmium. J Mar Biol Assoc UK 53:839–857

Cain DJ, Luoma SN (1986) Effect of seasonally changing tissue weight on trace metal concentrations in the bivalve *Macoma balthica* in San Francisco Bay. Mar Ecol Prog Ser 28:209–217

Carpene E, George SG (1981) Absorption of cadmium by gills of *Mytilus edulis* (L.). Mol Phys 1:23–34

Coombs TL (1980) Heavy metal pollutants in the aquatic environment. In: Gilles R (ed) Animals and environmental fitness. Pergamon Press, Oxford New York, pp 283–302

Cossa D (1976) Sorption du cadmium par une population de la diatomée *Phaeodactylum tricornutum* en culture. Mar Biol 34:163–167

Cossa D, Bourget E, Piuze J (1979) Sexual maturation as a source of variation in the relationship between cadmium concentration and body weight of *Mytilus edulis* L. Mar Pollut Bull 10:174–176

Cristiani G, Gassend R, Augier H (1980) Etude de la contamination expérimentale de la phanérogame marine *Posidonia oceanica* (L.) Delile par les composés mercuriques. Partie 1. Modalités de la contamination par le chlorure mercurique. Environ Pollut 1:153–162

Davies AG (1976) An assessment of the basis of mercury tolerance in *Dunaliella tertiolecta*. J Mar Biol Assoc UK 56:39–57

Decadt G (1985) Contribution to the study of the biochemical cycle of mercury in the Southerm Bight of the North Sea. Doct Thes. Vrije Univ Brussel, Belgium

Delbeke K, Joiris C (1985) Ecotoxicology of organochlorine residues in marine ecosystems. In: Grieken R van , Wollast R (eds) Progress in Belgian oceanographic research. Antwerp Univ, Antwerpen, pp 358 – 365
Delbeke K, Joiris C (1986) Transfer and accumulation of organochlorine residues in the North Sea ecosystem. Mar Environ Res (submitted)
Delbeke K, Joiris C (1987) Accumulation mechanisms and geographical distribution of PCBs in the North Sea. Int Conf Environ Protect North Sea, London, 24 – 27 March 1987, in press
Donkin P, Widdows J (1986) Scope for growth as a measurement of environmental pollution and its interpretation using structure-activity relationships. Chem Ind 21:732 – 737
Duinker J (1986) The role of small, low density particles on the partition of selected PCB congeners between water and suspended matter (North Sea area). Neth J Sea Res 20:229 – 238
Duncan DA, Klaverkamp JF (1983) Tolerance and resistance to cadmium in White Suckers (*Catostomus commersoni*) previously exposed to cadmium, mercury, zinc or selenium. Can J Fish Aquat Sci 40:128 – 138
Duursma EK, Nieuwenhuize J, Liere JM van, Hillebrand MTJ (1986) Partitioning of organochlorines between water, particulate matter and some organisms in estuarine and marine systems of the Netherlands. Neth J Sea Res 20:239 – 251
Eisler R, Gardner GR (1973) Acute toxicology to an estuarine teleost of mixtures of Cd, Cu and Zn salts. J Fish Biol 5:131 – 142
Elskens I, Goeyens L (1985) Transport, fate and recycling of heavy metals in sea-water ecosystems. In: Nürnberg HW (ed) Pollutants and their ecotoxicological significance. Wiley, New York, pp 239 – 253
Essink K (1985) Monitoring of mercury pollution in Dutch coastal waters by means of the teleostean fish *Zoarces viviparus*. Neth J Sea Res 19:177 – 182
Fischer H (1983) Shell weight as an independent variable in relation to cadmium content of molluscs. Mar Ecol Prog Ser 12:59 – 75
Fischer H (1986) Influence of temperature, salinity and oxygen on the cadmium balance of mussels, *Mytilus edulis*. Mar Ecol Prog Ser 32:265 – 278
Fisher NS, Bohe M, Teyssie J-L (1984) Accumulation and toxicity of Cd, Zn, Ag and Hg in four marine phytoplankters. Mar Ecol Prog Ser 18:191 – 200
Funk AE, Day FA, Brady FO (1987) Displacement of zinc and copper from copper-induced metallothionein by cadmium and by mercury: in vivo and ex vivo studies. Comp Biochem Physiol 86C:1 – 6
George SG (1983) Heavy metal detoxication in *Mytilus* kidney – an in vitro study of Cd- and Zn-binding to isolated tertiary lysosomes. Comp Biochem Physiol 76C:59 – 65
George SG, Coombs TL (1977, Effects of high-stability iron complexes on the kinetics of iron accumulation and excretion in *Mytilus edulis* (L.). J Exp Mar Biol Ecol 28:133 – 140
George SG, Carpene E, Coombs TL (1978) The effect of salinity on the uptake of cadmium by the common mussel, *Mytilus edulis* (L.). In: McLusky DS, Berry AJ (eds) Physiology and behaviour of marine organisms. Proc 12th Eur Symp Mar Biol, Stirling, Scotland, Sept 1977, Pergamon Press, Oxford New York, pp 189 – 193
George SG, Coombs TL, Pirie BJS (1982) Characterization of metal-containing granules from the kidney of the common mussel, *Mytilus edulis*. Biochim Acta 716:61 – 71
George SG, Pirie BJS, Calabrese A, Nelson DA (1986) Biochemical and ultrastructural observations of long-term silver accumulation in the mussel, *Mytilus edulis*. Mar Environ Res 18:255 – 265
Geyer J, Freitas D, Korte F (1984) Polychlorinated biphenyls (PCBs) in the marine environment, particularly in the Mediterranean. Ecotoxicol Environ Safety 8:129 – 151
Gray JS (1982) Effects of pollutants on marine ecosystems. Neth J Sea Res 16:424 – 443
Goldberg ED, Bowen VT, Farrington FW, Harvey G, Martin JH, Parker PL, Risebrough RW, Robertsen W (1978) The mussel watch. Environ Cons 5:101 – 125
Harding GC (1986) Organochlorine dynamics between zooplankton and their environment, a reassessment. Mar Ecol Prog Ser 33:167 – 191
Harding GC, Vass WP, Drinkwater KF (1981) Importance of feeding, direct uptake from seawater, and transfer from generation to generation in the accumulation of an organochlorine (p,p'-DDT) by the marine planktonic copepod *Calanus finmarchicus*. Can J Fish Aquat Sci 38:101 – 119
Holt D, Magos L, Webb M (1980) The interaction of cadmium-induced rat renal metallothionein with bivalent mercury in vitro. Chem Biol Interact 32:125 – 136

Jarnelov A, Lann H (1971) Mercury accumulation in food chains. Oikos 22:403–406

Jayasekara S, Brown DB, Sharma RP (1986) Tolerance to cadmium and cadmium-binding ligands in Great Salt Lake brine shrimp (*Artemia salina*). Ecotoxicol Environ Safety 11:23–30

Joiris C, Delbeke K (1985) The contamination by organochlorine pesticides and PCBs of the Belgian raptors, their eggs and their food. In: Nürnberg HW (ed) Pollutants and their ecotoxicological significance. Wiley, New York, pp 403–414

Joiris C, Billen G, Lancelot C, Daro MH, Mommaerts JP, Bertels A, Bossicart M, Nijs J, Hecq J-H (1982) A budget of carbon cycling in the Belgian coastal zone: relative roles of zooplankton, bacterioplanktion and benthos in the utilization of the primary production. Neth J Sea Res 16:260–275

Jorgensen SE (1983) In: Application of ecological modelling in environmental management , part A, Elsevier, Amsterdam New York, pp 455–483

Kantor YI (1984) Pseudohermaphroditism in *Buccinum undatum* (Gartropoda, Prosobranchia). Zool Zh 63:1256–1258 (in Russian)

Karin M (1985) Metallothioneins: proteins in search of function. Cell 41:9–10

Kayser H (1982) Cadmium effects in food chain experiments with marine planktion algae (Dinophyta) and benthic filter feeders (Tunicata). Neth J Sea Res 16:444–454

Knauer GA, Martin JH (1973) Seasonal variations of cadmium, copper, manganese, lead and zinc in water and phytoplankton in Monterey Bay, California. Limnol Oceanogr 18:597–604

Latouche YD, Mix MC (1981) Seasonal variation in soft tissue weights and trace metal burdens in the bay mussel, *Mytilus edulis*. Bull Environ Contam Toxicol 27:821–828

Lock RAC, Overbeeke AP van (1981) Effects of mercuric chloride and methylmercuric chloride on mucus secretion in rainbow trout, *Salmo gairdneri* Richardson. Comp Biochem Physiol 69C:67–73

Luoma SN, Cain DJ (1979) Fluctuations of copper, zinc and silver in tellenid clams as related to freshwater discharge in South San Francisco Bay. In: San Francisco Bay: the urbanized estuary. Pac Div, AAAS Publ

Luten JB, Bouquet W, Burggraaf MM, Rauchbaar AB, Rus J (1986a) Trace metals in mussels (*Mytilus edulis*) from the Waddenzee, coastal North Sea and the estuaries of Ems, western and eastern Scheldt. Bull Environ Contam Toxicol 36:770–777

Luten JB, Bouquet W, Burggraaf MM, Rus J (1986b) Accumulation, elimination and speciation of cadmium and zinc in mussels, *Mytilus edulis,* in the natural environment. Bull Environ Contam Toxicol 37:579–586

Martoja M, Martoja R (1984) La bioaccumulation de métaux, processus physiologique normal et conséquence de la pollution. Courr CNRS 54:32–37

Martoja M, Truchet M (1983) Données analytiques sur les concrétions du tissu conjonctif de quelques gastéropodes d'eau douce. Malacologia 23:333–349

Martoja M, Tan Tue VU, Elkaim B (1980) Bioaccumulation du cuivre chez *Littorina littorea* (L.) (Gastéropode prosobranche): signification physiologique et écologique. J Exp Mar Biol Ecol 43:251–270

Martoja M, Truchet M, Bouquegneau JM (1984) Accumulation naturelle du cadmium chez *Murex trunculus* et *Murex brandaris* (Prosobranches Néogastropodes): localisation histologique. CR Acad Sci 298 (16):461–466

Martoja M, Bouquegneau JM, Truchet M, Martoja R (1985) Recherche de l'argent chez quelques mollusques marins, dulcicoles et terrestres. Formes chimiques et localisation histologique. Vie Milieu 35 (1):1–13

Martoja R, Orcel L, Truchet M (1984) Accumulation de métaux et de metalloïdes dans différents viscères de poissons (foie, rate, rein). Ann Inst Oceanogr Paris 60(2):115–127

Mason AZ, Simkiss K (1983) Interactions between metals and their distribution in tissues of *Littorina littorea* (L.) collected from clean and polluted sites. J Mar Biol Assoc UK 63:661–672

McLean MW, Williamson FB (1977) Cadmium accumulation by marine red alga *Porphyra umbilicalis*. Physiol Plant 41:268–272

Miller ER, Pondick JS (1984) Heavy metal levels in *Nucella lapillus* (Gastropoda: Prosobranchia) from sites with normal and penis-bearing females from New England. Bull Environ Contam Toxicol 33:612–620

Moriarty F (1986) Measurement and prediction of bioaccumulation in aquatic organisms. Chem Ind 21:737–740

Morris AW (1971) Trace metal variarions in sea water of the Menai Straits caused by a bloom of *Phaeocystis*. Nature (London) 233:427 – 428

Nelson A, Donkin P (1985) Processes of bioaccumulation: the importance of chemical speciation. Mar Poll Bull 16:164 – 169

Noel-Lambot F (1981) Presence in the intestinal lumen of marine fish of corpuscles with high cadmium-, zinc- and copper-binding capacity: a possible mechanism of heavy metal tolerance. Mar Ecol Prog Ser 4:175 – 181

Noel-Lambot F, Bouquegneau JM, Frankenne F, Disteche A (1978) Le rôle des métallothioneines dans le stockage des métaux lourds chez les animaux marins. Rev Int Oceanogr Med XLIX:13 – 20

Noel-Lambot F, Bouquegneau JM, Disteche A (1980a) Some mechanisms promoting or limiting bioaccumulation in marine organisms. ICES CM 1980/E:39. Mar Environ Qual Comm

Noel-Lambot F, Bouquegneau JM, Frankenne F, Disteche A (1980b) Cadmium, zinc and copper accumulation in limpets (*Patella vulgata*) from the Bristol Channel with special reference to metallothioneins. Mar Ecol Prog Ser 2:81 – 89

Orren MJ, Eagle GA, Hennig HF-KO, Green A (1980) Variations in trace metal content of the mussel *Choromytilus meridionalis* (Kr.) with season and sex. Mar Pollut 1:253 – 257

Overnell J (1983) Protein and oxalate in mineral granules from the kidney of *Pecten maximus*. J Exp Mar Biol Ecol 52:173 – 183

Pentreath RJ (1975) Radiobiological studies with marine fish. In: Design of radiotracer experiments in marine biological systems. IAEA Tech Rep Ser no 167, Vienna, pp 137 – 170

Phillips DJH (1976a) The common mussel *Mytilus edulis* as an indicator of pollution by zinc, cadmium, lead and copper. I. Effects of environmental variables on uptake of metals. Mar Biol 38:59 – 69

Phillips DJH (1976b) The common mussel *Mytilus edulis* as an indicatior of pollution by zinc, cadmium, lead and copper. II. Relationships of metals in the mussel to those discharged by industry. Mar Biol 38:71 – 80

Prosi F (1979) Heavy metals in aquatic organisms. In: Forstner U, Wittmann CTW (eds) Metal pollution in the aquatic environment. Springer Berlin Heidelberg New York, pp 271 – 323

Radoux D, Bouquegneau JM (1979) The uptake of mercuric chloride from sea water by *Serranus cabrilla*. Bull Environ Contam Toxicol 22:771 – 778

Ramade F (1977) Ecotoxicologie. Masson, Paris, 228 pp

Ravera O (1984) Cadmium in freshwater ecosystems. Experientia 40:2 – 14

Ray S (1984) Bioaccumulation of cadmium in marine organisms. Experientia 40:14 – 23

Renfro WC, Fowler SW, Heyraud M, La Rosa J (1975) Relative importance of food and water in long-term zinc-65 accumulation by marine biota. J Fish Res Board Can 32:1339 – 1345

Richardot M (1979) Calcium cells and groove cells in calcium metabolism in the freshwater limpet *Ferrissia wautieri* (Basommatophora Ancylidae). Malacol Rev 12:67 – 78

Roch M, McCarter JA, Matheson AT, Clark MJR, Olafson RW (1982) Hepatic metallothionein in rainbow trouth (*Salmo gairdneri*) as an indicator of metal pollution in the Campbell River system. Can J Fish Aquat Sci 39:1569 – 1601

Roesijadi G (1981) The significance of low molecular weight, metallothionein-like proteins in marine invertebrates: current status. Mar Environ Res 4:167 – 179

Roesijadi G (1983) Enhanced tolerance in *Mytilus edulis*: influence of Cu, Zn and Cd pre-exposure and relationshop to metal-binding proteins. Abstr 5th Conf Eur Soc Comp Physiol Biochem, Sept 5 – 8, Taormina (Italy), pp 102 – 103

Roesijadi G, Drum AS, Thomas JM, Fellingham GW (1982) Enhanced mercury tolerance in marine mussels and relationship to low molecular weight mercury binding protein. Mar Pollut Bull 13:250 – 253

Sawhney BL, Frink CR, Glowa W (1981) PCBs in the Housatonic River: determination and distribution. J Environ Qual 10:444 – 448

Schneider R (1981) Polychlorinated biphenyls (PCBs) in cod tissues from the western Baltic: significance of equilibrium partitioning and lipid composition in the bioaccumulation of lipophilic pollutants in gill-breathing animals. Meeresforschung 29:69 – 79

Schultz-Baldes M, Lewin RA (1976) Lead uptake in two marine phytoplankton organisms. Biol Bull 150:118 – 127

Sheehan PJ, Miller DR, Butler GC, Bourdeau P (eds) (1984) Effects of pollutants at the ecosystem level. SCOPE 22. Wiley, New York, 443 pp

Sheehan P. Korte F, Klein W, Bourdeau P (eds) (1985) Appraisal of tests to predict the environmental behaviour of chemicals. SCOPE 25. Wiley, New York, 380 pp

Simkiss K (1984) Effects of metal ions or respiratory structures. In: Bolis L, Zadunaisky J, Gilles R (eds) Toxins, drugs and pollutants in marine animals. Springer, Berlin Heidelberg New York Tokyo, pp 137 – 146

Simkiss K, Taylor M (1981) Cellular mechanisms of metal ion detoxication and some new indices of pollution. Aquat Toxicol 1:279 – 290

Sternlieb I, Goldfischer S (1976) In: Dingle JT, Dean RT (eds) Lysosomes in biology and pathology, vol 5. Elsevier North Holland, Amsterdam New York, pp 185 – 197

Stone H, Overnell J (1985) Non-metallothionein cadmium binding proteins. Comp Biochem Physiol 80C:9 – 14

Strong CR, Luoma SN (1981) Variations in the correlation of body size with concentrations of Cu and Ag in the bivalve *Macoma balthica*. Can J Fish Aquat Sci 38:1059 – 1064

Spaargaren DH, Ceccaldi HJ (1984) Some relations between the elementary chemical composition of marine organisms and that of sea water. Oceanol Acta 7:63ı76

Tohyama C, Himeno S-I, Watanabe C, Suzuki T, Morita M (1986) The relationship of the increased level of metallothionein with heavy metal levels in the tissue of the harbor seal (*Phoca vitulina*). Ecotoxicol Environ Safety 12:85 – 94

Vandorpe B, Smigielski F (1978) Présence de métaux dans les nodules et nécroses de poissons (morue et pleuronectides). Rev Int Oceanogr Med 49:29 – 35

Veith GD, Macek KJ, Petrocelli SR, Carroll J (1980) In: Aquatic toxicology. Proc 3rd Annu Symp Aquat Toxicol, ASTM, Philadelphia, pp 116 – 129

Vernberg WB, Coursey PJ de, O'Hara J (1974) Multiple environmental factor effects on physiology and behavior of the fiddler crab, *Uca pugilator*. In: Vernberg FJ, Vernberg WB (eds) Pollution and physiology of marine organisms. Academic Press, London New York, pp 381 – 425

Vernberg WB, Coursey PJ de, Kelly M, Johns DM (1977) Effects of sublethal concentrations of Cd on adult *Palaemonetes pugio* under static and flow-through conditions. Bull Environ Contam Toxicol 17:16 – 24

Verschuren K (1983) Handbook of environmental data on organic chemicals. Van Nostrand Reynold, New York, 1310 pp

Viarango A, Palmero S, Zanicchi G, Capelli R, Vaissiere R, Orunesu M (1985) Role of metallothioneins in Cu and Cd accumulation and elimination in the gill and digestive gland cells of *Mytilus galloprovincialis* Lam. Mar Environ Res 16:23 – 36

Weis JS, Kim J (1986) Specificity of acquired cadmium tolerance in the fiddler crab, *Uca pugilator*. Bull Environ Contam Toxicol 37:362 – 368

Wright DA, Brewer CC (1979) Cd turnover in the shore crab *Carcinus maenas*. Mar Biol 50:151 – 156

Young JS, Roesijadi G (1983) Reparatory adaptation to copper-induced injury and occurrence of a copper-binding protein in the polychaete *Eudistylia vancouveri*. Mar Pollut Bull 14:30 – 32

Subject Index

GPSR Compliance

The European Union's (EU) General Product Safety Regulation (GPSR) is a set of rules that requires consumer products to be safe and our obligations to ensure this.

If you have any concerns about our products, you can contact us on ProductSafety@springernature.com

In case Publisher is established outside the EU, the EU authorized representative is:

Springer Nature Customer Service Center GmbH
Europaplatz 3
69115 Heidelberg, Germany

Batch number: 10620881

Printed by Printforce, the Netherlands